Hot Quarks 2010: Workshop for Young Scientists on the Physics of Ultrarelativistic Nucleus-Nucleus Collisions

Journal of Physics: Conference Series Volume 270

La Londe-les-Maures, Cote d'Azur, France
21 – 26 June 2010

ISBN: 978-1-61782-372-5
ISSN: 1742-6588

Printed from e-media with permission by:

Curran Associates, Inc.
57 Morehouse Lane
Red Hook, NY 12571

Some format issues inherent in the e-media version may also appear in this print version.

Printed by Curran Associates, Inc. (2011)

For permission requests, please contact the Institute of Physics
at the address below.

Institute of Physics
Dirac House, Temple Back
Bristol BS1 6BE UK

Phone: 44 1 17 929 7481
Fax: 44 1 17 920 0979

techtracking@iop.org

Additional copies of this publication are available from:

Curran Associates, Inc.
57 Morehouse Lane
Red Hook, NY 12571 USA
Phone: 845-758-0400
Fax: 845-758-2634
Email: curran@proceedings.com
Web: www.proceedings.com

TABLE OF CONTENTS

BULK OBSERVABLES IN PP AT THE LHC

JETS AND PARTICLES AT HIGH TRANSVERSE MOMENTUM

COLLECTIVE MOTION IN NUCLEAR COLLISIONS

THE PHASE DIAGRAM AND THE CRITICAL POINT

THE INITIAL STATE IN HEAVY-ION COLLISIONS

VISCOSITY AND NON-FLOW EFFECTS

PREFACE

The 4th Workshop for Young Scientists on the Physics of Ultrarelativistic Nucleus–nucleus Collisions (Hot Quarks 2010) was held in La Londe-Les-Maures, France, from June 21–26, 2010. Following the traditions of the conference, this meeting gathered more than 70 participants in the first years of their scientific careers. The present issue contains the proceedings of this workshop.

The articles published in this volume clearly show the presence of a dynamic new generation of physicists interested in the different aspects of high energy nuclear collisions. The newest results from RHIC at Brookhaven and SPS at CERN were presented, as well as the latest results from the proton–proton programme from the LHC at CERN, while waiting for the data of the lead–lead collisions only available some months after the meeting. Along with these experimental findings, the corresponding theoretical research was also extensively discussed as well as the new perspectives for future facilities like FAIR, EIC and LHeC.

We wish to thank the sponsors of the Hot Quarks 2010 Conference, who supported the authors of this volume: IN2P3/CNRS (France), EMMI (Germany), Institut Pluridisciplinaire Hubert Curien (France), National Science Foundation (USA), CERN (Switzerland), Helmholtz International Center for FAIR (Germany), Xunta de Galicia (Spain) and the *Journal of Physics G*.

Markus Bleicher (Frankfurt (HIC4FAIR), Germany)
Helen Caines (Yale University, USA)
Manuel Calderon de la Barca Sanchez (UC Davis, USA)
Alessandro de Falco (Cagliari/INFN, Italy)
Rainer Fries (Texas A & M University, USA)
Raphael Granier de Cassagnac (Ecole Polytechnique, France)
Boris Hippolyte (IPHC, Strasbourg, France)
Andre Mischke (Utrecht University, The Netherlands)
Marzia Nardi (Torino/INFN, Italy)
Carlos A Salgado (Universidade de Santiago de Compostela, Galicia, Spain)

Physics at the LHC: a short overview

David d'Enterria
ICREA & ICC-UB, Universitat de Barcelona, 08028 Barcelona, Catalonia
CERN, PH Department, 1211 Geneva, Switzerland

Abstract. The CERN Large Hadron Collider (LHC) started operation a few months ago. The machine will deliver proton-proton and nucleus-nucleus collisions at energies as high as $\sqrt{s} = 14$ TeV and luminosities up to $\mathcal{L} \sim 10^{34}$ cm^{-2}s^{-1}, never reached before. The main open scientific questions that the seven LHC experiments – ATLAS, CMS, ALICE, LHCb, TOTEM, LHCf and MOEDAL – aim to solve in the coming years are succinctly reviewed.

1. Introduction

The LHC [1] is the ultimate particle collider in terms of centre-of-mass (c.m.) energies and luminosity. The machine together with its 7 international experiments at the CERN laboratory in Geneva, have been built to study the frontier of our knowledge about the particles, interactions, and space-time structure of the Universe. Their design goal addresses, in particular, at least 6 fundamental questions that still remain unsolved today in high-energy physics:

(i) Mass generation problem: How do the elementary particles of the Standard Model (SM) acquire their bare masses ? Is it via their coupling to the Higgs boson – the last predicted missing piece of the SM – or through other mechanisms ?

(ii) Hierarchy / fine-tuning problem: What mechanism stabilizes the Higgs boson mass up to the next known physics scale at Planck energies (10^{16} orders-of-magnitude above) ? Supersymmetry (SUSY) ? Higgs boson compositeness ? new space dimensions ?

(iii) Dark matter (DM) problem: What weakly-interacting particle accounts for one fourth of the (invisible) content of the universe ? Can the lightest sparticle or other new massive particles (lightest technihadron or Kaluza-Klein tower, axions, sterile ν's, ...) explain DM ?

(iv) Flavour problem: Why do we observe a matter-antimatter asymmetry in the Universe ? Why are there so many types of matter particles and they mix the way they do ?

(v) Strong interaction in non-perturbative regime: Why quarks are confined in hadrons ? What is the energy evolution of the total hadronic cross sections ? Can one experimentally test the conjectured duality between gauge and string theories (AdS/CFT) ?

(vi) Origin/Nature of the highest-energy cosmic-rays (CRs): What are the sources and type of particles constituting CRs at energies up to 10^{20} eV ?

Of course, the solutions to these open problems need to be consistent, if not directly connected, among each other. The two large multipurpose detectors, ATLAS [2] and CMS [3], have the experimental capabilities to address the full set of questions. The two mid-size experiments, LHCb [4] and ALICE [5], are mostly optimised for problems (iv) and (v) respectively, although they can cover a subset of the rest of research topics. The smallest ones, TOTEM [6], LHCf [7] and MOEDAL [8], aim at studying particular aspects of questions (v), (vi) and (iii) respectively.

2. Preface: Rediscovering the Standard Model (SM)

The SM is a renormalizable quantum-field-theory which – unifying quantum mechanics and special relativity – explains the fundamental interactions (except gravity) among elementary particles via a local $SU_C(3) \times SU_L(2) \times U_Y(1)$ gauge-symmetry group[1]. The three gauge-symmetry terms give rise to the strong, weak and electromagnetic interactions. The particles fall into different representations of these groups. The SM Lagrangian (without neutrino masses) contains 19 free parameters: 3 gauge couplings, 6 quark masses, 3 lepton masses, 3 mixing-angles, 2 CP phases, and 2 Higgs-boson couplings, to be determined experimentally. The SM (except the Higgs sector) has been verified to high precision in many measurements and the LHC will allow one to test it in a regime of energies up to 7 times higher than those probed before. Prior to looking for new physics signals one needs to confirm that we have a good experimental and theoretical control of processes which have been already measured at lower energies.

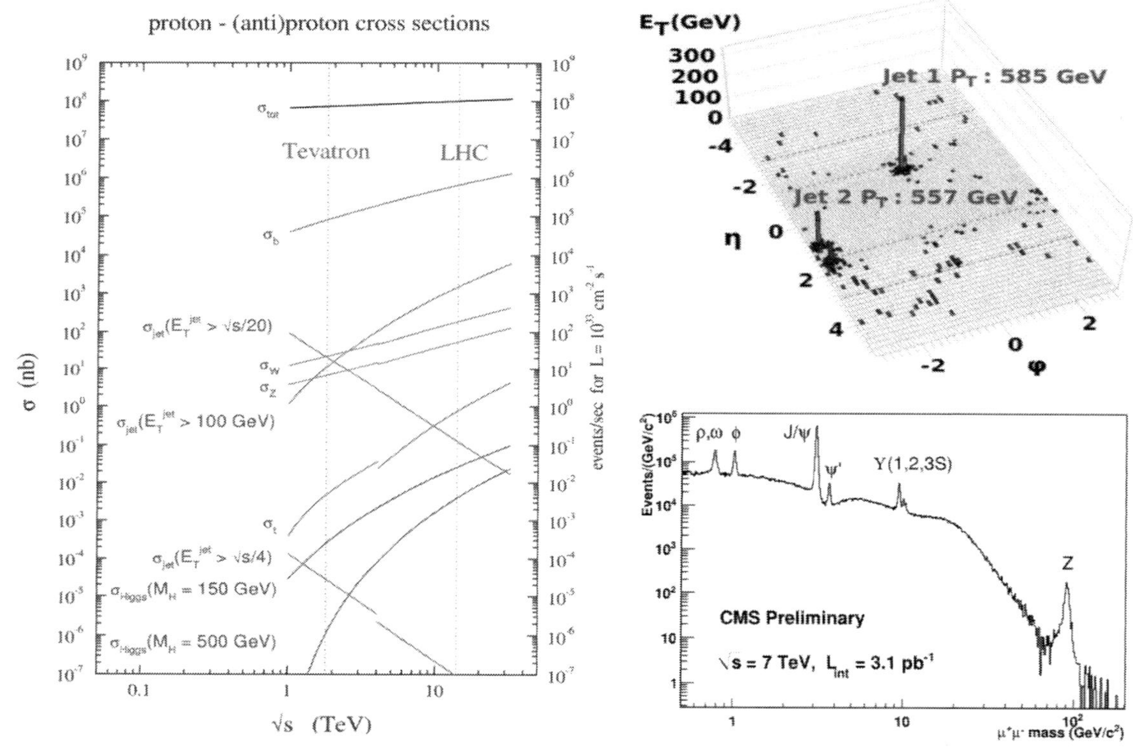

Figure 1. Left: Cross sections (and rates for $\mathcal{L} = 10^{34}$ cm^{-2}s^{-1}) as a function of $\sqrt{s}$ for various SM processes in p-p(p-$\bar{p}$) collisions [9]. Right: First measurements in p-p at 7 TeV: dijet event with $m_{jj} = 2.13$ TeV/c^2 (top), and mass spectrum of all known dimuon resonances (bottom) [10].

The processes with the largest cross sections in hadronic collisions are mediated by the strong force (upper curves in Fig. 1, left) and thus the first LHC measurements in p-p at $\sqrt{s} = 7$ TeV are linked to QCD observables such as those shown in Fig. 1 right. Perturbative QCD calculations agree well with preliminary data such as high-p_T hadrons, jets, and heavy-flavours measured by ATLAS [11], CMS [10], ALICE [12] and LHCb [13]. Next in importance in terms of cross sections are the electroweak processes: the first W and Z bosons (highest-mass dimuon peak in Fig. 1 right) have been detected with the first hundred nb^{-1} integrated by CMS and ATLAS.

[1] The subindices indicate the conserved color and weak-hypercharge and the action on left-handed fermions.

3. Problem I: Mass generation problem – the Higgs boson

The LHC together with the ATLAS and CMS experiments have been designed primarily to be able to solve the last missing element of the SM: the mechanism of generation of the elementary particle masses. Indeed, the electroweak (EWK) sector of the theory suffers from two problems:

- The $\mathrm{SU_L(2)\times U_Y(1)}$ symmetry imposes zero-masses for the gauge bosons and fermions in striking contradiction, in particular, with the large observed masses for W and Z.
- Without a mechanism to generate the vector-boson masses, the longitudinal WW scattering cross sections grow quadratically with energy and break unitarity at energies $\mathcal{O}(1\ \mathrm{TeV})$.

A simple and elegant solution to these problems, was proposed by Englert-Brout-Higgs-Guralnik-Hagen-Kibble [14] who demonstrated that the W and Z bosons can acquire a mass while preserving the $\mathrm{SU_L(2)\times U_Y(1)}$ symmetry of the EWK Lagrangian, if the vacuum is filled with a field with a non-zero expectation value which couples to the (massless) vector-bosons. The electroweak symmetry is preserved in the Lagrangian but it is broken spontaneously by the ground-state actually realized in nature. Such a mechanisms involves adding new terms, an extra scalar-doublet and associated couplings, to the SM Lagrangian. Three of the four degrees of freedom of this extra field mix with the W,Z bosons to provide them with mass, while the other one becomes the Higgs boson, a new scalar particle. In a similar manner, Higgs bosons "lurking virtually" in the vacuum drag on all fermions to give them mass. The quadratic rise of the WW,ZZ cross sections is thus damped by new diagrams involving Higgs boson exchange.

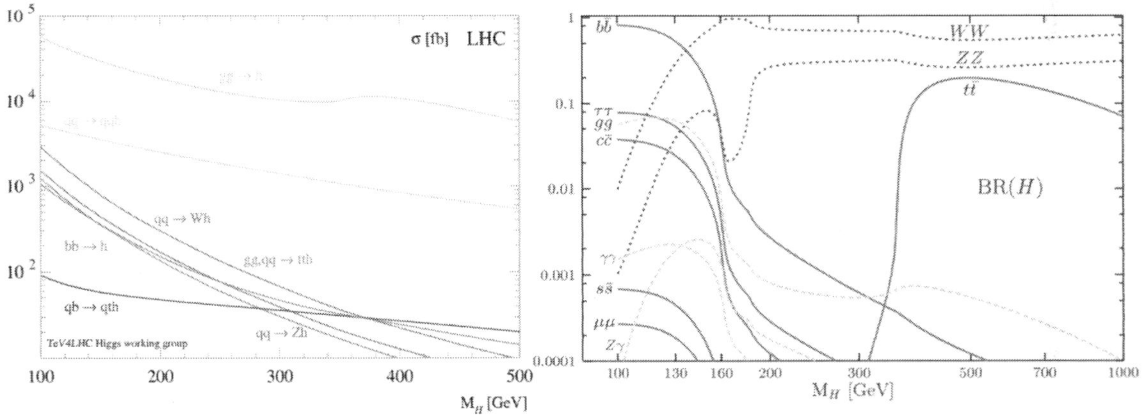

Figure 2. SM Higgs boson: LHC production channels (left) [15] and branching ratios (right) [16].

The mass of the Higgs boson is a free parameter of the SM. Direct searches at LEP [17] and Tevatron [18] have constrained its value at 95% confidence level (CL) to the range 114.4 – 158 GeV/c^2 and $m_{\mathrm{H}} > 175$ GeV/c^2, whereas indirect constraints from global fits to precision EWK data (accounting for virtual Higgs contributions e.g. to the W and top-quark masses) exclude $m_{\mathrm{H}} > 185$ GeV/c^2 at 95% CL [19]. At the LHC, several Higgs production processes and decay channels are accessible (Fig. 2). Depending on its mass, different production/decay modes can be exploited for a 5σ-discovery integrating a few tens of fb^{-1} in p-p at 14-TeV [2, 3]:

(i) For $m_{\mathrm{H}} < 135$ GeV/c^2, the preferred channels involve $\gamma\gamma$ and $\tau\tau$ decays, in particular in vector-boson-fusion production processes accompanied by forward-backward jets.

(ii) For $m_{\mathrm{H}} > 135$ GeV/c^2, the $H \to WW^{(*)}, ZZ^{(*)}$ modes, characterized by four (two) high-p_T electrons or (and) muons in the final-state, provide a relatively clean signal.

The determination of the Higgs couplings to all SM particles and its quantum numbers will likely require a more precise e^+e^- collider or resort to other more suppressed production channels [20].

4. Problem II: Hierarchy/fine-tuning – New symmetries at high energies ?

The hierarchy problem of the SM is related to the "uncontrollable" running of the Higgs boson mass with energy. Indeed, (i) the $m_{\rm H}$ value is not predicted by the theory, nor protected by any internal SM symmetry; and (ii) being a scalar field, the loop radiative corrections make $m_{\rm H}$ increase *quadratically*[2] up to next physics scale known today, where gravity becomes as strong as the gauge interactions. Thus, if $m_{\rm H}$ is imposed from a symmetry at the Planck scale, then its value has to be fine-tuned to 1 part in $m_{\rm EW}/m_{\rm Planck} \approx G_{\rm F}^{-1/2}/G_{\rm N}^{-1/2} \approx 10^{16}$!

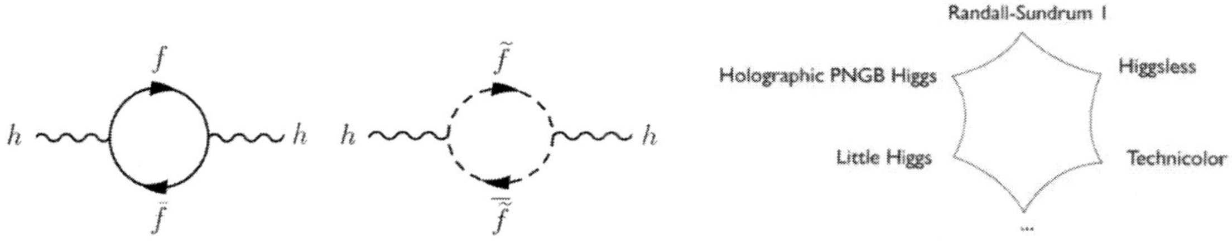

Figure 3. Left: Virtual contributions to the Higgs mass from SM fermions and SUSY sfermions. Right: Alternative (non-SUSY) models beyond the SM [21].

There are three general theoretical solutions to the hierarchy problem. All of them imply physics beyond the Standard Model (BSM), and the existence of new symmetries and associated new particles at the TeV scale:

(i) SUSY: The existence of new SUSY partners, differing by 1/2 unit of spin, for every SM particle provides a simple way to stabilize the Higgs potential, as their amplitude in the quantum corrections come with opposite sign and cancel the SM contributions (Fig. 3, left).

(ii) Non-Standard "Higgs" models: The W and Z masses can also be generated *dynamically* via a Goldstone boson corresponding to a spontaneously broken global symmetry of a new strongly-interacting (QCD-like) sector at some higher mass scale Λ, as done e.g. in Technicolour [22] or little-Higgs models. In the previous case, the Higgs boson is not an elementary scalar but a condensate of techni-fermions. In the latter, the Higgs is a pseudo-Goldstone boson and its mass is protected from acquiring quadratically-divergent loop corrections by the contributions of new particles (heavy-top, W', Z', ...).

(iii) Low-energy quantum gravity: The huge hierarchy between the EWK and Planck scales can be solved if one considers that the apparent relative weakness of gravitation is not real but only due to the fact that it expands over extra (hidden) spatial dimensions, whereas the other interactions are confined to our visible 3-D space. Theories with extra dimensions (flat as in ADD [23] or warped as in RS [24] models) predict also new particles at the TeV-scale (Kaluza-Klein towers, radion, mini-blackholes, ...).

5. Problem III: Dark matter – New heavy particles ?

The existence of dark matter (DM), accounting for one-fourth of the content of the Universe has been confirmed by several experimental evidences such as (i) the fact that the galactic rotation speed curves do not follow the expected Copernican fall-off with distance from the center, (ii) the structure of the power spectrum of the temperature fluctuations of the cosmic microwave background, (iii) the separation in the distribution of matter observed by gravitational-lensing and by radiation in the collision of two clusters of galaxies. The current DM signatures favour

[2] In contrast, the radiative corrections for the fermion masses are only logarithmic with energy.

a Weakly-Interacting Massive Particle (WIMP) which is sensitive only to the weak-interaction and gravitation, and which is a stable heavy relic from the early Universe. All BSM theories mentioned in the previous section naturally contain DM candidates such as:

(i) the lightest SUSY Particle (LSP), such as neutralinos or gravitinos;
(ii) the lightest technibaryon in technicolor models;
(iii) the lightest Kaluza-Klein tower (resonances arising from quantized waves in the extra dimension), gravitons from adjacent branes, or radions; in ADD or RS approaches.

In addition, axions or heavy right-handed (sterile) neutrinos, among others have also been proposed as DM particle candidates.

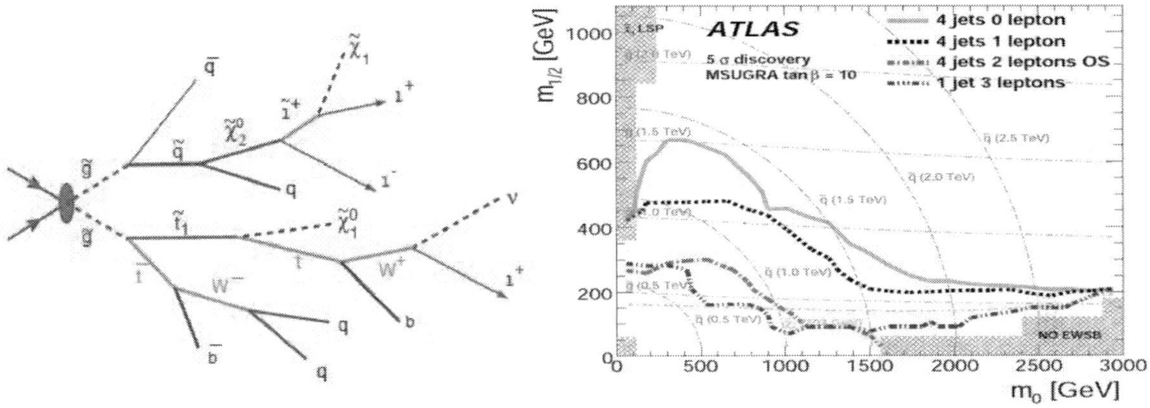

Figure 4. Left: Cascade decay of a pair of gluinos in a *p-p* collision. Right: mSUGRA discovery contours in jets+leptons+$\not\!\!E_T$ channels (*p-p* 14 TeV, 1 fb^{-1}) as a function of m_0 and $m_{1/2}$ [2].

By its own characteristics, any heavy WIMP produced at the LHC will appear as (large) missing transverse energy in a *p-p* event. In the SUSY case[3] the LSP – e.g. the neutralino (a neutral-charge mixture of superpartners of the SM gauge bosons) or the gravitino – is a paradigmatic stable heavy candidate for DM. LSPs are produced in events whose experimental event topologies (Fig. 4, left) are characterized by:

- missing transverse energy ($\not\!\!E_T$) from the invisible WIMP at the end of a decay cascade,
- multi-jets: from intermediate sparticle decays into heavy SM particles (top, W, Z),
- same-sign leptons/multi-photons: from decay of the intermediate sparticles.

An example of the discovery potential of a particular class of SUSY models (mSUGRA, $\tan\beta = 10$) based on such experimental signatures, is shown in Fig. 4 (right) as a function of two of the free parameters of the theory (the common scalar m_0 and fermion $m_{1/2}$ masses).

In addition, in many BSM approaches the next-to-lightest new particle is metastable (long-lived) and has a mass/charge ratio which makes it a highly ionizing particle when interacting with normal matter[4]. The MOEDAL experiment [8] (sharing the interaction point with LHCb) aims at measuring such highly-ionizing tracks in large plastic detectors.

[3] The LSP is stable, as it cannot decay anymore, if SUSY conserves R-parity (i.e. sparticles are produced in pairs and decay into another SUSY particle plus any number of normal particles).
[4] Typical SUSY highly-ionizing particles are NLSPs such as staus as well as R-hadrons (gluino-parton bound-states).

6. Problem IV: Matter-antimatter asymmetry – new virtual particles ?

The process of changing a quark flavour into another is controlled by the (charged-current) weak interaction. The observation that kaons and B-mesons can transform into their anti-particles with different probabilities in each direction and that they can decay into a state with different Charge-Parity (CP) quantum numbers, implies that the weak force is not invariant under C- (particles and antiparticles exchange) and P- (changing a particle by its mirror image) transformations. In the SM, weak interactions involving W exchange act exclusively on left-(right-)handed (anti)particles, and the coupling among quarks is done in terms of mixed-flavour objects which do not correspond to the quark mass-eigenstates (to which QCD and QED couple) but to a superposition of them. The relationship between the mass- and weak- eigenstates is implemented via the Cabibbo-Kobayashi-Maskawa (CKM) matrix V_{ij}, which describes how quarks mix/decay among each others. CP violation is then incorporated via a CKM matrix complex phase, constrained by the condition $V_{ud}V_{ub} + V_{cd}V_{cb} + V_{td}V_{tb} = 0$, which is often displayed as a unitarity triangle with angles $\alpha = f(V_{td,tb}, V_{ud,ub})$, $\beta = f(V_{cd,cb}, V_{td,tb})$ and $\gamma = f(V_{ud,ub}, V_{cd,cb})$ obtained from kaon and B-meson measurements (Fig. 5, left).

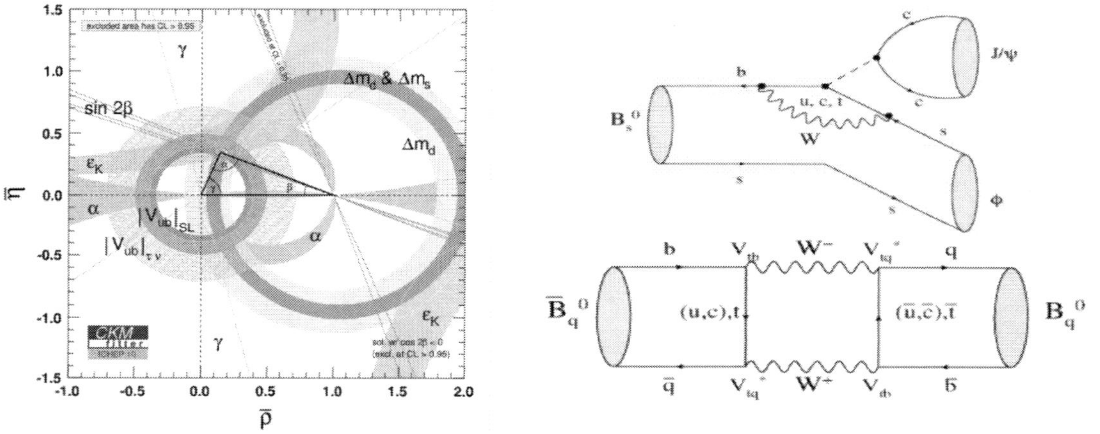

Figure 5. Left: Unitarity triangle fit [25]. Right: Examples of penguin (top) and box (bottom) diagrams for B-meson decay and oscillations in the SM.

The currently known differences between particles and antiparticles (i.e. the amount CP-violation in the SM) are however way too small to explain the matter-antimatter imbalance observed today, $(n_B - n_{\bar{B}})/n_\gamma \approx 10^{-9}$, and new particles/CP-phases are needed to explain how baryon dominance appeared in the Universe (baryogenesis). CP-violation studies at the LHC involve indirect searches of new virtual particles contributing to higher-order (Penguin or box) loops in flavour-changing neutral current processes (Fig. 5, right). The LHCb experiment [4] focuses on measurements in the bottom sector (e.g. $b \to s$ transitions) which are less constrained by current data and can access higher energy scales than the kaon system, such as:

(i) detailed B-meson studies of rare decay rates, branching ratios, decays asymmetries, oscillation frequencies, and lifetimes;

(ii) over-constraints of the unitarity triangle via improved precision of CKM angles and sides, and cross-checks of the same quantity via various measurements.

In all cases, any possibly found inconsistency (in the phases of the couplings, their absolute values, and/or their Lorentz structure) can be a sign of new physics.

Journal of Physics: Conference Series **270** (2011) 012001 doi:10.1088/1742-6596/270/1/012001

7. Problem V: Strongly-interacting matter – QGP, CGC, AdS/CFT, σ_{tot}

The strong interaction among quarks and gluons is described by QCD, a quantum field theory with a very rich dynamical content including asymptotic freedom, infrared slavery, (approximate) chiral symmetry, non trivial vacuum topology (instantons), strong CP problem, $U_A(1)$ axial-vector anomaly, ... All these properties translate into a diverse many-body phenomenology at various limits (Fig. 6, left). Interestingly, QCD is the only SM sector whose full *collective* behaviour – phase diagram, (deconfinement and chiral) phase transitions, thermalization of fundamental fields – is accessible to scrutiny in the laboratory, via high-energy heavy-ion collisions. The study of strongly-interacting matter in extreme conditions of temperature, density and small parton momentum fraction (low-x) can be carried out by measuring different observables in nucleus-nucleus (A-A) collisions which are sensitive to the underlying QCD medium properties (Fig. 6, right) [26].

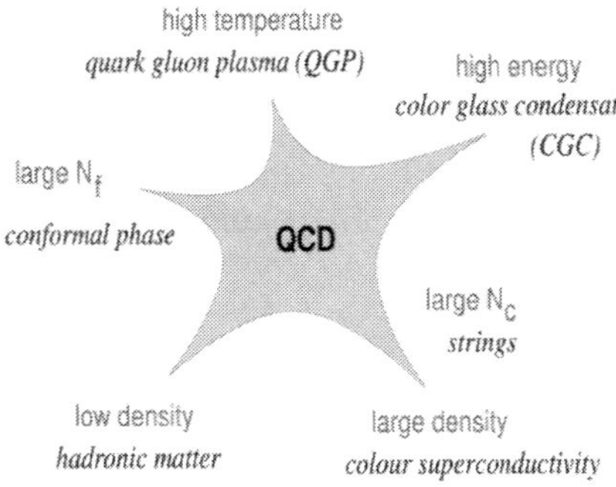

Heavy-ion measurement	Experimental observable	QCD medium property	Theoretical tool
$dN_h/d\eta$ (soft)	multiplicity	gluon $xG(x,Q^2)$	CGC
dN_h/dp_T (soft)	radial flow	Eq. of State $p(T)$	Hydro+lattQCD
$dN_h/d\phi$ (soft)	elliptic flow	shear viscosity η	Hydro+AdS/CFT
dN_γ/dp_T (semihard)	thermal γ	temperature T_{crit}	Hydro+lattQCD
$Q\bar{Q}$ (hard)	color screening	energy density ε_{crit}	Hydro+lattQCD
$dN_{h,jet}/dp_T$ (hard)	jet quenching	transport coeff. $\langle\hat{q}\rangle$	Hydro+pQCD

Figure 6. Left: Many-body QCD at various limits [27]. Right: Experimental and theoretical tools in high-energy A-A collisions (dN_h stands for differential hadron distributions).

The collisions of lead nuclei at the LHC ($\sqrt{s_{NN}}$ = 5.5 TeV, 30 times higher than those attained previously) will allow one to study quark-gluon matter at unprecedented values of energy density ($\varepsilon_{\text{Bjorken}} \approx 10$ GeV/fm^3 at times $\tau_0 = 1$ fm/c) using pQCD probes produced with cross sections 10 to 10^4 higher than at the Relativistic Heavy-Ion Collider. The fractional momenta of the colliding partons will be as low as $x \approx p_T/\sqrt{s_{NN}}\, exp(-\eta) = \mathcal{O}(10^{-5})$, where gluon saturation effects, as described in the Color-Glass-Condensate (CGC) approach [28], are expected to dominate the parton dynamics. A part from the standard searches of the formation of a deconfined and chirally-symmetric Quark-Gluon-Plasma (QGP) via quarkonia suppression or jet quenching, one of the important measurements will be that of the azimuthal anisotropies of bulk hadron production with respect to the reaction plane. Such "explosive" anisotropies, known under the name of "elliptic flow", have been found to be very sensitive to the viscosity/entropy ratio of the produced medium, according to advanced relativistic fluid-dynamics calculations [29] including the QCD equation-of-state computed in the lattice. The field of heavy-ion physics has attracted lots of theoretical interest as testbed for the application of the Anti-de-Sitter/Conformal-Field-Theory (AdS/CFT) duality between weakly-coupled gravity and strongly-coupled gauge theories [30]. Applications of such a formalism have led to the determination of transport properties such as the viscosity [31] of Supersymmetric-Yang-Mills (SYM) plasmas, from simpler black-hole thermodynamics calculations. The large differences between SYM theory and QCD (extra SUSY degrees of freedom, no running-coupling, no confinement, ...) seem to "wash out" at finite temperature [32].

The measurements of the total and elastic p-p cross sections are also part of the LHC physics programme, providing a valuable test of fundamental quantum mechanics relations such as the Froissart bound $\sigma_{tot} < Const \ln^2 s$, the optical theorem $\sigma_{tot} \sim \mathrm{Im} f_{el}(t = 0)$, and dispersion relations $\mathrm{Re} f_{el}(t = 0) \sim \mathrm{Im} f_{el}(t = 0)$ [33]. The current extrapolations of the total p-p cross section at the LHC ($\sigma_{tot} = 90 - 140$ mb), of which the elastic contribution accounts for about one fourth, suffer from large uncertainties due to disagreeing measurements at the Tevatron and uncertain extractions from cosmic-ray collisions with air nuclei (Fig. 7, left). The main goal of the TOTEM experiment [6] is to obtain a measurement of the total and elastic p-p cross sections, with an uncertainty of about 1%, over a large range of 4-momentum transfers from $-t \approx 2 \cdot 10^{-3}\,\mathrm{GeV}^2$ to $8\,\mathrm{GeV}^2$. TOTEM Roman-pots detectors will measure the elastically scattered protons inside the LHC tunnel area adjacent to the CMS collision point.

8. Problem VI: Origin and nature of the highest-energy cosmic-rays

The origin and nature of cosmic rays (CRs) with energies between 10^{15} eV and the Greisen-Zatsepin-Kuzmin (GZK) cutoff at about 10^{20} eV, measured by the HiRes [34] and recently confirmed by the Auger [35] experiments (Fig. 7, right), remains a central open question in high-energy astrophysics. One key to solving this question is the determination of the elemental composition of cosmic rays in this energy range. The candidate particles, ranging from protons to Fe ions, are measured with surface detectors that detect the "extended air-showers" (EAS) generated in the CR interactions with air nuclei when entering the Earth's atmosphere [36].

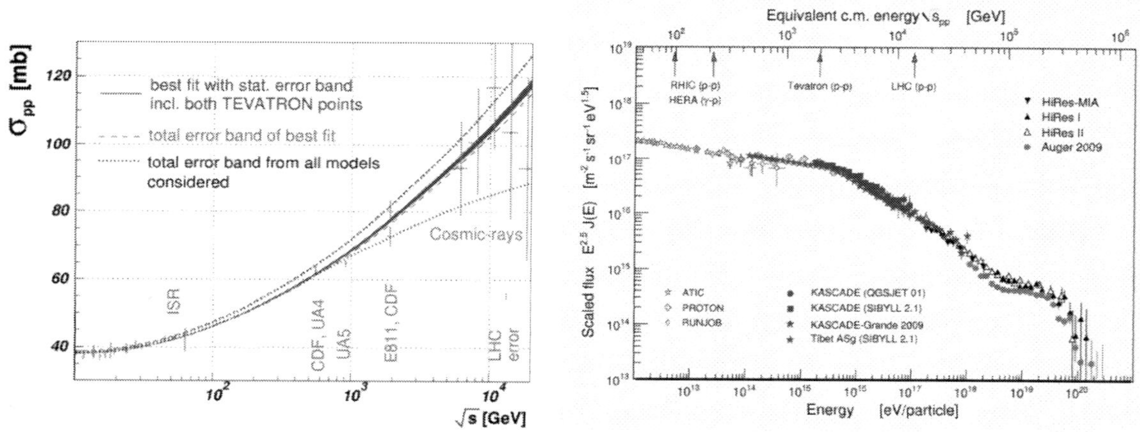

Figure 7. Left: COMPETE fits [37] of σ_{tot} as a function of $\sqrt{s}$. Right: Measured (scaled) cosmic ray flux as a function of E_{lab} (the equivalent $\sqrt{s}$ of different colliders is also shown) [36].

The determination of the primary energy and mass relies on hadronic Monte Carlo (MC) codes which describe the interactions of the primary cosmic-ray with air nuclei (N, O). The bulk of the primary particle production is dominated by forward and soft QCD interactions, modeled commonly in Regge-Gribov+pQCD approaches with parameters tuned to reproduce the pre-LHC collider data ($E_{lab} \lesssim 10^{15}$ eV). When extrapolated to energies around the GZK-cutoff, the current MCs predict energy and multiplicity flows differing by factors as large as three with significant inconsistencies in the forward region. The LHCf experiment [7] has installed scintillator/silicon calorimeters inside the tunnel area 140 m away of the ATLAS interaction point to detect neutral particles (photons, π^0, neutrons) close to the beam-rapidity ($|\eta| \gtrsim 8.3$). Measurement of forward particle production in p-p, p-Pb, and Pb-Pb collisions at LHC energies ($E_{lab} \approx 10^{17}$ eV) will provide strong constraints on these models and allow for more reliable determinations of the CR energy and composition at the highest energies.

Acknowledgments
Support by the 7th EU Framework Programme contract FP7-ERG-2008-235071 is acknowledged.

References

[1] L. Evans and P. Bryant, JINST **3** (2008) S08001.
[2] G. Aad *et al.* [ATLAS Collaboration], arXiv:0901.0512.
[3] A. de Roeck (ed.) [CMS Collaboration], J. Phys. G: Nucl. Part. Phys. **34** (2007) 995.
[4] B. Adeva *et al.* [LHCb Collaboration], arXiv:0912.4179.
[5] B. Alessandro *et al.* [ALICE Collaboration], J. Phys. G: Nucl. Part. Phys. **32** (2006) 1295.
[6] V. Berardi *et al.* [TOTEM Collaboration], CERN/LHCC-2004-002 (2004).
[7] Technical Design Report of the LHCf experiment, CERN/LHCC-2006-004 (2006).
[8] Technical Design Report of the MOEDAL experiment, CERN/LHCC-2009-006 (2009).
[9] J. M. Campbell, J. W. Huston and W. J. Stirling, Rept. Prog. Phys. **70** (2007) 89
[10] G. Tonelli [CMS Collaboration], ICHEP 2010.
[11] F. Gianotti [ATLAS Collaboration], ICHEP 2010.
[12] J. Schukraft [ALICE Collaboration], ICHEP 2010.
[13] A. Golutvin [LHCb Collaboration], ICHEP 2010; R. Aaij *et al.* [LHCb Collaboration], arXiv:1009.2731.
[14] Englert F and Brout R 1964 *Phys. Rev. Lett.* **13** 321; Higgs P W 1964 *Phys. Lett.* **12** 132; Guralnik G S, Hagen C R and Kibble T W B 1964 *Phys. Rev. Lett.* **13** 585.
[15] U. Aglietti *et al.*, hep-ph/0612172.
[16] A. Djouadi, Phys. Rept. **457**, 1 (2008).
[17] R. Barate *et al.* [LEP-Higgs Working Group], Phys. Lett. B **565**, 61 (2003).
[18] CDF and D∅ Collaborations, arXiv:1007.4587.
[19] LEP-Tevatron-SLD Electroweak Working Group; arXiv:0811.4682.
[20] D. d'Enterria, arXiv:0905.4307.
[21] H. C. Cheng, arXiv:0710.3407.
[22] S. Weinberg, Phys. Rev. D **19** (1979) 1277; L. Susskind, Phys. Rev. D **20** (1979) 2619.
[23] N. Arkani-Hamed, S. Dimopoulos and G. R. Dvali, Phys. Lett. B **429** (1998) 263.
[24] L. Randall and R. Sundrum, Phys. Rev. Lett. **83** (1999) 3370.
[25] J. Charles *et al.* [CKMfitter Group], Eur. Phys. J. C **41** (2005) 1.
[26] See e.g. D. d'Enterria, J. Phys. G **34** (2007) S53.
[27] Schaefer T 2005 "HUGS 2005 Lectures" World Scientific *Preprint* hep-ph/0509068
[28] F. Gelis, E. Iancu, J. Jalilian-Marian and R. Venugopalan, arXiv:1002.0333.
[29] M. Luzum and P. Romatschke, Phys. Rev. C **78** (2008) 034915 [Erratum-ibid. C **79** (2009) 039903]
[30] J. M. Maldacena, Adv. Theor. Math. Phys. **2**, 231 (1998) [Int. J. Theor. Phys. **38**, 1113 (1999)]; E. Witten, Adv. Theor. Math. Phys. **2**, 505 (1998).
[31] P. Kovtun, D. T. Son and A. O. Starinets, Phys. Rev. Lett. **94**, 111601 (2005)
[32] K. Kajantie, T. Tahkokallio and J. T. Yee, JHEP **0701**, 019 (2007).
[33] C. Bourrely, N. N. Khuri, A. Martin, J. Soffer and T. T. Wu, arXiv:hep-ph/0511135 (2005).
[34] R. Abbasi *et al.* [HiRes Collaboration], Phys. Rev. Lett. **100**, 101101 (2008).
[35] T. Yamamoto [Pierre Auger Collaboration], arXiv:0707.2638 [astro-ph].
[36] J. Bluemer, R. Engel and J. R. Hoerandel, Prog. Part. Nucl. Phys. **63** (2009) 293.
[37] J. R. Cudell *et al.* [COMPETE Collaboration], Phys. Rev. Lett. **89**, 201801 (2002).

Journal of Physics: Conference Series **270** (2011) 012002 doi:10.1088/1742-6596/270/1/012002

Measurement of charged-particle pseudorapidity density with ALICE at LHC

Maria Nicassio for the ALICE Collaboration

University of Bari and INFN Sezione di Bari, Via Orabona 4, 70126 Bari, Italy

E-mail: maria.nicassio@ba.infn.it

Abstract. The ALICE experiment, designed primarily for studying heavy-ion collisions at the Large Hadron Collider, has several features in its apparatus that allow it to contribute significantly to the first proton–proton physics. ALICE started proton–proton data taking at the end of 2009: data collected at a centre-of-mass energy of 0.9, 2.36 and 7 TeV allowed several first physics studies to be carried out. In this contribution, the first measurements of the charged-particle pseudorapidity density at the three energies will be presented. These measurements have been obtained using mainly the two innermost layers of the inner tracking system. The reconstruction algorithm and the analysis procedure are described and the results are compared to other measurements and model predictions.

1. Introduction

The ALICE experiment [1] has been taking data since November 2009 when LHC started colliding proton beams. With the data collected at the three energies explored so far, several studies have been carried out. The first measurement is the multiplicity of primary charged particles and its pseudorapidity density distribution. ALICE results have been compared with those from CMS experiment [2] and, at the lowest energy, also with the existing results for p-$\overline{\text{p}}$ obtained at Sp$\overline{\text{p}}$S [3]. At high energies the multiplicity measurement will constrain the hadroproduction models allowing the Monte Carlo generators to be correctly tuned and the energy dependence of the charged-particle density to be determined.

2. The pixel detector and the tracklet algorithm

The ALICE Silicon Pixel Detector (SPD) consists of two cylindrical layers equipped with hybrid pixels and placed at 3.9 and 7.6 cm from the beam line. The pseudorapidity coverage is $|\eta| < 2$ for the inner layer and $|\eta| < 1.4$ for the outer layer.

Using the reconstructed hits in the two layers and exploiting their spatial correlations, the position of the primary interaction vertex can be calculated. Then, charged tracks crossing the detector can be reconstructed using the vertex as the origin and looking for pairs of reconstructed points, aligned with the vertex within predefined tolerances, so called "tracklets". The algorithm consists of an iterative procedure that associates the pairs with the minimum distance in the φ-θ plane as described in [4]. The multiplicity is estimated counting tracklets and the pseudorapidity is calculated using the θ angle given by the point in the inner layer and the primary vertex. There are several advantages in using the SPD to reconstruct tracks compared to the measurement based on tracks reconstructed in the full tracking system: the SPD has a larger acceptance coverage both in pseudorapidity and transverse momentum than the rest of the central tracking system and relies on faster alignment and calibration procedures.

3. Data taking conditions and data samples

The first proton–proton collisions at $\sqrt{s}$=0.9 TeV took place on the 23rd of November 2009. The first physics publication of the LHC is based on 284 events collected that day with the ALICE detector [5]. Subsequently, in the data taking of December 2009 and March-April 2010, higher statistics samples have been collected at $\sqrt{s}$=0.9, 2.36 and 7 TeV with magnetic field B=0.5 T. At $\sqrt{s}$=0.9 and 7 TeV the trigger used for data acquisition was based on the SPD and the V0 detector: essentially this trigger selects events with at least one charged particle in $|\eta|$ <8. The SPD can contribute to the Level 0 (L0) trigger of the experiment with its digital Fast-Or signal [1]. The V0 detector consists of arrays of scintillators placed at either side of the nominal interaction point around the beam-pipe and allows collisions and beam–gas interactions to be distinguished using its timing signals with respect to beam crossing time. The trigger requirement was at least one SPD Fast-Or or at least one hit in either one of the two V0 arrays. The signals were required to be in coincidence with passing bunches. Control triggers to remove beam-induced and accidental background were used during data taking as described in [6]. At $\sqrt{s}$=2.36 TeV the V0 detector was switched off, so the trigger condition was looser and based on SPD signals only, in coincidence with passing bunches. This results in a higher systematic error.

4. Data analysis

The event classes chosen for the analysis are the following: inelastic (INEL), non-single-diffractive (NSD) and events with at least one charged particle, i.e. at least one tracklet, reconstructed in $|\eta| < 1$ (INEL>0) [7]. The latter has been chosen for the 7 TeV data analysis, where no information about the cross section of diffractive processes is available yet, to reduce the dependency of the measurements on the models used to calculate the corrections and hence the systematic uncertainty.

Offline selections were applied to reject beam-induced background. In addition, at 0.9 TeV a subset of the triggered sample has been extracted for the NSD analysis: in order to reduce the fraction of single-diffractive events, signals in both V0 counters have been required. The vertex range of events accepted in the analysis has been limited to $|z_{vtx}|$ <10 cm since for larger values the efficiency in reconstructing the vertex decreases due to the SPD acceptance in z. For the INEL>0 class the range has been further restricted to $|z_{vtx}|$ <5.5 cm to have the pseudorapidity range used to define the event class fully within the SPD acceptance.

Primary particles are defined as prompt particles produced in the collision and all decay products, except products from weak decays of strange particles. To obtain the number of primaries, the number of reconstructed tracklets has been corrected for the following effects: contaminations from secondary interactions, long-lived particle decay products and gamma conversions, transverse momentum cut-off (approximately 50 MeV/c), detector acceptance and efficiency, reconstruction efficiency and combinatorial background. The corrections are calculated as a function of η and the longitudinal position of the primary vertex. The number of events has also been corrected for the trigger and selection efficiencies.

Monte Carlo samples have been generated to calculate the corrections and to study the systematic uncertainties. The combinatorial background has been assessed from data by rotating the SPD inner layer by 180°, thereby destroying real correlations of hits. The event generators used are PhoJet [8] and several tunes of PYTHIA (D6T [9], ATLAS-CSC [10] and Perugia-0 [11]).

Table 1. Measured $dN_{ch}/d\eta$ at central pseudorapidity in $|\eta|$ <0.5 for INEL and NSD classes at $\sqrt{s}$=0.9 and 2.36 TeV [6] and in $|\eta|$ <1 for INEL>0 class at the three centre-of-mass energies [7]. PYTHIA has been used to calculate the corrections. The first uncertainty is statistical and the second is systematic. The asymmetric systematic uncertainties indicate that there are differences when calculating corrections with PhoJet.

	0.9 TeV	2.36 TeV	7 TeV
INEL	$3.02\pm0.01^{+0.08}_{-0.05}$	$3.77\pm0.01^{+0.25}_{-0.12}$	
NSD	$3.58\pm0.01\pm0.12$	$4.43\pm0.01^{+0.17}_{-0.12}$	
INEL>0	$3.81\pm0.01\pm0.07$	$4.70\pm0.01^{+0.11}_{-0.08}$	$6.01\pm0.01^{+0.20}_{-0.12}$

5. Results

In Tab. 1 the values of the pseudorapidity density at central pseudorapidity at the three energies and for different event classes are quoted.

In Fig. 1 the $dN_{ch}/d\eta$ as a function of pseudorapidity measured at $\sqrt{s}$=0.9 TeV (left panel) and 2.36 TeV (right panel) are shown for the INEL and NSD events. ALICE results at 0.9 TeV are compared to UA5 p-$\overline{\text{p}}$ results for both event classes and to CMS distribution for NSD events. At 2.36 TeV the NSD results are compared with the CMS measurement and with PYTHIA tune D6T and PhoJet predictions (which are the lowest and the highest values for INEL events respectively). As for the systematic uncertainty shown as shaded areas in the figures, it can be observed that at 2.36 TeV the errors are higher (at 0.9 TeV 2.5% and 3.3% for INEL and NSD respectively and at 2.36 TeV 6.7% and 3.7% for INEL and NSD events). The systematic uncertainties are dominated by the uncertainty in the cross sections of diffractive processes and their kinematics. CMS did not include charged leptons in its definition of primary charged particles so the measured values are expected to be 1.5% lower that the ALICE ones.

In Fig. 2 the energy dependence of pseudorapidity density at central pseudorapidity is shown for ALICE and CMS data, as well as for existing measurements of ISR, Sp$\overline{\text{p}}$S, RHIC and Tevatron experiments. In Fig. 3 the relative increase of the $dN_{ch}/d\eta$ predicted by PhoJet and the three tunes of PYTHIA considered is shown for the INEL>0 class. These predictions are compared

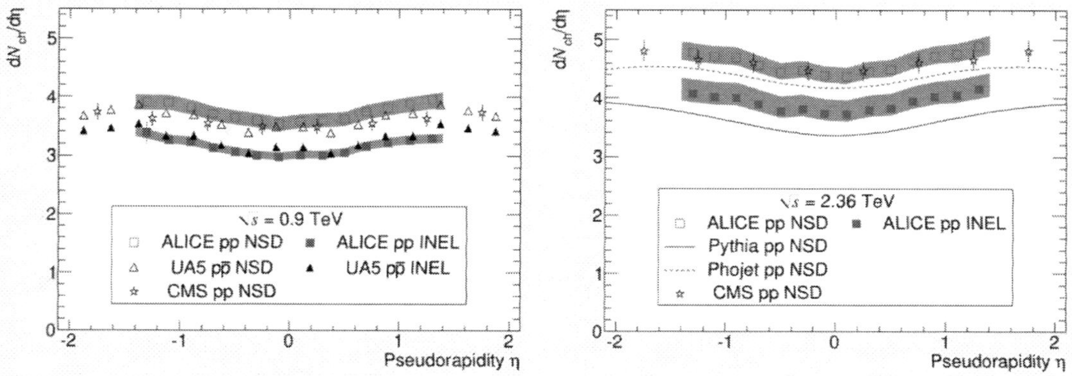

Figure 1. $dN_{ch}/d\eta$ distribution measured by ALICE and compared with other measurements and model predictions. The statistical error is not visible and the systematic uncertainty is represented by the shaded areas. The error bars for CMS data points are the sum in quadrature of systematic and statistical uncertainties. Figures reproduced from [6].

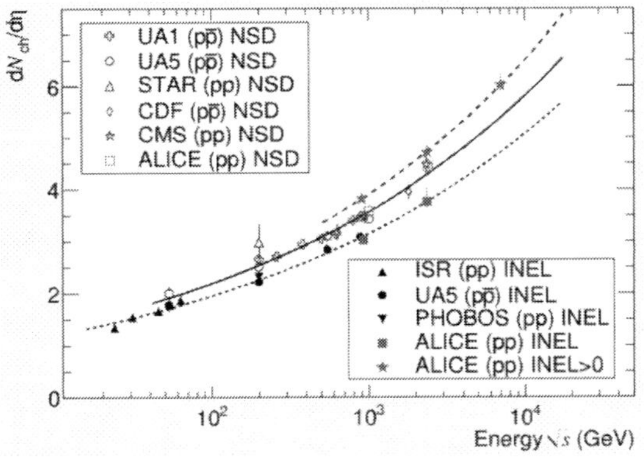

Figure 2. $dN_{ch}/d\eta$ at central pseudorapidity ($|\eta| < 0.5$) as a function of the centre-of-mass energy. For the INEL>0 class the pseudorapidity range considered is $|\eta| < 1$. The data for the three event classes have been fitted using a power law (curves). Figure reproduced from [7].

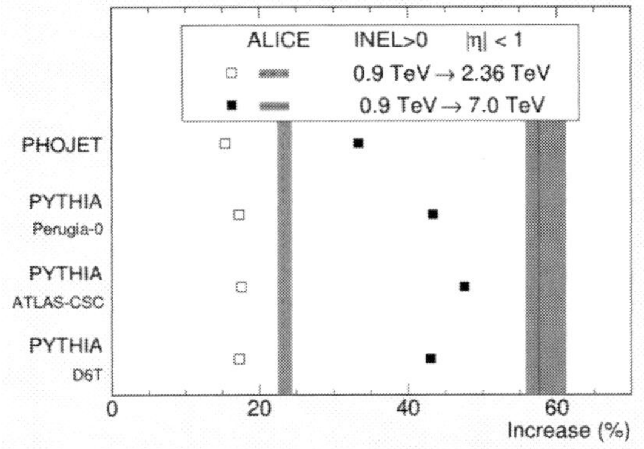

Figure 3. Predictions of various models for the relative increase of the $dN_{ch}/d\eta$ in the (INEL>0) event class at central pseudorapidity from 0.9 to 2.36 TeV (open symbols) and from 0.9 to 7 TeV (full symbols). The increase measured by ALICE in both cases is represented by the two lines where the width of the grey regions is the sum in quadrature of the statistical and systematic uncertainties. Figure reproduced from [7].

to the increase measured by ALICE that is $57.6\% \pm 0.4\%^{+3.6}_{-1.8}\%$ between the 0.9 and 7 TeV data.

6. Conclusions

ALICE has measured the pseudorapidity density distribution using data provided mainly by the SPD and collected in December 2009 and March-April 2010. The values are consistent with previous UA1 measurements and the new CMS experiment results. Model predictions at the highest energy significantly underestimate the value of $dN_{ch}/d\eta$ at central pseudorapidity, except for the PYTHIA tune ATLAS-CSC. Studying the dependence of the $dN_{ch}/d\eta$ on the centre-of-mass energy, it is observed that it increases much faster than predicted by the models considered.

References

[1] ALICE Collaboration Aamodt K *et al* 2008 JINST 3 S08002
[2] CMS Collaboration Khachatryan V *et al* arXiv:1006.0948v2
[3] UA5 Collaboration Alner G J *et al* 1986 Z. Phys. C 33 1
CMS Collaboration Khachatryan V *et al* arXiv:1005.3299v2
[4] Elia D *et al* 2009 ALICE-INT-2009-021
[5] ALICE Collaboration Aamodt K *et al* 2010 Eur. Phys. J. C 65 111
[6] ALICE Collaboration Aamodt K *et al* 2010 Eur. Phys. J. C 68 89
[7] ALICE Collaboration Aamodt K *et al* arXiv:1004.3514 accepted by Eur. Phys. J. C [doi:10.1140/epjc/s10052-010-1350-2]
[8] Engel R, Ranft J and Roesler S 1995 Phys. Rev. D 52 1459
[9] Albrow M G *et al* 2006 arXiv:hep-ph/0610012
[10] ATLAS Collaboration Moraes A 2009 ATL-COM-PHYS-2009-119
[11] Skands P Z 2009 arXiv:0905.3418[hep-ph]

Journal of Physics: Conference Series **270** (2011) 012003 doi:10.1088/1742-6596/270/1/012003

Early Measurements of Bulk Observables in CMS

Edward Wenger, for the CMS Collaboration
Massachusetts Institute of Technology, Cambridge, Massachusetts, United States
E-mail: edwenger@mit.edu

Abstract. With the large acceptance and excellent granularity of its tracking system, the CMS experiment is well-equipped to explore the bulk properties of the quark-gluon plasma with the first Pb+Pb collisions recorded at the LHC. Given the twenty-fold increase in collision energy compared to previous experiments, CMS measurements will be a powerful test of parton saturation, jet quenching, and hydrodynamic models, which have been successful in describing many features of particle production at RHIC. In preparation for early measurements of charged multiplicity, transverse-momentum spectra, and two-particle angular correlations, the corresponding measurements performed on p+p collision data are reviewed.

1. Introduction

The large increase in collision energy from existing measurements at RHIC ($\sqrt{s}_{_{NN}} = 0.2$ TeV) to future measurements at the LHC ($\sqrt{s}_{_{NN}} = 2.76$ TeV) ensures that the bulk properties of the produced QGP, accessible from the very first Pb+Pb collisions, will be exciting and perhaps unexpected. In these proceedings, three specific early Pb+Pb measurements are addressed: charged particle multiplicity, transverse momentum spectra, and two-particle correlations. For each, the corresponding first CMS measurements, performed on 10 - 200k p+p collisions at $\sqrt{s} = 0.9$, 2.36, and 7 TeV, are presented and the implications for Pb+Pb running discussed.

These bulk measurements of charged particle distributions are primarily reliant on the silicon pixel and strip detectors [1], which reside in a 3.8 T magnetic field and cover the pseudorapidity range $|\eta| < 2.4$ with full azimuthal acceptance. The performance of the CMS tracking system has been well-understood since the first 0.9 TeV collisions in 2009, showing excellent agreement with detector simulations [2].

A common minimum bias event selection was applied for the first CMS p+p measurements of bulk QCD properties (details in Ref. [3]). The detector readout was triggered by a single hit paddle in the Beam Scintillator Counters (BSC), coincident with the passing of a proton bunch indicated by the Beam Pick-up Timing for the eXperiments (BPTX) devices. To preferentially select non-single-diffractive (NSD) events, at least one forward calorimeter (HF) tower with $E > 3$ GeV in each of the forward and backward hemispheres was required. Further selections were made on the timing of BSC hits to reject beam backgrounds. Finally, a valid primary vertex reconstructed from pixel tracks was required.

2. Charged Particle Multiplicity

The $dN_{ch}/d\eta$ distributions for p+p collisions at $\sqrt{s} = 0.9$, 2.36, and 7 TeV [3, 4] were obtained using three methods based on the counting of (1) reconstructed clusters in the pixel barrel detector; (2) pixel tracklets composed of pairs of clusters in different pixel barrel layers; and (3) tracks reconstructed in the full tracker volume, combining the pixel and strip hits. The three different methods, which are sensitive to different systematic effects, are compared in Fig. 1 for the three collision energies. The results of the three methods were then weighted by their uncorrelated errors and averaged (not shown).

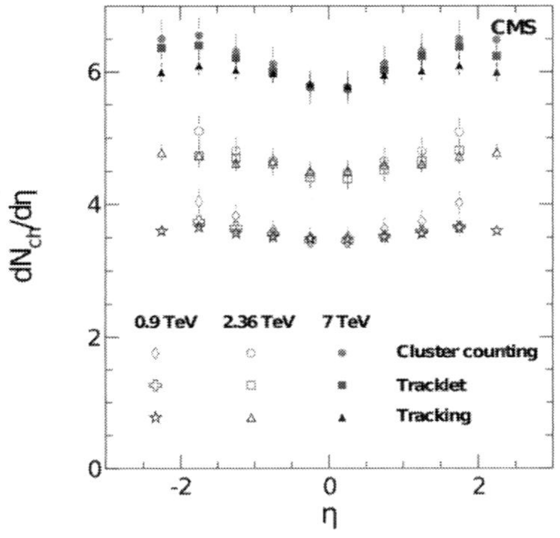

Figure 1. Reconstructed $dN_{ch}/d\eta$ distributions obtained from the cluster counting, tracklet and tracking methods, in p+p collisions at 0.9, 2.36 and 7 TeV. The error bars include systematic uncertainties, excluding those common to all the methods.

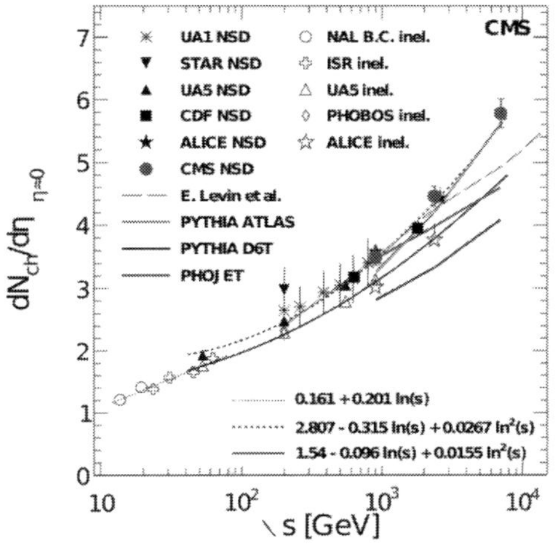

Figure 2. Charged-hadron pseudorapidity density in the central region as a function of centre-of-mass energy in p+p and p+$\bar{p}$ collisions, together with lower energy data. Lines represent various empirical fit parameterizations and predictions from PYTHIA tunes and a Pomeron model [5].

In Fig. 2, the collision energy dependence of the midrapidity yield is shown compared to lower energy data and the predictions of a number of models. The $dN_{ch}/d\eta$ values measured at the LHC energies are consistent with the trend, but the increase with energy is significantly steeper than predicted by most models. The steeper-than-expected energy dependence in p+p collisions has interesting implications for Pb+Pb collisions, where predictions of the multiplicity vary by as much as a factor of three [6]. Unless the ratio of Pb+Pb to p+p multiplicities is significantly smaller than the trend observed at lower energies, the new LHC multiplicity measurements suggest that the lowest Pb+Pb estimates ($dN_{ch}/d\eta \approx 1000$ at 5.5 TeV) are disfavored.

3. Transverse Momentum Spectra

In addition to the charged multiplicity measurements, the transverse momentum (p_T) spectra have been measured by CMS at all three LHC collision energies [3, 4]. The spectra shown in Fig. 3 are well-described up to $p_T = 6$ GeV/c by the empirical Tsallis fit functions, which simultaneously capture the exponential behavior at low p_T and the power-law behavior at high p_T. As expected, the spectra are significantly harder at the highest collision energy. Preliminary results from the high-statistics spectra analysis (p_T out to 140 GeV/c), which were mentioned at the conference, have since been made public [7]. The resulting spectra are consistent with expectations from leading-order QCD and x_T-scaling and show a significant power to discriminate between various MC tunes that vary among themselves by as much as a factor of two.

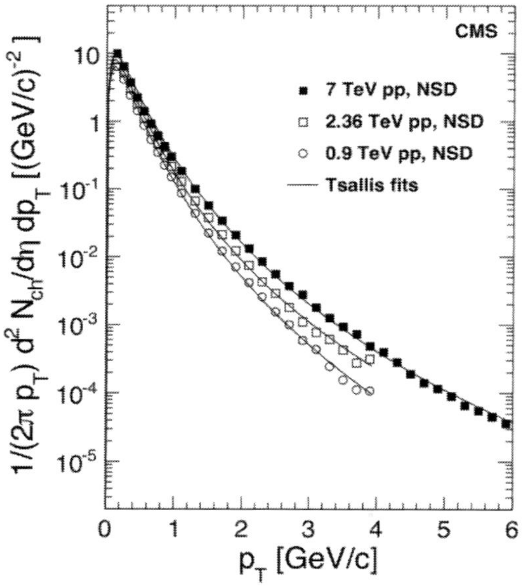

Figure 3. Measured yield of charged hadrons for $|\eta| < 2.4$ with systematic uncertainties (symbols) at $\sqrt{s} = 0.9$, 2.36 and 7 TeV. Lines represent empirical fits based on the Tsallis function described in Ref. [4].

The charged particle spectrum in p+p collisions is an important reference for future CMS measurements of high-p_T particle suppression in the dense QCD medium produced in Pb+Pb collisions. At RHIC, the sizable suppression of high-p_T spectra, compared to the superposition of a corresponding number of p+p collisions, was one of the first indications of strong final-state medium effects on particle production. It is similarly expected to be one of the first measurements at the LHC, where first-year statistics should allow measurements to 50-60 GeV/c. Again, expectations on the scale and the p_T dependence of the suppression vary by large factors [8]. By interpolating between the 0.9, 2.36, and 7 TeV measurements, the systematics of the 2.76 TeV p+p reference can be significantly better constrained, improving the discriminating power of the Pb+Pb measurement to various jet quenching scenarios.

4. Two-particle Correlations

Angular correlations between final-state particles are related to the hadronization process responsible for their production. From studying various MC models, it is apparent that QCD string fragmentation results in a roughly Gaussian $\Delta\eta$ correlation that is similar for all $\Delta\phi$ and peaked at $\Delta\eta = 0$. Conversely, dijets result in a characteristic 'near-side' peak centered at $\Delta\phi = \Delta\eta = 0$ and an 'away-side' structure around $\Delta\phi = \pi$ that is swept out over the observable range of $\Delta\eta$. The correlations between particles with $0.1 < p_T < 5.0$ GeV/c, measured by CMS in minimum bias collisions at all three p+p energies [9], are less 'jet-like' than PYTHIA (shown for 7 TeV in Figs. 4 and 5) but not as 'string-like' as, e.g., HERWIG++ (not shown).

As was done at lower energies [10], the strength and spatial extent of the $\Delta\eta$ correlation structure can be characterized in terms of an effective cluster size and width. The new CMS measurements are consistent with the trend of slowly increasing cluster size with collision energy, significantly exceeding what is seen in PYTHIA.

In heavy-ion collisions at RHIC, two-particle correlations have yielded a wealth of information on the physics of jet quenching, elliptic flow, and much more. The large $\Delta\eta$ acceptance of the CMS tracker will be particularly advantageous for exploring correlations in heavy ions. For example, based on the same two-particle correlation technique discussed above, it is possible to

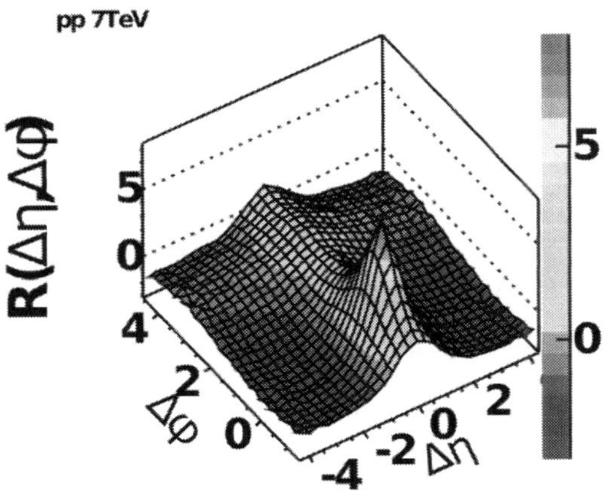

Figure 4. Two-particle angular correlation function in $\Delta\eta$ and $\Delta\phi$ for p+p collisions at 7 TeV.

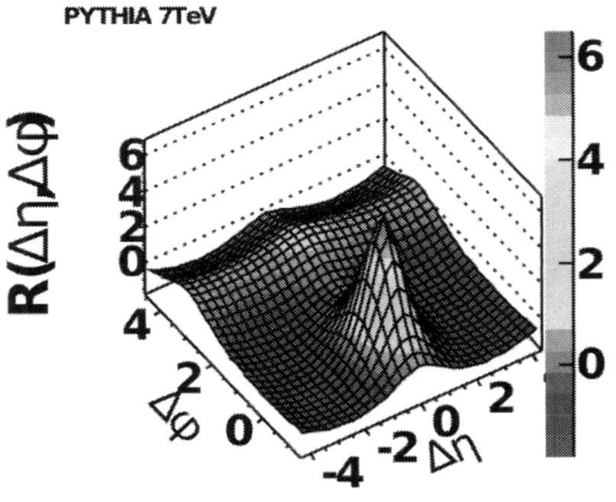

Figure 5. Two-particle angular correlation function in $\Delta\eta$ and $\Delta\phi$ for PYTHIA at 7 TeV.

extract the relative contributions of 'flow' and 'non-flow' in two-particle correlations [11], for many years the dominant uncertainty on elliptic flow.

5. Summary

CMS is an excellent detector for studying minimum bias QCD and heavy-ion physics. First p+p measurements have now been published at $\sqrt{s} = 0.9$, 2.36, and 7 TeV. These results will provide an important reference for future heavy-ion measurements of bulk observables, which will already be possible with only a few tens of thousands of events (or about fifteen minutes of colliding beams).

References

[1] **CMS** Collaboration, "The CMS experiment at the CERN LHC", *JINST* **3** (2008) S08004. doi:10.1088/1748-0221/3/08/S08004.
[2] **CMS** Collaboration, V. Khachatryan et al., "Tracking and Vertexing Results from First Collisions", *CMS PAS* **TRK-10-001** (2010).
[3] **CMS** Collaboration, V. Khachatryan et al., "Transverse momentum and pseudorapidity distributions of charged hadrons in pp collisions at sqrt(s) = 0.9 and 2.36 TeV", *JHEP* **02** (2010) 041, arXiv:1002.0621.
[4] **CMS** Collaboration, V. Khachatryan et al., "Transverse-momentum and pseudorapidity distributions of charged hadrons in pp collisions at sqrt(s) = 7 TeV", *Phys. Rev. Lett.* **105** (2010) 022002, arXiv:1005.3299.
[5] E. Gotsman, E. Levin, and U. Maor, "QCD motivated approach to soft interactions at high energy: inclusive production", *Phys. Rev.* **D81** (2010) 051501, arXiv:1001.5157.
[6] N. Armesto, "Predictions for the heavy-ion programme at the Large Hadron Collider", arXiv:0903.1330.
[7] **CMS** Collaboration, "Jet-triggered charged particle transverse momentum spectra in pp collisions at 7 TeV", *CMS PAS* **QCD-10-008** (2010).
[8] D. d'Enterria, "Jet quenching", arXiv:0902.2011.
[9] **CMS** Collaboration, V. Khachatryan et al., "Observation of Long-Range Near-Side Angular Correlations in Proton-Proton Collisions at the LHC", *JHEP* **09** (2010) 091, arXiv:1009.4122.
[10] **PHOBOS** Collaboration, B. Alver et al., "Cluster properties from two-particle angular correlations in p + p collisions at s**(1/2) = 200-GeV and 410-GeV", *Phys. Rev.* **C75** (2007) 054913, arXiv:0704.0966.
[11] **PHOBOS** Collaboration, B. Alver et al., "Non-flow correlations and elliptic flow fluctuations in gold-gold collisions at $\sqrt{s_{NN}} = 200$ GeV", *Phys. Rev.* **C81** (2010) 034915, arXiv:1002.0534.

π^0 measurement with ALICE electromagnetic calorimeters in p+p collisions at LHC

Renzhuo WAN [1,2] for the ALICE collaboration

[1]Key Laboratory of Quark & Lepton Physics (Huazhong Normal University), Ministry of Education, Wuhan, China

[2]Institut Pluridisciplinaire Hubert Curien, Université de Strasbourg, Strasbourg, France

E-mail: wanrz@iopp.ccnu.edu.cn

Abstract.
The electromagnetic calorimeters, PHOS (PHOton Spectrometer) and EMCAL (ElectroMagnetic CALorimeter) have been designed for photon detection and jet measurement in ALICE experiment at LHC. We discuss the strategy adopted for calorimetric π^0 measurements via the $\pi^0 \rightarrow 2\gamma$ channel in ALICE and present the two-photon invariant mass spectra obtained in p+p collisions at $\sqrt{s_{NN}} = 7\ TeV$.

1. Introduction

The goal of heavy-ion physics at ultra high energy is to explore the QCD phase diagram and probe the nature of QCD matter at extreme temperature and energy density. At LHC energies, the initial state temperature is expected to be well above the deconfinement transition temperature at which the Quark Gluon Plasma (QGP) is expected to be formed. Photon detection is one of the most promising methods to study the medium properties at the earliest phase of the collision thanks to electromagnetic calorimetry. Through the measurement of inclusive photon p_T spectra it is essential for the achievement of γ_{direct}, π^0, η and $\omega(782)$ distributions. All together, these measurements provide tests of perturbative QCD, investigation of m_T and x_T scaling and studies of the interaction of partons with the medium. In addition, γ and π^0 triggered jet measurements are sensitive tomographic probes of the hot and dense medium formed in heavy-ion collisions. Among this program of measurements, the light neutral meson π^0 measurement over a wide momentum range will play an important role in ALICE at the LHC.

2. ALICE electromagnetic calorimeters: PHOS and EMCAL

The ALICE experiment at LHC [1, 2] was specifically designed to cope with the high multiplicity environment in heavy-ion collisions at $\sqrt{s_{NN}} = 5.5\ TeV$. PHOS [3] is a high resolution photon calorimeter designed to measure photons over a large momentum range from 0.5 GeV/c to 100 GeV/c. It is placed at 4.6 m from the interaction point and covers $\Delta\phi = [220°, 320°]$ in azimuthal angle and $|\Delta y| < 0.12$ in rapidity. It is made of $PbWO_4$ [4] crystals with a fine granularity of $\Delta\phi \times \Delta\eta = 4.8 \cdot 10^{-3} \times 4.3 \cdot 10^{-3}$ to allow precise identification and measurement of electromagnetic showers in a high multiplicity environment. The PHOS beam test measured an energy resolution of $\frac{3\%}{\sqrt{E}} \oplus 1.1\%$. EMCAL [5], a lead-scintillator sampling electromagnetic calorimeter, covers a much larger acceptance over $\Delta\phi = [80°, 190°]$ and $|\Delta y| < 0.7$ with a granularity of $\Delta\phi \times \Delta y = 14 \cdot 10^{-3} \times 14 \cdot 10^{-3}$ and allows precise jet measurements in addition to those of electrons and photons. The beam test of EMCAL module prototype has shown that the energy resolution corresponds to $\frac{11\%}{\sqrt{E}} \oplus 1.7\%$ [6]. These two different calorimeter designs offer complementary methods to cross-check measurements and better estimate systematic uncertainties.

3. Event reconstruction

The raw data taken with the ALICE experiment are reconstructed by using the ALICE offline software package AliRoot [7]. They are then decoded to provide energy and timing information. An electromagnetic cluster is defined as the total energy deposition in adjacent cells fired by a same incident particle. However, several clusters may overlap in high multiplicity context induced by heavy-ion collisions. The details of the cluster finding and unfolding algorithm can be found in [3, 5]. After the clustering procedure, the position and energy of the incident particles can be obtained with additional parameters related to the shower topology. To minimize charged particle contamination of the photon spectra, track matching in the (x, z) plane of the calorimeters is employed, using information from the central tracking detectors. By combining information from particle time of flight, track position matching and the shower shape parameters, particle identification can be performed as studied in simulation [8]. We also employ the Bayesian method which assigns a PID weight on an event-by-event basis for each reconstructed particle in order to improve the particle identification efficiency in the pp and AA environments (see detailed study in [9]).

4. Strategy of π^0 spectra measurement

To measure the π^0 p_T spectra to hundred GeV/c, we employ a multi-component method based on invariant mass analysis at low p_T, shower shape analysis at intermediate p_T and an isolation cut method at high p_T to enhance the extraction of the π^0 signal and suppress the background for direct photon measurement.

4.1. Invariant mass analysis

π^0 are identified via the $\pi^0 \to 2\gamma$ channel with a branching ratio of 98.8%. We assume that the two decay photons come from the collision point. At low p_T, < 20 GeV/c for PHOS and < 8 GeV/c for EMCAL, the two decay photons are well separated and thus can be reconstructed using the invariant mass formula

$$m_{\gamma\gamma} = \sqrt{\vec{p_1}^2 + \vec{p_2}^2} = \sqrt{2E_1E_2(1 - \cos\phi_{12})} , \quad (1)$$

where $(\vec{p_1}, E_1)$ and $(\vec{p_2}, E_2)$ are the momentum and energy of the two photon candidates, ϕ_{12} is the opening angle in the laboratory frame. The uncorrelated photon pairs, which are not originating from a parent π^0, will produce a combinatorial background that we will subtract via an event mixing technique. The energy asymmetry cut $\alpha_{\gamma\gamma} = |E_1 - E_2|/(E_1 + E_2)$ and distance cut $\Delta R = \sqrt{(\phi_1 - \phi_2)^2 + (\eta_1 - \eta_2)^2}$ could be used to improve the signal-to-background ratio especially in heavy-ion collisions. For more technical details, refer to some of the RHIC publications [10, 11, 12].

4.2. Shower shape analysis

Going to higher p_T, > 20 GeV/c for PHOS and > 8 GeV/c for EMCAL, the two decay photons start to overlap. A cluster unfolding algorithm becomes thus essential to separate the two clusters. With such overlapping clusters, the shower eccentricity becomes measurable evolving

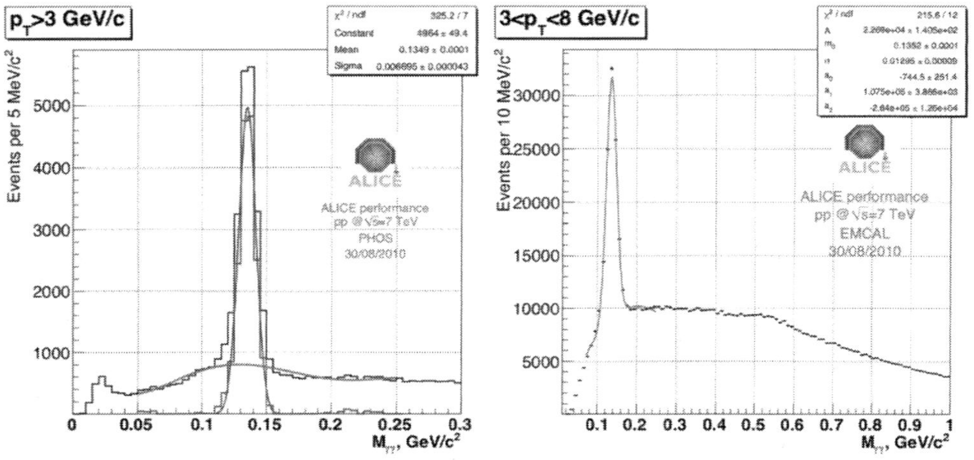

Figure 1. Two-photon invariant mass distributions with PHOS (left) at $p_T > 3.0\ GeV/c$ and EMCAL (right) with p_T bins of [3.0, 8.0] GeV/c from p+p collisions at $\sqrt{s_{NN}} = 7\ TeV$.

from a circular shape when a single photon presents to an elliptic shape when a π^0 is produced. Measurement of shower eccentricity enables us to extend π^0 PID capability to 50 GeV/c for PHOS and 30 GeV/c for EMCAL.

4.3. Isolation cut method

However, when going to higher p_T, the two photons from π^0 decay are completely overlapped to a single cluster and this becomes a challenge for π^0 identification. The isolation cut method [13, 14] originates from topology consideration induced by the initial hard processes at leading order of Compton scattering ($q(\overline{q}) + g \rightarrow \gamma + g$) and annihilation ($q + \overline{q} \rightarrow \gamma + g$) for direct photon identification. We thus can extract high p_T π^0 by subtracting the direct photon contribution from the inclusive photon spectrum. Although the next-to-leading order processes will spoil the isolation in both p+p and heavy-ion collisions, the starting point is to study the hadron and photon kinematics within a cone of the photon candidate $R = \sqrt{(\phi - \phi_\gamma)^2 + (\eta - \eta_\gamma)^2}$ and its surrounding. The combination of the isolation cut method and shower shape analysis simultaneously allows us to enhance the γ/π^0 separation at high p_T [15].

5. π^0 extraction

The early LHC data runs beginning in Dec. 2009 with p+p collisions at 900 GeV, 2.36 TeV and 7 TeV will be used as reference data for the future pA and AA collisions. During these runs, the ALICE central barrel detectors ITS, TPC, TOF, 7/18 of TRD, 3/5 of PHOS (without CPV) and 4/10 EMCAL were installed. After several iterations made for data reconstruction using the OCDB (Offline Conditions Data Base) where the calibration, alignment and reconstruction parameters are stored, the raw data have been converted to the final physics data to be analyzed by different physics working groups. The two-photon invariant mass spectra have been then constructed. In each p_T bin, the combinatorial background in real events is subtracted using the mass spectra from mixed events. The π^0 signals are extracted on a statistical basis within its nominal mass window.

From the data with pp collisions at 7 TeV, clear peaks around the π^0 nominal mass ($m_{\pi^0} = 0.135\ GeV/c^2$) have been observed up to 20 GeV/c. Fig. 1 shows the $m_{\gamma\gamma}$ spectra for PHOS at $p_T > 3.0\ GeV/c$ and EMCAL for p_T bins of [3.0, 8.0] GeV/c. The π^0 peaks are extracted by a fitting of Gaussian+Polynomial function. The mass resolutions of 4~7 MeV/c^2 for PHOS and 8~14 MeV/c^2 for EMCAL correspond to the expected detector performance. We are now at the stage of collecting additional data for analysis, while improving the calibration to a level allowing full use of the calorimeters. The goal of the first analysis is the invariant cross section measurement. It will also be essential to estimate the systematic uncertainties and compare the final spectra with the existing experimental data to cross-check.

6. Conclusion

The π^0 analysis plays an important role itself in QCD physics and also in photon and jet physics in the new TeV energy region at LHC. We have presented the π^0 analysis strategy in ALICE using PHOS and EMCAL calorimeters. These detectors joined the global data-taking and have collected $\sim 3 \times 10^8 M$ events for physics analysis through the end of August, 2010. The clear peaks around the π^0 nominal mass have been observed up to 20 GeV/c. The analysis is in a good progress moving toward the final physics results.

Acknowledgement

We thank the ALICE collaboration and the ALICE funding agencies (the same as the acknowledgement in [16]). The work is supported partly by the NSFC (10875051, 10635020, 10975061 and 11020101060), the Key Project of Chinese Ministry of Education (306022 and IRT0624), and the Program of Introducing Talents of Discipline to Universities of China: B08033 and CCNU09C01002 (Key Project) of China.

References

[1] ALICE Collaboration, *J. Phys. G: Nucl. Part. Phys.* **30** (2004)1517
[2] ALICE Collaboration, *J. Phys. G: Nucl. Part. Phys.* **32** (2006)1295
[3] ALICE Collaboration, *CERN-LHCC-99-4*
[4] Aleksandrov D. V. et al., *Nucl. Instrum. Meth. A* **550** (2005) 169
[5] ALICE Collaboration, *CERN-LHCC-2008-014*
[6] Allen J. et al., *arXiv: 0912.2005*
[7] *http://aliweb.cern.ch/Offline/AliRoot/Releases.html*
[8] Conesa G. et al., *ALICE-INT-2005-053*
[9] Conesa G. et al., *ALICE-INT-2005-016*
[10] Adler S. S. et al. PHENIX collaboration, *Phys. Rev. C* **76** (2007) 034904
[11] Abelev B. I. et al. STAR collaboration, *Phys. Rev. C* **80** (2009) 044905
[12] Abelev B. I., et al. STAR collaboration, *arXiv: hep-ex/0912.3838v2*
[13] Aaltonen T. et al. CDF collaboration, *Phys. Rev. D* **80** (2009) 111106
[14] Abazov V. M. et al. D0 collaboration, *Phys. Lett. B* **639** (2006) 151
[15] Conesa G., *Nucl. Instrum. Meth. A* **580** (2007) 1446
[16] K. Aamodt et. al ALICE collaboration, *Eur. Phys. J. C* **68** (2010) 89

Baryon number transport at LHC energies with the ALICE experiment

Panos Christakoglou[1,2] **for the ALICE Collaboration**
[1]NIKHEF, Science Park 105, 1098 XG Amsterdam, The Netherlands
[2]Utrecht University, Institute of Subatomic Physics, PO Box 80000, 3508 TA Utrecht, The Netherlands

E-mail: Panos.Christakoglou@cern.ch

Abstract. This article presents the ratio of the yields of antiprotons to protons in pp collisions as measured by the ALICE experiment at $\sqrt{s} = 0.9$ and 7 TeV. The ratio is measured to be $R_{|y|<0.5} = 0.957 \pm 0.006(stat.) \pm 0.014(syst.)$ at 0.9 TeV and $R_{|y|<0.5} = 0.991 \pm 0.005(stat.) \pm 0.014(syst.)$ at 7 TeV independent of both rapidity and transverse momentum. The results are compared with theoretical predictions.

1. Introduction

The ratios of particle yields in hadronic interactions are important indicators of the collision dynamics. The ratio of the yields of antiprotons to protons, is of particular interest since it provides information about the transport of the baryon number (BN). The latter has been debated theoretically for some time [1, 2, 3, 4, 5, 6]. Experimental data on soft hadronic processes are well understood with the Quark-Gluon String Model (QGSM) [2], within which the baryon is considered to be a bound diquark–quark state. Within this framework the mechanism responsible for baryon-number transport is the break-up of this configuration [2]. The diquark in general retains a large fraction of the proton momentum and therefore stays close to beam rapidity, typically within one or two units. Additional processes have been proposed to transport the baryon number over larger distances in rapidity, in particular via purely gluonic exchanges, where the proton breaks up into three quarks. In this case, the baryon number resides in a non-perturbative configuration of gluon fields, the so-called "baryon string junction", which connects the valence quarks [1]. In this picture, the baryon-number transport is suppressed exponentially with the rapidity interval Δy, proportional to $\exp[(\alpha_J - 1)\,\Delta y]$, where α_J is the intercept of the junction. In one other approach [3], based on a modified version of the string junction exchange model described above, the junction trajectory is assumed to have unit intercept and thus lead to an energy independent uniform rapidity component.

The $\overline{\mathrm{p}}/\mathrm{p}$ ratio has been reported by several experiments [7, 8, 9] at lower energies. The LHC opens the possibility to investigate the baryon transport over very large rapidity intervals by measuring the antiproton-to-proton production ratio at midrapidity, $R = N_{\overline{\mathrm{p}}}/N_{\mathrm{p}}$. In this article, we describe the measurement of the $\overline{\mathrm{p}}/\mathrm{p}$ ratio at midrapidity in non-diffractive pp collisions at center-of-mass energies $\sqrt{s} = 0.9$ TeV and 7 TeV ($\Delta y \approx 6.9$–8.9), with the ALICE experiment at the LHC [10].

2. Data Analysis

Data from 2.8 ($\sqrt{s} = 0.9$ TeV) and 4.2 ($\sqrt{s} = 7$ TeV) million pp collisions, recorded during the first LHC runs (December 2009 and March–April 2010) were used for this analysis. The events were recorded with both magnetic field polarities for each energy. The trigger required a hit in one of the VZERO counters or in the SPD detector [11], in coincidence with the signals from two beam pick-up counters, one on each side of the interaction region, indicating the presence of passing bunches. Measurements of momentum and particle identification are performed using information from the TPC detector, which measures the ionization in the TPC gas and the particle trajectory. The phase space of the analysis was restricted to the rapidity and momentum range of $|y| < 0.5$ and $0.45 < p < 1.05$ GeV/c, respectively.

Corrections that do not cancel out in the ratio were applied. Corrections for absorption in detector material were extracted using a complete Monte Carlo model, simulating the detector response with GEANT3. In addition, a detailed study using GEANT3 and FLUKA was performed to extract the scaling factors for the p($\overline{\text{p}}$)–A inelastic cross-sections. Corrections for background (i.e. protons originating from the interaction of particles with the material) and feed–down (i.e. weak decay of hyperons) were estimated by parameterizing the distribution of the distance of closest approach (dca) of the tracks to the primary vertex from data. Finally, the effect of the differences in the efficiencies of the analysis cuts for the different charges, resulting from the corresponding differences between p–A and $\overline{\text{p}}$–A elastic cross-sections, was taken also into account. For more details about the corrections and the estimation of the systematic uncertainty, see [11, 12].

3. Results

The final, feed-down corrected $\overline{\text{p}}$/p ratio R integrated within our rapidity and p_t acceptance rises from $R_{|y|<0.5} = 0.957 \pm 0.006(stat.) \pm 0.014(syst.)$ at $\sqrt{s} = 0.9$ TeV to $R_{|y|<0.5} = 0.991 \pm 0.005(stat.) \pm 0.014(syst.)$ at $\sqrt{s} = 7$ TeV. Within statistical errors, the measured ratio R shows no dependence on transverse momentum (Fig. 1), rapidity or multiplicity [12]. The ratio is also independent of momentum and rapidity in our acceptance for all theoretical models considered, with the exception of HIJING/B, which predicts a decrease with increasing transverse momentum for the lower energy.

The data are compared with various model predictions for pp collisions [5, 6, 13] in Tab. 1 (integrated values) and Fig. 1. The QGSM model does not predict the p_t dependence and is therefore not included in Fig. 1. For both energies, two of the PYTHIA tunes [13] (ATLAS-CSC and Perugia-0) as well as the version of QGSM with the value of the string junction intercept $\alpha_J = 0.5$ [5] describe the experimental values well. On the other hand QGSM without string junctions ($\epsilon = 0$, ϵ is a parameter proportional to the probability of the string-junction exchange) is slightly above the data. HIJING/B [6] underestimates the experimental results, in particular at the lower LHC energy. Also, QGSM with a value of the junction intercept $\alpha_J = 0.9$ [5] predicts a smaller ratio, as does the Perugia-SOFT tune of PYTHIA, which also includes enhanced baryon transfer.

Figure 3 shows a compilation of central rapidity measurements of the ratio R in pp collisions as a function of center-of-mass energy (upper axis) and the rapidity interval Δy (lower axis). The ALICE measurements correspond to $\Delta y = 6.87$ and $\Delta y = 8.92$ for the two energies, with the lower energy data points taken from [7, 8, 9]. The lower ALICE point indicates that there is still a small but significant excess of protons over antiprotons. On the other hand, the ratio at $\sqrt{s} = 7$ TeV is consistent with unity, which sets a stringent limit on the amount of baryon transport over 9 units in rapidity.

The Δy dependence of the ratio R can be parameterized with the help of different diagrams describing the particle production of both p and $\overline{\text{p}}$ based on the Regge model. The baryon pair production at very high energy is governed by Pomeron exchange and baryon transport by string-junction exchange [4]. Following this formulation the p/$\overline{\text{p}}$ ratio can be described by the simple form $1/R = 1 + C \exp[(\alpha_J - \alpha_P)\Delta y]$ [11, 12]. The value for the Pomeron intercept is chosen to be $\alpha_P = 1.2$ in accordance with the energy dependence of the rapidity density [14]

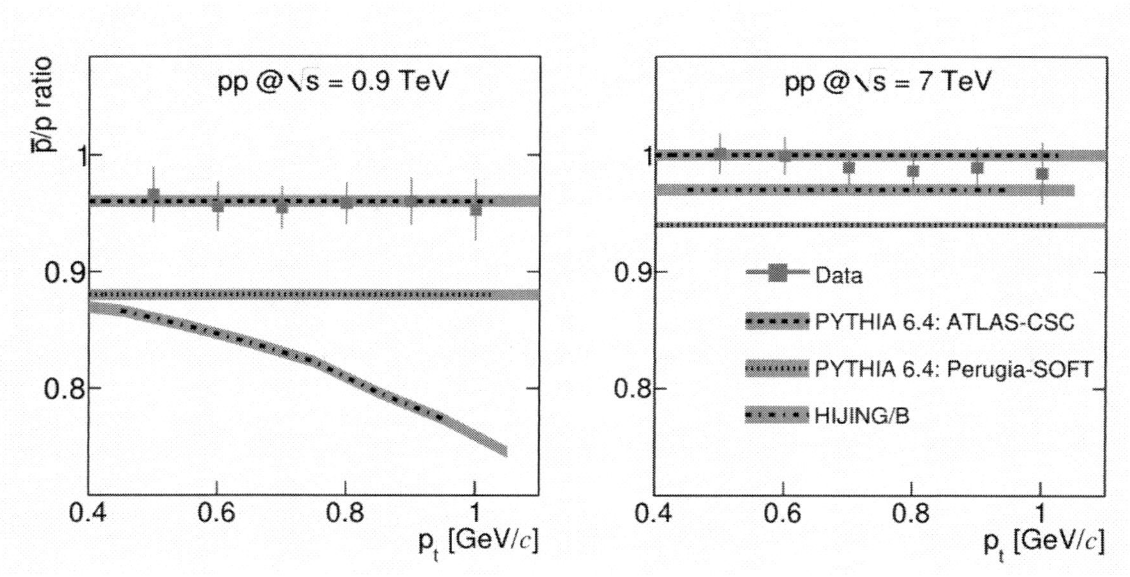

Figure 1. The p_t dependence of the $\overline{\rm p}$/p ratio integrated over $|y| < 0.5$ for pp collisions at $\sqrt{s} = 0.9$ TeV (left) and $\sqrt{s} = 7$ TeV (right).

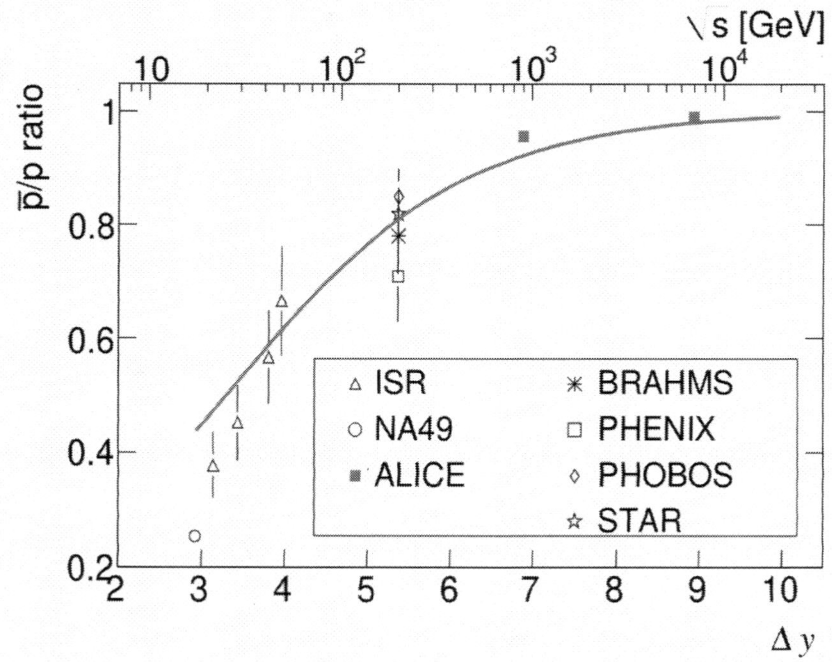

Figure 2. Central rapidity $\overline{\rm p}$/p ratio as a function of the rapidity interval Δy (lower axis) and center-of-mass energy (upper axis). Error bars correspond to the quadratic sum of statistical and systematic uncertainties for the RHIC and LHC measurements and to statistical errors otherwise.

Table 1. The measured central rapidity $\overline{\mathrm{p}}/\mathrm{p}$ ratio compared to the predictions of different models (the statistical uncertainties in the models are less than 0.005). The quoted errors for the ALICE points are the quadratic sum of statistical and systematic uncertainties.

	Energy [TeV]	0.9	7
ALICE		0.957 ± 0.015	0.991 ± 0.015
	ATLAS-CSC Tune (306)	0.96	1.0
PYTHIA	Perugia-0 Tune (320)	0.95	1.0
	Perugia-SOFT Tune (322)	0.88	0.94
	$\epsilon = 0$	0.98	1.0
QGSM	$\epsilon = 0.076,\ \alpha_J = 0.5$	0.96	0.99
	$\epsilon = 0.024,\ \alpha_J = 0.9$	0.89	0.95
HIJING/B		0.83	0.97

and $\alpha_J = 0.5$ (intercept of the Reggeon). The parameter C, which determines the relative contributions of the two diagrams, is adjusted to the measurements from ISR, RHIC, and LHC. The fit, shown in Fig. 3, gives a reasonable description of the data with only one free parameter (C), except at lower energies, where contributions of other diagrams (exchange of two junctions at both vertices) cannot be neglected [4]. The contribution of a second string junction diagram with a larger intercept [3], i.e., $1/R = 1 + C\exp[(\alpha_J - \alpha_P)\Delta y] + C'\exp[(\alpha_{J'} - \alpha_P)\Delta y]$ with $\alpha_{J'} = 1$, is compatible with zero ($C \approx 10$, $C' \approx -0.1 \pm 0.1$).

4. Summary

In summary, we have measured the ratio of antiproton to proton production at $\sqrt{s} = 0.9$ and $\sqrt{s} = 7$ TeV. The obtained values are $R_{|y|<0.5} = 0.957 \pm 0.006(stat.) \pm 0.014(syst.)$ at 0.9 and $R_{|y|<0.5} = 0.991 \pm 0.005(stat.) \pm 0.014(syst.)$ at 7 TeV. The $\overline{\mathrm{p}}/\mathrm{p}$ ratio is independent of both rapidity and transverse momentum and the results are consistent with standard models of baryon-number transport over very large rapidity intervals in pp collisions.

References

[1] G.C. Rossi and G. Veneziano, Nucl. Phys. **B123**, (1977) 507.
[2] A. Capella *et al.* Phys. Rep. **236**, 225 (1994); A.B. Kaidalov and K.A. Ter-Martirosyan, Sov. J. Nucl. Phys. **39**, 1545 (1984).
[3] B.Z. Kopeliovich, Sov. J. Nucl. Phys. **45**, 1078 (1987).
[4] D. Kharzeev, Phys. Lett. **B378**, 238 (1996).
[5] C. Merino *et al.* Eur.Phys.J. **C54** 577 (2008); C. Merino, M.M. Ryzhinskiy, Yu.M. Shabelski, arXiv:0906.2659.
[6] S. E. Vance and M. Gyulassy, Phys. Rev. Lett. **83**, 1735 (1999).
[7] T. Anticic *et al.* (NA49 Collaboration), Eur. Phys. J. **C65**, 9 (2010).
[8] A.M. Rossi *et al.* Nucl. Phys. **B84**, 269 (1975); M. Aguilar-Benitez *et al.* Z. Phys. **C50**, 405 (1991).
[9] B.I. Abelev *et al.* (STAR Collaboration), Phys. Rev. **C79**, 034909 (2009); I.G. Bearden *et al.* (BRAHMS Collaboration), Phys. Lett. **B607**, 42 (2005); B.B. Back *et al.* (PHOBOS Collaboration), Phys. Rev. **C71**, 021901 (2005); S.S. Adler *et al.* (PHENIX Collaboration), Phys. Rev. **C69**, 034909 (2004).
[10] K. Aamodt *et al.* (ALICE Collaboration), JINST **3**, S08002 (2008).
[11] K. Aamodt *et al.* (ALICE Collaboration), Phys. Rev. Lett. **105**, 072002 (2010), arXiv:1006.5432.
[12] P. Christakoglou, ALICE Internal Note ALICE-INT-2010-006, 2010.
[13] T. Sjostrand, P. Skands, Eur. Phys. J. **C39**, 129 (2005); P. Skands, arXiv:1005.3457 [hep-ph] (2010), Perugia-0 (320) and Perugia-SOFT (322) tunes; A. Moraes (ATLAS Collaboration), ATLAS Note ATL-COM-PHYS-2009-119, 2009.
[14] A.B. Kaidalov, L.A. Ponomarev and K.A. Ter-Martirosyan, Sov. J. Nucl. Phys., **44**, 468 (1986).

Journal of Physics: Conference Series **270** (2011) 012006 doi:10.1088/1742-6596/270/1/012006

Measurement of identified charged hadron spectra with ALICE using the ITS in proton-proton collisions at the LHC

Marek Chojnacki for the ALICE Collaboration
Utrecht University Princetonplein 5 3584CC Utrecht the Netherlands
E-mail: Marek.Chojnacki@cern.ch

Abstract. The method used by ALICE to measure identified charged hadron spectra in the Inner Tracking System (ITS) is presented. The analysis results are compared with measurements using the Time Projection Chamber and Time-of-Flight detector at larger transverse momentum.

1. Introduction

The measurement of identified charged hadron spectra is among the first measurements that are performed with the ALICE experiment in proton-proton collisions at $\sqrt{s} = 900$ GeV collected in 2009. The results provide basic insights into particle production processes. They also give the possibility to test and tune commonly used event generators at LHC energies. In addition, they are the baseline for measurements in heavy-ion collisions.

The ALICE experiment features multiple particle identification systems, including: the Time Projection Chamber (TPC), the Inner Tracking System (ITS), a dedicated time-of-flight system (TOF) and a Ring-imaging Cherenkov detector. With these detector systems, ALICE can identify charged hadrons in the momentum range from 100 MeV/c up to 5 GeV/c. More details about the ALICE experiment can be found in [1, 2, 3].

The ITS is the detector of ALICE placed closest to the interaction point. It is built out of three subdetectors. Each subdetector contains two concentric silicon layers. The innermost part of the ITS is the SPD (Silicon Pixel Detector), the middle part is the SDD (Silicon Drift Detector) and the outer one is the SSD (Silicon Strip Detector). The ITS provides high resolution tracking for the identification of secondary vertices of heavy flavour weak decays and improved momentum resolution at high transverse momentum (p_T). In addition, the ITS is used for standalone tracking to measure tracks at very low p_T which do not reach the TPC.

The measurement of the identified charged hadron spectra in the ITS is based on particle energy loss in 4 (SSD and SDD) out of 6 silicon layers. This measurement is performed for global tracks (TPC+ITS) and for ITS standalone tracks. Global tracks are reconstructed using information from the TPC, the ITS and other ALICE tracking detectors (e.g TOF). Standalone tracks are reconstructed using only information from the ITS (6 layers). The p_T spectra obtained from both methods are compared to p_T spectra obtained from the TPC and the TOF.

Before the p+p data taking in 2009, the ITS was aligned using the cosmic ray data collected by ALICE in 2008. The corresponding procedure is described in [4]. The status of the ITS energy loss calibration for 2009 data is shown in Figures 1 and 2. The SDD was calibrated at the module level in 2009. For the SSD, which has more modules than the SDD, the amount of collected data was not sufficient to perform the calibration at the module level. As can be seen from Figure 2, the module to module fluctuations of the most propable value of the charge deposition are at the per cent level, which is small compared to the intrinsic resolution.

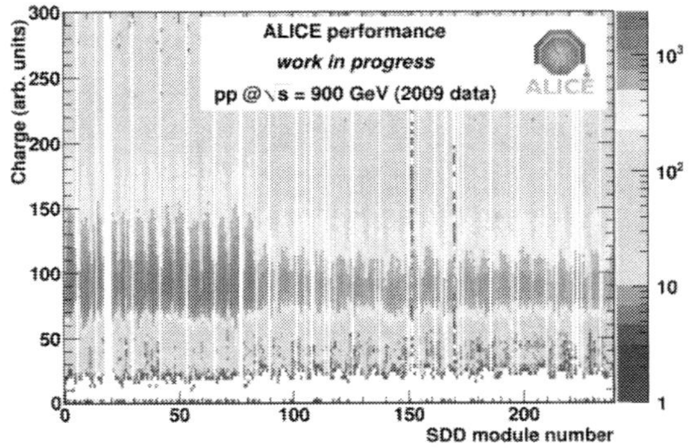

Figure 1. The SDD calibration in 2009 data. Distribution of measured cluster charge per unit path length in each module of the SDD.

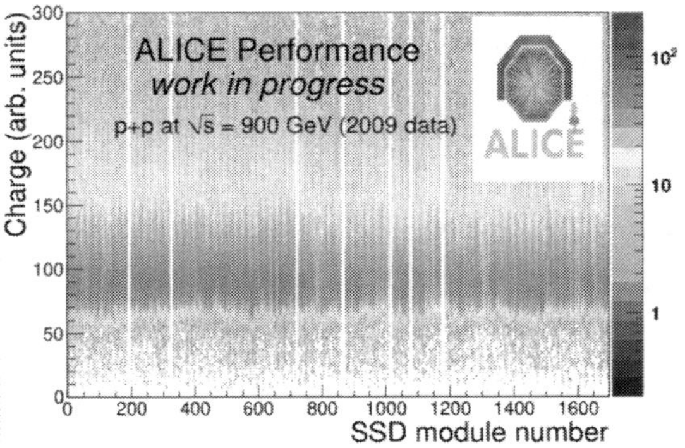

Figure 2. The SSD calibration in 2009 data. Distribution of measured cluster charge per unit path length in each module of the SSD.

2. Particle identification in the ITS

Only tracks with 3 or 4 measured points in the SDD and SSD are used in the analysis. The charge deposit measured in each layer is normalised to the energy loss per 300 μm (thickness of the SDD and SSD modules) of silicon using the local track angles as reconstructed by the tracking algorithm. A truncated mean is calculated using the two lowest normalised charge signals. Figure 3 shows the measured energy loss signal in the ITS (value of the truncated mean) as function of the total momentum for global tracks, compared to the parametrized response for pions, kaons and protons (black lines). This parametrization is based on a formula from [5], with an additional polynomial correction for the low momentum part.

For each track fulfilling the track quality requirements, the rapidity is calculated using three different mass hypotheses: pion, kaon and proton. Tracks with $|y| < 0.5$ are selected. For each mass hypothesis in all p_t bins, a histogram is filled with the difference between the logarithm of the measured energy loss signal in the ITS and the expected value calculated using the parametrization $\ln[\mathrm{d}E/\mathrm{d}x]_{meas} - \ln[\mathrm{d}E/\mathrm{d}x]_{calc}$. This procedure gives a series of three histograms for each charged particle in all p_t bins. Each histogram has one, two or three maxima depending on the momentum. Figure 4 shows an example of these distributions for the kaon mass hypothesis. The area below the peak centered at zero is the raw yield of particles from the mass hypothesis in the given p_t bin. The raw yield is determined by fitting the histograms with the function show in Figure 4: a sum of three asymmetric Gaussians for the ITS standalone analysis and a sum of a Gaussian with an exponential tail and the tail of the peak at lower $\mathrm{d}E/\mathrm{d}x$ for the TPC+ITS analysis. The raw yield is then corrected for the contamination from secondary particles (products of weak decays and particles coming from interactions with the detector material) and the efficiency. Pions are also corrected for the contamination of electrons and muons which can not be distinguished from pions based on the measured energy loss signal in the ITS. All corrections are calculated from the simulation using the PYTHIA 6.4 event generator [6] (tune D6T) [7] and GEANT3 [8]. The contamination corrections will be calculated

Journal of Physics: Conference Series **270** (2011) 012006 doi:10.1088/1742-6596/270/1/012006

from the data for the final result.
Figure 5 shows the efficiencies for both analyses. The standalone analysis has a larger efficiency at low p_t, where tracks do not reach the TPC. In addition, the efficiency for kaons is larger over the full p_t range, because the combined tracking rejects a large fraction of the kaons that decay inside the TPC.
The final result is compared to spectra measured in the TPC and the TOF. This is shown in Figure 6.

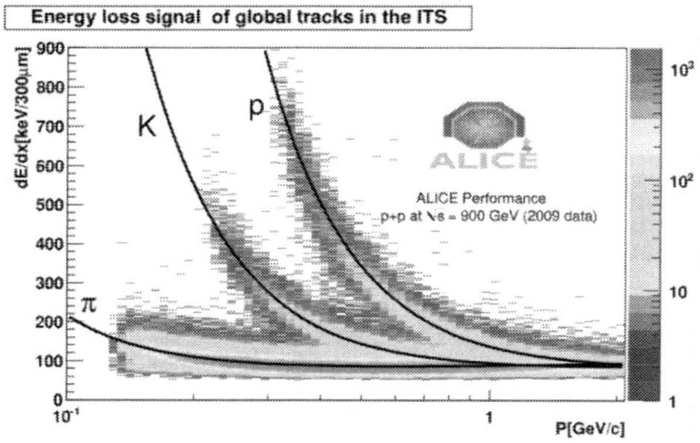

Figure 3. Energy loss signal as a function of momentum for global tracks. The solid lines show the expected signal for pions, kaons and protons.

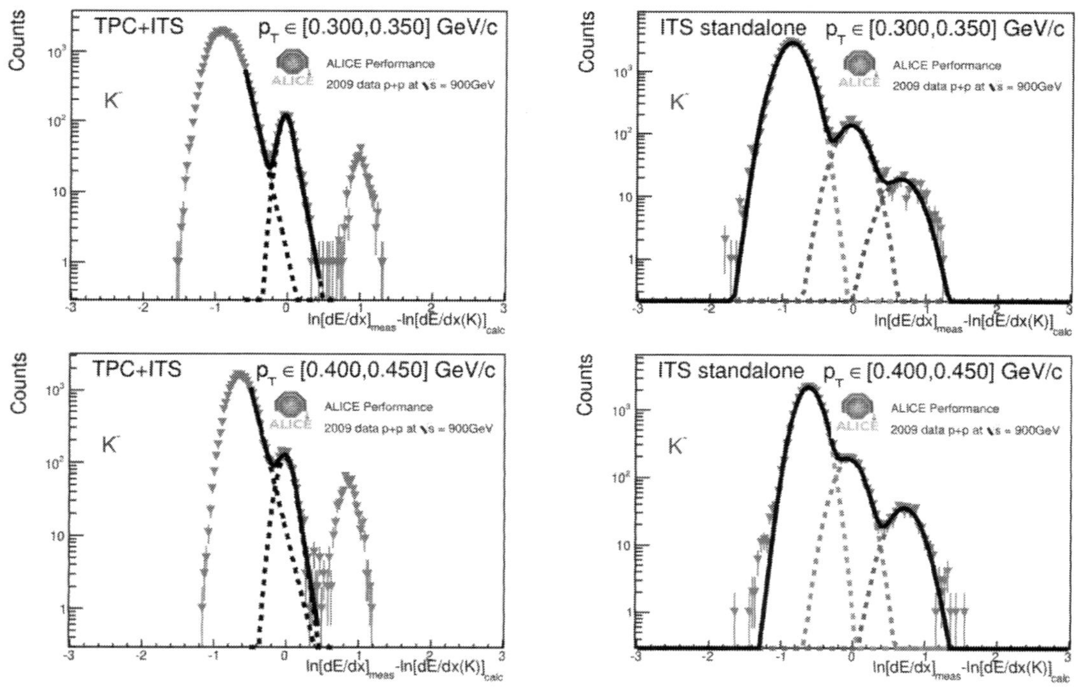

Figure 4. Distributions of the difference between the logarithm of the measured energy loss signal in the ITS and the expected value for Kaon, for a global (TPC+ITS, left) and a standalone (right) track sample. The solid curves show the fit used to determine the raw yield of Kaons.

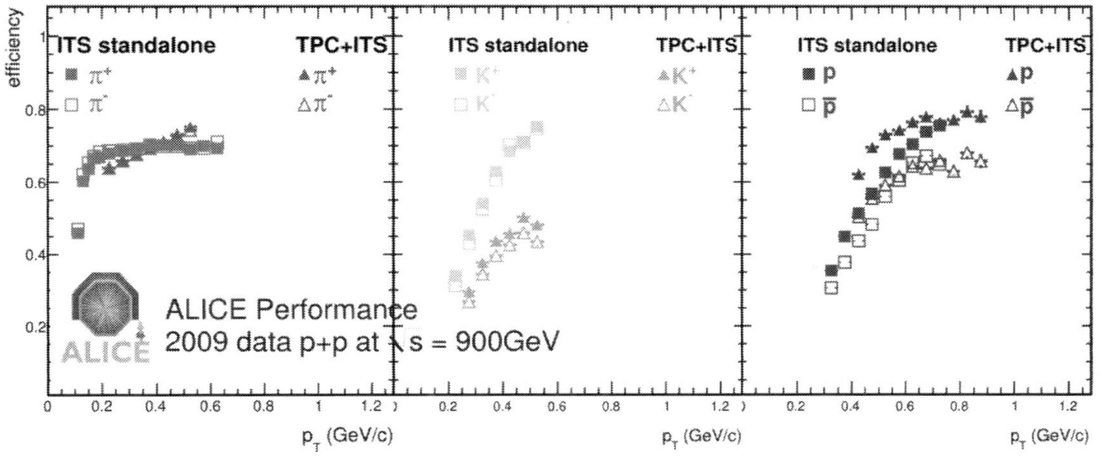

Figure 5. Efficiency as function of p_t of global (TPC+ITS) and standalone tracks for each particle species.

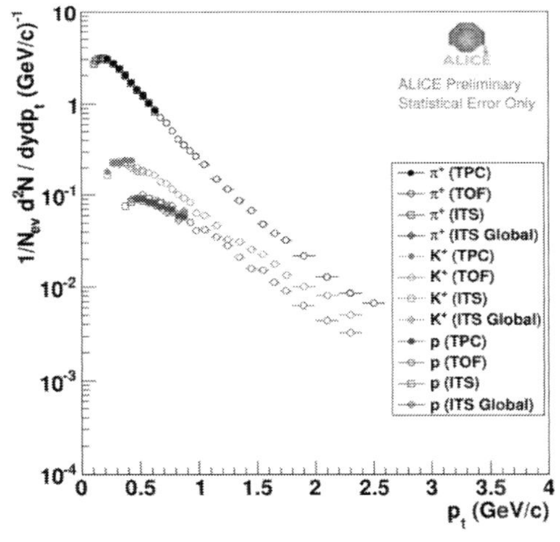

Figure 6. Transverse momentum spectra of identified positive hardons in p+p collisions at 900 GeV measured in different ALICE detectors. Two ITS spectra are shown: a standalone (ITS) and a global (ITS Global)

3. Summary

The identified charged hadron spectra measured in the ITS are in a good agreement with the measurements performed in other ALICE detectors.

References

[1] B. Alessandro *et al.* [ALICE Collaboration], *J. Phys.* G **32** (2006) 1295
[2] F. Carminati *et al.* [ALICE Collaboration], *J. Phys.* G **30** (2004) 1517
[3] K. Aamodt *et al.* [ALICE Collaboration], *JINST* **3** (2008) S08002
[4] K. Aamodt *et al.* [ALICE Collaboration], *JINST* **5** (2010) P03003
[5] B.B. Back *et al.* [PHOBOS Collaboration], *Phys. Rev.* C **75** 2007 024910
[6] T.Sjostrand, S.Mrenna, P.Z Skands, *JHEP* **05**, 026 2006, hep-ph/0603175
[7] M.G Albrow *et al.*, (Tev4LHC QCD Working group), arXiv:hep-ph/0610012
[8] R.Brun, F. Carminati, S. Giani CERN-W5013

Analysis of ϕ resonance in pp collisions at 900 GeV with the ALICE detector

A Pulvirenti[1,2,*], for the ALICE Collaboration
[1] INFN, Sezione di Catania
[2] Università di Catania
[3] Centro Studi e Ricerche e Museo Storico della Fisica "Enrico Fermi"
[*] Corresponding author (email: alberto.pulvirenti@ct.infn.it)

Abstract. ALICE is the LHC experiment mainly dedicated to the study of hot and high energy density nuclear matter created in heavy-ion collisions. ALICE has also developed a detailed proton-proton physics programme, in order to exploit its capabilities to investigate the novel energy regime made available by LHC, which turns out to be quite interesting in itself besides being important as a baseline for the heavy-ion data. Results will be presented of the ϕ resonance measurements done with the first data sets taken at LHC between 2009 and 2010, in pp collisions at 900 GeV. ϕ transverse momentum spectra and global yields will be shown.

1. Introduction
The study of identified particle production in pp collisions provides a fundamental baseline for properly tuning the QCD-inspired models in order to make predictions at higher energies. In the low-p_T region this helps especially in understanding the nature of the soft part of the underlying event. The ALICE detector [1], thanks to its low operating magnetic field (0.5 T) and to its excellent particle identification (PID) capabilities, is well suited for the study of low-p_T particle production at mid-rapidity. The ϕ resonance is of great interest in this respect and has already been widely studied at lower energies (see for example [2]-[5]).

Studying the hadronic resonances in pp collisions provides also a baseline for a better understanding of heavy-ion collisions, where they are useful probes of the hot and dense matter created there [6].

A measurement of the ϕ production has been carried out with the first LHC commissioning runs (November-December 2009), consisting in almost 250k events at the center of mass energy of 900 GeV. dN/dy at mid-rapidity and the average transverse momentum were measured. Section 2 will briefly illustrate the most important aspects of the ALICE detector exploited for this study. Section 3 will present the analysis details and section 4 will show the results obtained. Conclusions will be given in section 5.

2. Experimental setup
ALICE [1] is the LHC experiment most specifically devoted to the study of heavy-ion collisions. It is designed to guarantee a very complete reconstruction and identification of the particles produced in those collisions (for details, refer to [1]).

The contribution of several detectors was used in this analysis. Charged particles were tracked by the Time Projection Chamber (TPC) and the Inner Tracking System (ITS). The Silicon

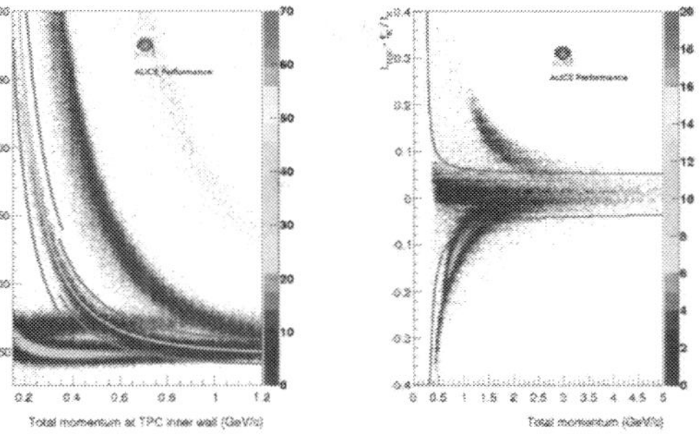

Figure 1. (Left) TPC signal vs. total momentum at the inner TPC wall. The central continuous curve is the Bethe-Bloch estimation, and the two broken curves define the momentum-dependent fiducial compatibility window. (Right) The relative difference between TOF measured time and Kaon integrated time, vs. total momentum at the vertex; the curves define the fiducial window.

Pixel Detectors (SPD) in the two innermost ITS layers also contributed to the primary vertex reconstruction. The TPC returns a dE/dx measurement with a resolution $\sigma_{TPC} = 6\%$, useful for particle identification (PID). At larger momenta (above $\sim$0.7 GeV/c) PID is also supported by the Time of Flight detector (TOF), which reached a time resolution of almost 170 ps. At the trigger level the two VZERO counters helped in rejecting most of the beam-gas interactions.

The first LHC data-taking period consisted essentially in low luminosity (10^9 p/bunch) pp beams at $\sqrt{s}$ = 900 GeV. The online minimum bias trigger was based on the signals from VZERO counters, the Fast-OR trigger coming from the SPD (see [1] for details) and from the LHC beam pickup signals.

For a better quality of the analysis, events with longitudinal primary vertex position $Z_V \geq 10$ cm were rejected, in order to avoid biases due to the detector acceptance. After the whole set of online and offline selections, the analyzed sample consisted in $\sim$250k events.

3. Analysis details

3.1. Track selection

The ϕ resonance was reconstructed through its K^+K^- decay channel (branching ratio = 49.2%).

Tracks used for analysis were required to have been reconstructed by both TPC and ITS, with at least 80 clusters in the TPC and a total χ^2 (normalized to the number of TPC points) smaller than 4, in order to ensure a reasonable track quality. Since resonance daughters are indistinguishable from primaries, tracks were also required to have a distance of closest approach to primary vertex smaller than 0.5 cm (3 cm) in the transverse (longitudinal) direction.

A PID request was also made, in the form of a 'compatibility cut': a track was accepted if its associated TPC and TOF signals were close to the expected one for a Kaon with the same total momentum, within a fiducial window. The TPC fiducial window was chosen to be $3\,\sigma_{TPC}$ ($5\,\sigma_{TPC}$) if the track total momentum at the TPC inner wall was larger (smaller) than 350 MeV/c (see fig. 1 left). The TOF fiducial window on time difference was defined by two hyperbolas, as it is shown in figure 1 (right), to cope with the worsening resolution at small momenta (for $p_T \leq 260$ MeV/c the TOF is never used). Since the TPC-TOF matching efficiency is almost 60%, whenever a TOF signal was absent in the track, the TPC PID cut only was used.

3.2. Signal extraction and correction

The measurement was done in four p_T bins between 0.7 and 3 GeV/c and in a rapidity window $|y| \leq 0.6$, where the efficiency was found to be almost flat. An invariant mass distribution was

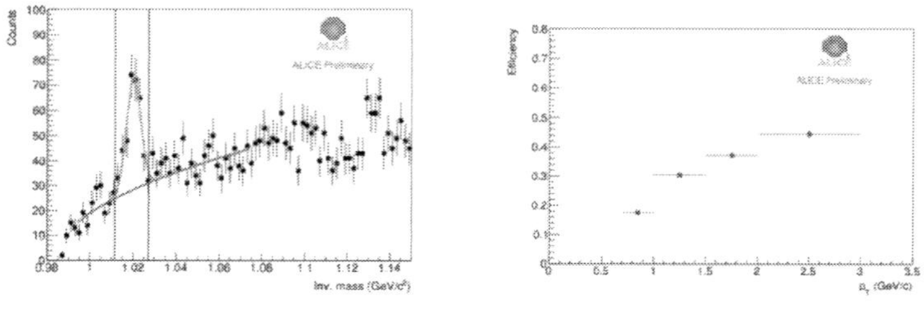

Figure 2. (Left) K^+K^- invariant mass spectrum in the p_T bin between 1 and 1.5 GeV/c. The curve is the combined fit of square-root function + Gaussian to the points, and the vertical lines delimit the region where the subtraction is computed. (Right) Efficiency of reconstructed ϕ computed in the PYTHIA sample cited in the text, in the four measured p_T bins.

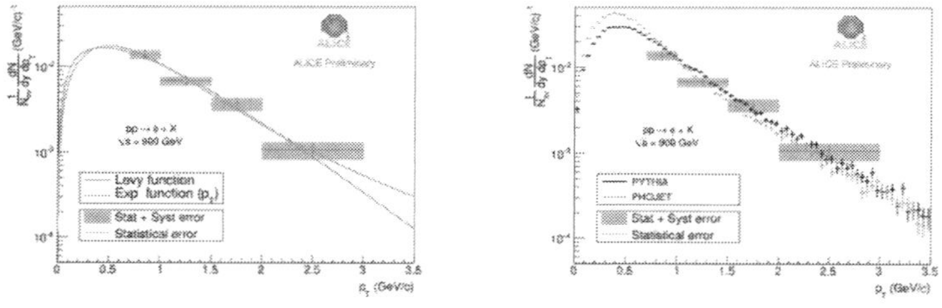

Figure 3. (Left) Corrected ϕ spectrum as a function of p_T. Shaded boxes represent the sum in quadrature of statistical and systematic errors, and the bars indicate the statistical errors only. The curves are the fit to the functions described in the text. (Right) Comparison of corrected spectrum from data with the corresponding spectrum of ϕ generated in a PYTHIA and in a PHOJET sample, normalized to the same integral in the measured p_T range (0.7 to 3 GeV/c).

computed with all unlike-sign charged track pairs (see fig 2, left) and fitted with a square-root function ($A\sqrt{m - m_0}$) plus a Gaussian, to reproduce the peak plus the background. Due to the small size of the measured sample and the consequently large statistical errors in the fit function, the bin counting method was preferred to compute the number of ϕ. This was then obtained from the subtraction of the background function integral from the total number of entries counted in a fixed invariant mass range. The chosen counting range was $m_\phi \pm 4\sigma$, where m_ϕ is the nominal mass of the ϕ resonance [7] and σ is computed from the full-width at half maximum (FWHM) of a Gaussian, expressed in terms of its σ parameter ($\sigma = FWHM/2.35$), and then assuming as FWHM the nominal width Γ_ϕ [7] of the ϕ resonance. This results in a range of $m_\phi \pm 1.7/\Gamma_\phi$.

To correct the measured counts, the ϕ reconstruction efficiency was estimated through a PYTHIA simulation made with a realistic implementation of the ALICE detector (see fig. 2, right). Finally, the corrected counts were normalized to the total number of inelastic collisions [8].

Two main sources of systematic error were identified: one was related to the background subtraction, and another is due to TPC-related cuts (dE/dx and number of clusters). They were estimated to range between 3.0 to 7.0% in the first case and from 0.9 to 6% in the second one. Anyway, they are smaller than the statistical error which ranges between 10% and 20% due to the sample size.

Integral	Levy	%	Exponential	%
0 - 0.7 GeV/c	0.0090 ± 0.0007	45	0.009 ± 0.002	47
0.7 - 3 GeV/c	0.0104 ± 0.0007	52	0.0104 ± 0.0007	52
3 - ∞ GeV/c	0.0006 ± 0.0001	3	0.0002 ± 0.0001	1
Total	0.020 ± 0.001	100	0.020 ± 0.002	100
$\langle p_T \rangle$ (GeV/c)	0.98 ± 0.07		0.88 ± 0.06	

Table 1. Computed dN/dy and mean p_T after the spectra fit (see text). First and third line come from function integration, second line comes from measured histogram integral.

4. Results

In order to extract the global integrated yields, the corrected p_T spectrum was fitted with a p_T-dependent exponential and with a Levy function:

$$\frac{d^2N}{dydp_T} = A \times p_T \times e^{-\frac{p_T}{T}} \tag{1}$$

$$\frac{d^2N}{dydp_T} = \frac{(n-1)(n-2)}{nT[nT+m(n-2)]} \times \frac{dN}{dy} \times p_T \times \left(1 + \frac{m_T - m}{nT}\right)^{-n} \tag{2}$$

where $m_T = \sqrt{m^2 + p_T^2}$. The full dN/dy distribution was computed by summing the integral fit functions outside the measured range, and the integral of measured bins inside it. In both cases the average transverse momentum was also computed.

Figure 3 (left) shows the measured points and the two used functions, and table 1 reports the computed yields and mean p_T. As expected, the two functions return different values for the latter, due to the very different contribution coming from the high momentum tail. Due to the exiguity of sample size, and the necessity to carry on the measurement on large p_T bins, it was not possible to discriminate between the two functions; then, these difference can give a preliminary estimate of a systematic uncertainty on the value. Figure 3 (right) reports a comparison of the measured points with PYTHIA and PHOJET normalized to the same integral in the measured region (0.7 to 3 GeV/c): while in this range they appear to be quite compatible, both generators give a larger expectation in the extrapolation region at low p_T, which reflects in a smaller mean p_T (0.55 for PHOJET and 0.77 for PYTHIA).

5. Conclusions

Preliminary results have been shown of the measurement of ϕ resonance in ALICE obtained with the first LHC data at 900 GeV. The p_T spectrum was measured between 0.7 and 3 GeV/c.

These results provide a useful baseline for comparisons with QCD-inspired particle production models and will be a reference for future measurements in heavy-ion collisions at LHC. Moreover, despite the rather low statistics (250 k events), this study demonstrates the excellent capability of ALICE for resonance measurements which will be exploited in pp collisions at higher energies and in heavy-ion collisions.

6. Bibliography

[1] ALICE Collaboration, *JINST* **3** (2008)S08002
[2] UA5 Collaboration, *Z. Phys. C* **41** (1988), 179; *Nucl. Phys. B* **328** (1989), 36
[3] E735 Collaboration, *Z. Phys. C* **67** (1995), 411
[4] AFS Collaboration, *Nucl. Phys. B* **203** (1982), 27
[5] STAR Collaboration, *Phys. Lett. B* **612** (2005), 181; Phys. Rev. C 79 (2009), 064903
[6] Markert C et al, *Phys. Lett. B* **669** (2008), 92 and references therein
[7] Nakamura K et al, *J. Phys. G* **37** (2010), 075021
[8] Floris M, *proceedings in this book*

Journal of Physics: Conference Series **270** (2011) 012008 doi:10.1088/1742-6596/270/1/012008

Tsallis-Pareto like distributions in hadron-hadron collisions

G. G. Barnaföldi[1], **K. Ürmössy**[1,2], **T. S. Biró**[1]

[1]MTA KFKI RMKI, Konkoly-Thege Miklós út 29-33, H-1121 Budapest, Hungary

[2]Eötvös University, Pázmány Péter sétány 1/A, H-1117 Budapest, Hungary

E-mail: `bgergely@rmki.kfki.hu`

Abstract. Non-extensive thermodynamics is a novel approach in high energy physics. In high-energy heavy-ion, and especially in proton-proton collisions we are far from a canonical thermal state, described by the Boltzmann – Gibbs statistic. In these reactions low and intermediate transverse momentum spectra are extremely well reproduced by the Tsallis-Pareto distribution, but the physical origin of Tsallis parameters is still an unsettled question.

Here, we analyze whether Tsallis-Pareto energy distribution do overlap with hadron spectra at high-p_T. We fitted data, measured in proton-proton (proton-antiproton) collisions in wide center of mass energy range from 200 GeV RHIC up to 7 TeV LHC energies. Furthermore, our test is extended to an investigation of a possible $\sqrt{s}$-dependence of the power in the Tsallis-Pareto distribution, motivated by QCD evolution equations. We found that Tsallis-Pareto distributions fit well high-p_T data, in the wide center of mass energy range. Deviance from the fits appears at $p_T > 20-30$ GeV/c, especially on CDF data [7]. Introducing a p_T-scaling *ansatz*, the fits at low and intermediate transverse momenta still remain good, and the deviations tend to disappear at the highest-p_T data.

1. Introduction

Heavy ion physics is one possible testbed for non-extensive thermodynamics, due to the fact that high-energy nuclear reactions are few-body and non-thermalized processes. Furthermore, strong correlations between particles, especially at the final state or during the hadronization, rule out the Boltzmann – Gibbs and rather lead to non-extensive statistics [1].

Non-extensive thermodynamics is a novel approach to high-energy heavy ion collisions, it is particularly successfull for describing the intermediate and high transverse momentum regimes. Based on the idea in Refs. [2, 3] we found that the hadron production in $AuAu$ collisions at RHIC energies can be well fitted by the Tsallis – Pareto distribution [4].

Recent experimental data measured at the LHC in proton-proton collisions [5, 6] have also shown Tsallis – Pareto-like distribution, but the fitted parameters are not satisfactorily interpreted. Furthermore, fits for earlier Tevatron data [7] result in a deviance from the parameterization at the highest transverse momentum region, as it was pointed out in Refs. [8, 9].

In this short contribution we report on various measured spectra from RHIC to LHC energies. We test a particular, scale evolution *ansatz* for the power in the Tsallis – Pareto distribution, suitable to fit the highest momentum data. We compare our results to several existing experimental data, the parameters are estimated up to the highest LHC energies.

2. Tsallis – Pareto fit on various experimental data

We investigated the neutral pion (π^0) and charged hadron ($h^{\pm}$) production spectra in proton-proton (pp) and proton-antiproton ($p\bar{p}$) collisions in a wide center of mass energy ($\sqrt{s}$) range from 200 GeV (RHIC,UA1) up to 7 TeV (LHC). In all cases we selected data with similar experimental acceptance, taken at around midrapidity ($y, \eta \approx 0$) in symmetric (pseudo)rapidity regions. We plotted the data as a function of logarithmic transverse momenta, on the *left panel* of Fig. 1. Points are: PHENIX data [10] *blue full triangles*, CERN UA1 data [11, 12] *blue open triangles*, CDF data [7, 13] *red full squares*, ATLAS data [14] *magenta full stars*, earlier CMS data [5] *green full circles*, while the latest CMS (upscaled) data [6] *green full triangles*. The p_T bins and errors are also indicated on the data points with bars on all panels of Fig. 1.

We assumed that the spectra measured in pp or $p\bar{p}$ collisionis are Tsallis – Pareto-like. They are determined by two essential parameters: the *effective temperature*, T refers to the average slope and the *parameter* q determines the tail of the distribution. For the Boltzmann – Gibbs-distribution, when $q \to 1$, $1/T$ gives the logarithmic slope of the curve, while for non-unity q values this effective temperature is p_T-dependent, especially at higher momenta. This behavior modifies the shape of the whole distribution-curve to a power-law tailed one as p_T increases. Since fits to existing proton-(anti)proton data well agree with $q \neq 1$, one concludes that these spectra are Tsallis – Pareto-distributions.

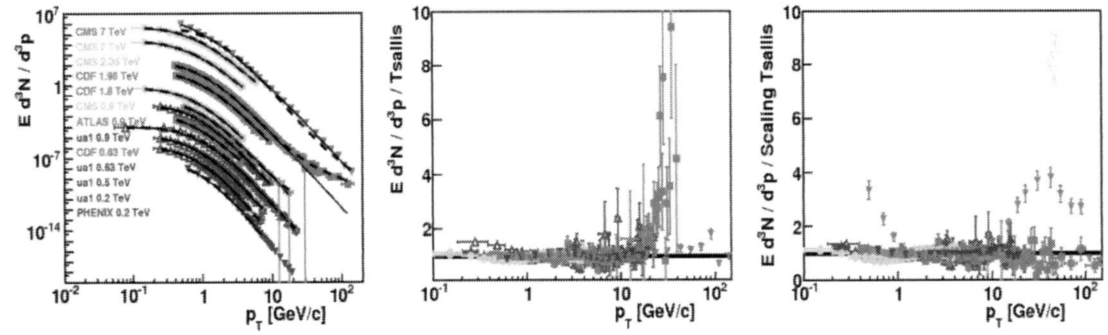

Figure 1. Tsallis fits on experimental data plotted on a logarithmic plot on the *left panel*. 'Data/Tsallis' best fit for the Tsallis – Pareto distribution on the *middle panel*. On the *right panel* 'Data/Scaling Tsallis' points are shown. (Color online.)

The data above were fitted by using the least-square method. We used the standard parameterization of the Tsallis – Pareto distribution given by:

$$E\frac{\mathrm{d}N}{\mathrm{d}p_T} = A \cdot m_T \left(1 + \frac{1}{n \cdot T} m_T\right)^{-n}, \tag{1}$$

where $m_T = \sqrt{m^2 + p_T^2}$ is the *transverse mass* of the hadrons and the power is $n = 1/(q-1)$. Values of T and q are obtained via the best fit. The results of the fits are drawn by *solid curves* on Fig. 1. (We note, some of the distributions were upscaled to avoid multiple overlaps.)

These Tsallis fits are in a remarkably good agreement with the experimental data, except CDF at 1.96 TeV. It shows up to 3 – 4 orders of magnitude deviations beyond $p_T \approx 20 - 30$ GeV/c. For a better comparison we plot the ratio of the experimental data to the fitted Tsallis distribution on the *middle panel* of Fig. 1. Here the 3 – 4 orders of magnitude deviations are clearly seen (all points above 30 GeV/c are off scale).

Motivated by the QCD scale evolution idea we investigate a similar distribution ('Scaling Tsallis') as given by Eq. (1), but with a p_T-dependent power:

$$n \longrightarrow n(p_T) = \frac{1}{q-1} - 2 \cdot \log\log\left(1 + \kappa \cdot p_T\right) \quad , \tag{2}$$

with a parameter $\kappa = 0.013$ (GeV/c)$^{-1}$. The obtained 'Data/Scaling Tsallis' points are plotted on the *right panel* of Fig. 1. Here, all experimental data are in close agreement with the 'Scaling Tsallis' formula, except the latest CMS result. It has a factor of $2-4$ deviance for low ($p_T < 1$ GeV/c) and at the highest ($p_T > 15$ GeV/c) transverse momenta. Based on this latter panel, introduction of a scaling may help on the earlier data especially for Tevatron. On the other hand latest CMS results explicitly contradict to this behavior at high-p_T.

3. Analyzing Tsallis parameters

We investigated the center of mass energy dependence of the parameters fitted to the Tsallis – Pareto-distributions in Fig. 1. We summarize all of these properties on Fig. 2.

On the *left panel* of Fig. 2 we plot the obtained $q(\sqrt{s})$ relation with *blue* signs for the standard 'Tsallis' and with *red* ones for the scaling Tsallis – Pareto-distribution. For both cases, the values of the Tsallis-q parameter are slightly increasing then saturating with $\sqrt{s}$, within the values $1.07 < q < 1.18$. Generally the Tsallis-q parameters display the same function of the $\sqrt{s}$ for both fits.

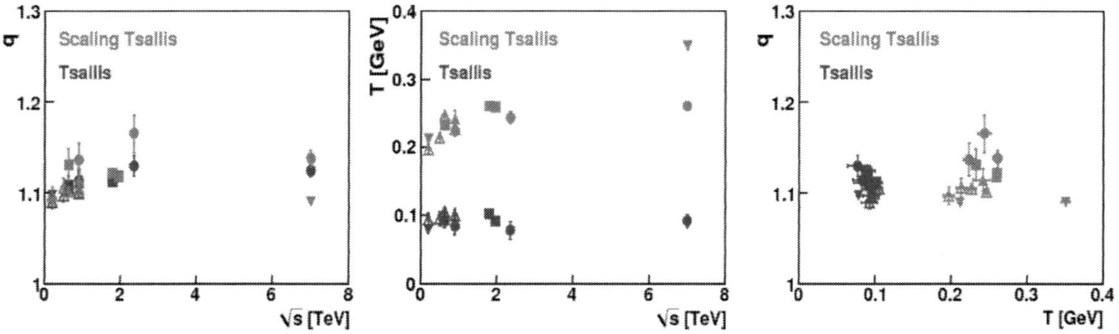

Figure 2. The $\sqrt{s}$ dependence and the parameter space of fitted Tsallis parameters T and q.

We indicate the $T(\sqrt{s})$ dependence on the *middle panel* of Fig. 2. Here the two parameterizations deviate from each other. The value of the effective temperature in case of 'Tsallis' is around ≈ 0.1 GeV – almost independently of $\sqrt{s}$ (see *blue points*). Meanwhile, after introducing scaling by Eq. (2) into Eq. (1), an increase can be seen on the T value starting from ≈ 0.2 GeV, and may saturate around 0.3 GeV between $2.5 - 3.5$ GeV for the highest energies, $\sqrt{s} > 1$ TeV (see *red points*). These latter results are closer to the physical picture of high-energy collisions stating that a higher energy density reflects hotter matter. To make an estimate for the values of the effective temperature at higher $\sqrt{s}$ values we use the following two parameterizations: (i) a power-like: $T(\sqrt{s})[\text{GeV}] = (0.24 \pm 0.006) \cdot (\sqrt{s})^{0.12 \pm 0.02}$ and (ii) a logarithmic one: $T(\sqrt{s})[\text{GeV}] = (0.029 \pm 0.005) \cdot \log(\sqrt{s}) + (0.242 \pm 0.006)$.

The parameter space for the standard 'Tsallis' and for the 'Scaling Tsallis' are summarized on the *right panel* of Fig. 2. Since both parameters have explicit dependence on the $\sqrt{s}$, the evolution of the parameters can be inspected clearly on the plot. As already noted, the

Tsallis-q values are similar for both cases, but due to the different effective temperature finally two deviating trajectories appear. They cover well separated regions in the parameter space. However, due to the inclusion of the above mentioned low-p_T CMS data with huge errors, a given degree of uncertainty remains. Based on the scattering of the fitted values around $q \leq 1.22$, we surmise that the Tsallis-q distribution is not yet reached within the present statistic. This may correspond to a strong correlation of partons or to the limited number of steps during hadronization, as applied in Monte Carlo generators Ref. [15].

4. Conclusions and outlook

Based on various experimental data from RHIC to LHC energies, we fitted Tsallis – Pareto distributions to transverse momentum spectra measured in pp and $p\bar{p}$ collisions. Two sets of Tsallis parameterization were applied: 'standard' and 'scaling' – this latter with p_T-dependent power *ansatz*. Parameters T and q were obtained by using the least square method.

We investigated the $\sqrt{s}$ dependence on the parameter values for the original and for a modified, scaling Tsallis – Pareto distribution. Tsallis-q parameters were found to be almost $\sqrt{s}$ independent distributed around $q = 1.12 \pm 0.6$. This value seems to be constant (or slowly increasing) beyond recent LHC energies, which supports the idea that these experiments are just getting to reach Tsallis-q distribution with $q = 1.22$. It may then be reached at the highest energies of the LHC.

The effective temperature, T behaves differently in the cases of two parameterizations. Standard Tsallis has an almost constant value $T = 0.8 \pm 0.2$ GeV as a function of $\sqrt{s}$. With this parameterization no further changes are expected as increasing the energy. By using the 'scaling Tsallis – Pareto' ansatz, defined by Eqs. (2) and (1), values of T increase with $\sqrt{s}$. We speculate that this results is physically reliable, since increasing the $\sqrt{s}$ may result in a higher energy density and higher temperature. Assuming both a power like and a logarithmic function, we estimated a common value of $T \approx 0.32 - 0.33$ for pp collisons at 14 TeV LHC energy.

Based on the experimental data, the Tsallis – Pareto distribution seems to be a solid and invariant assumption for high-energy hadron spectra both in proton-(anti)proton and heavy-ion collisions. The possible scaling property of the power, however, needs to be investigated in more detail in the light of forthcoming data.

Acknowledgments

This work was supported by the Hungarian OTKA (PD73596, K68108) and Eötvös University. One of the authors (GGB) thanks to the János Bolyai Research Scholarship of the HAS.

References

[1] C. Tsallis, J. Stat. Phys. **52** 479 (1988); Eur. Phys. J. **A40** 257 (2009)
[2] T. S. Biró, Europhys. Lett. **84**, 56003 (2008)
[3] G. Wilk and Z. Wlodarczyk, Phys. Rev. C **79**, 054903 (2009)
[4] T. S. Biró, K. Urmossy and G. G. Barnafoldi, J. Phys. G **35**, 044012 (2008)
[5] V. Khachatryan *et al.* [CMS Collaboration], JHEP 1002: 041, 2010.
[6] V. Khachatryan *et al.* [CMS Collaboration], CMS-QCD-10-006, CERN-PH-EP-2010-009, FERMILAB-PUB-10-170-CMS, e-Print: arXiv:1005.3299
[7] T. Aaltonen *et al.* [CDF Collaboration] Phys. Rev. D 79: 112005, 2009.
[8] R. Ichou and D. d'Enterria, Phys. Rev. D **82**, 014015 (2010)
[9] F. Arleo, D. d'Enterria and A. S. Yoon, JHEP **1006**, 035 (2010)
[10] A. Adare *et al.* [PHENIX Collaboration], arXiv:1005.3674 [hep-ex].
[11] C. Albajar *et al.* [UA1 Collaboration] Nucl. Phys. B 335: 261, 1990.
[12] G. Bocquet *et al.* [UA1 Collaboration], Phys. Lett. B 366: 434-440, 1996.
[13] F. Abe *et al.* [CDF Collaboration], Phys. Rev. Lett. 61: 1819, 1988.
[14] G. Aad *et al.* [ATLAS Collaboration] Phys. Lett. B 688: 21-42, 2010.
[15] T. Sjostrand, S. Mrenna and P. Z. Skands, JHEP **0605**, 026 (2006) [arXiv:hep-ph/0603175].

Journal of Physics: Conference Series **270** (2011) 012009 doi:10.1088/1742-6596/270/1/012009

On jet structure in the QGP

Mehtar-Tani Y
Departamento de Física de Partículas and IGFAE
Universidade de Santiago de Compostela
E-15782 Santiago de Compostela Spain
E-mail: ymehtar@gmail.com

Abstract. We investigate the gluon radiation pattern of a quark-antiquark pair traversing a dense QCD medium at high energy. We show that the medium-induced gluon radiation spectrum is logarithmically divergent and that the angular distribution exhibits an anti-angular ordering of soft-gluon emissions in contrast with angular ordering in in-vacuum radiation.

1. Introduction

The study of jets in the heavy-ion program at the LHC is of a great importance, as it provides us with probes of the quark-gluon plasma (QGP). However, whereas jets observables in the vacuum are well understood theoretically to a very high accuracy [1, 2, 3], medium-jet modification is still at its early stages [4, 5, 6, 7, 8, 9]. As an attempt toward understanding jets in the QGP from first principles we have proposed in a recent work [10] to study a simple process, namely gluon radiation off a pair quark-antiquark ($q\bar{q}$), in color singlet state, passing through a QCD medium. The aim is to gain insight on in-medium jet dynamics.

The $q\bar{q}$ pair, with momenta $p \equiv (E, \mathbf{p})$ and $\bar{p} \equiv (\bar{E}, \bar{\mathbf{p}})$, respectively, is assumed to be in a singlet state. In case of gluon radiation in the vacuum, the cross-section exhibits a soft and a collinear singularity. Namely, when the emitted gluon of momentum $k \equiv (\omega, \mathbf{k})$, is collinear to either the quark or the anti-quark and when its energy $\omega \to 0$. The collinear poles can be split into two terms: independent emissions from the quark and from the anti-quark, $dN \equiv dN_q + dN_{\bar{q}}$, with

$$dN_q = \frac{\alpha_s C_F}{2} \frac{d\omega}{\omega} \frac{\sin\theta d\theta}{1-\cos\theta} \Theta(\cos\theta - \cos\theta_{q\bar{q}}), \tag{1}$$

where θ is the angle between the gluon and the quark and $\theta_{q\bar{q}}$ is the angle formed by the pair, where the azimuthal angle has been integrated out. We see that gluons are coherently emitted inside the cone delimited by the pair. This coherence is responsible for the depletion of soft emissions leading to the so-called humpbacked plateau.

2. Classical Yang-Mills Equations (CYM)

Let us return now to the case of medium induced gluon radiation off the pair. In the classical Yang-Mills (CYM) equation framework [8], which holds for soft gluon radiation, the amplitude of emission of a gluon is related to the classical gauge field by the reduction formula

$$\mathcal{M}^a_\lambda(\mathbf{k}) = \lim_{k^2 \to 0} -k^2 A^a_\mu(k) \epsilon^\mu_\lambda(\mathbf{k}), \tag{2}$$

where $\epsilon_\lambda^\mu(\mathbf{k})$ is the gluon polarization vector. The gauge field, $A^\mu \equiv A^{\mu,a}t^a$, where t^a is the generator of SU(3) in the fundamental representation, is the solution of the CYM equations, $[D_\mu, F^{\mu\nu}] = J^\nu$, with $D_\mu \equiv \partial_\mu - igA_\mu$ and $F_{\mu\nu} \equiv \partial_\mu A_\nu - \partial_\nu A_\mu - ig[A_\mu, A_\nu]$. The covariantly conserved current, i.e., $[D_\mu, J^\mu] = 0$, describes the projectiles. Furthermore, we shall set our calculation in the light-cone gauge (LCG), $A^+ = 0$. The classical eikonalized current that describes the pair created at time $t_0 = 0$ inside the medium reads $J_p^\mu = J_q^\mu + J_{\bar{q}}^\mu$, where $J_q^\mu = g\frac{p^\mu}{E}\,\delta^{(3)}(\mathbf{x} - \frac{\mathbf{p}}{E}t)\,\Theta(t)\,\mathcal{C}_q^a t^a$ and analogously for the anti-quark, where $\mathcal{C}_q = -\mathcal{C}_{\bar{q}} = \mathcal{C}$ is the quark and anti-quark color charge, respectively. In Fourier space the total current reads

$$J_p^\mu(k) = ig\left(\frac{p^\mu}{p\cdot k + i\epsilon} - \frac{\bar{p}^\mu}{\bar{p}\cdot k + i\epsilon}\right)\mathcal{C}^a t^a. \tag{3}$$

We assume that the $q\bar{q}$ pair moves in the $+z$ direction and interacts with a medium moving in the opposite direction at nearly the speed of light. This approximation is valid as long as the pair opening angle $\theta_{q\bar{q}} \to 0$ and at asymptotic energies [4, 6, 8]. Also, we restrict our calculation, for simplicity, to one gluon exchange with the medium. To do so, the pair field is treated as a perturbation around the strong medium field A_0. In the asymptotic limit, the medium gauge field can be described by $A_0^-(x^+, \boldsymbol{x}_\perp)$ which is a solution to the Poisson equation $-\partial_\perp^2 A_0^-(x^+, \boldsymbol{x}_\perp) = \rho_0(x^+, \boldsymbol{x}_\perp)$, where the medium source density in treated as a Gaussian white noise, while $A_0^i = A_0^+ = 0$ [8]. In Fourier space it reads $A_0^-(q) = 2\pi\,\delta(q^+)\int dx^+\,\mathcal{A}_0(x^+, \boldsymbol{q}_\perp)\,e^{iq^- x^+}$. In the present approximation $t \approx z$, thus, $x^+ \approx \sqrt{2}z$. The medium average is defined as $\langle \mathcal{A}_0^a(x^+, \boldsymbol{q}_\perp)\mathcal{A}_0^{*b}(x'^+, \boldsymbol{q}'_\perp)\rangle \equiv \delta^{ab}\,n_0\,m_D^2\,\delta(x^+ - x'^+)(2\pi)^2\,\delta^{(2)}(\boldsymbol{q}_\perp - \boldsymbol{q}'_\perp)\mathcal{V}(\boldsymbol{q}_\perp)$, where $\mathcal{V}(\boldsymbol{q}_\perp) = 1/(\boldsymbol{q}_\perp^2 + m_D^2)^2$ is the Coulomb potential, m_D is the Debye mass and n_0 is the 1-dimensional density of scattering centers. At first order in opacity the continuity relation reads $\partial_\mu J_1^\mu = ig[A_0^-, J_p^+]$ which is solved by

$$J_1^\mu = ig\frac{p^\mu}{p\cdot\partial}\,[A_0^-, J_q^+] + ig\frac{\bar{p}^\mu}{\bar{p}\cdot\partial}\,[A_0^-, J_{\bar{q}}^+]. \tag{4}$$

Then, the transverse part of the CYM equations read $\Box A_1^i = 2ig\left[A_0^-, \partial^+ A_p^i\right] - \frac{\partial^i}{\partial^+}J_1^+ + J_1^i$. In fourier space we obtain

$$-k^2 A_1^i(k) = 2g\int\frac{d^4q}{(2\pi)^4}\left[A_0^-(q), (q-k)^+ A_p^i(k-q)\right] - \frac{k^i}{k^+}J_1^+(k) + J_1^i(k)\,. \tag{5}$$

After performing the q^+ and q^- integrals we obtain the amplitude for gluon radiation off the quark, via Eq. (2),

$$\mathcal{M}_{q,1}^{i,a} = ig^2 f^{abc}\mathcal{C}^c\int\frac{d^2\boldsymbol{q}_\perp}{(2\pi)^2}\int_0^{L^+} dx^+\mathcal{A}_0^b(x^+, \boldsymbol{q}_\perp)\left[\frac{\nu^i}{p\cdot v}\left(1 - e^{i\frac{p\cdot v}{p^+}x^+}\right) + \frac{\kappa^i}{p\cdot k}e^{i\frac{p\cdot v}{p^+}x^+}\right]e^{i(k-v)^- x^+} \tag{6}$$

where $v \equiv \left(v^+ = k^+, v^- = \boldsymbol{v}_\perp^2/2k^+, \boldsymbol{v}_\perp = \boldsymbol{k}_\perp - \boldsymbol{q}_\perp\right)$, $\boldsymbol{\nu}_\perp = p^+\boldsymbol{v}_\perp/k^+ - \boldsymbol{p}_\perp$ and $\boldsymbol{\kappa}_\perp = p^+\boldsymbol{k}_\perp/k^+ - \boldsymbol{p}_\perp$. Here, the subindex 1 in the l.h.s of Eq. (6) stands for the first order in the medium field A_0. $L^+ = \sqrt{2}L$ is the medium size. The amplitude for gluon radiation off the anti-quark is deduced from $\mathcal{M}_q$ by the substitution $p \to \bar{p}$ and $\mathcal{C} \to -\mathcal{C}$. The first term in Eq. (6) corresponds to the interaction of the emitted gluon with the medium, denoted $\mathcal{M}_q^g$, while the second term corresponds to the interaction of the quark before the bremsstrahlung emission of the gluon, denoted $\mathcal{M}_q^{\text{brem}}$. The cross-section is obtained after squaring $\mathcal{M}_1$, plus terms of the type $\mathcal{M}_2\mathcal{M}_0^*$ which simply lead to a redefinition of the potential such that $\mathcal{V}(\boldsymbol{q}_\perp) \to \mathcal{V}(\boldsymbol{q}_\perp) - \delta(\boldsymbol{q}_\perp)\int d^2\boldsymbol{q}'_\perp\mathcal{V}(\boldsymbol{q}'_\perp)$, they insure that the spectrum is finite in the limit $\boldsymbol{q}_\perp \to 0$. Here, $\mathcal{M}_2$ and $\mathcal{M}_0$ are the amplitudes at second and zeroth order in A_0, respectively.

Interestingly, the interference of bremsstrahlung gluons off the quark and anti-quark, given by $\mathcal{I}_{\text{brems}} = 2\ \text{Re}\ \mathcal{M}_q^{\text{brem}} \mathcal{M}_{\bar{q}}^{*\text{brem}}$, exhibits a soft divergency and thus will dominate at small ω,

$$\omega \frac{dN^{\text{soft}}}{d^3k} = \frac{8\pi C_A C_F\, \alpha_s^2\, n_0\, m_D^2}{(2\pi)^2} \frac{\boldsymbol{\kappa}_\perp \cdot \bar{\boldsymbol{\kappa}}_\perp}{(p\cdot k)(\bar{p}\cdot k)} \int_0^{L^+} dx^+ \cos\Omega_0\, x^+ \int \frac{d\boldsymbol{q}_\perp}{(2\pi)^2} \mathcal{V}(\boldsymbol{q}_\perp) \left(1 - \cos\Delta\Omega\, x^+\right), \quad (7)$$

where we have written the contact term explicitly, and where

$$\Omega_0 = \frac{p\cdot k}{p^+} - \frac{\bar{p}\cdot k}{\bar{p}^+}, \quad \Delta\Omega = \frac{\boldsymbol{p}_\perp \cdot \boldsymbol{q}_\perp}{p^+} - \frac{\bar{\boldsymbol{p}}_\perp \cdot \boldsymbol{q}_\perp}{\bar{p}^+}, \quad \boldsymbol{\kappa}_\perp \cdot \bar{\boldsymbol{\kappa}}_\perp = -p\cdot\bar{p} + \frac{p^+}{k^+}\bar{p}\cdot k + \frac{\bar{p}^+}{k^+} p\cdot k\,. \quad (8)$$

The soft divergency in Eq. (7) is manifest. Note that in the soft limit $\Omega_{q\bar{q}}^0 \to 0$ the integrals in Eq. (7) are straightforward, yiedling

$$\int_0^{L^+} dx^+ \int \frac{d\boldsymbol{q}_\perp}{(2\pi)^2} \mathcal{V}(\boldsymbol{q}_\perp) \left(1 - \cos \frac{\bar{\boldsymbol{p}}_\perp \cdot \boldsymbol{q}_\perp}{\bar{p}^+} x^+\right) \approx \frac{L^+ r_\perp^2}{24\pi} \left[\ln\left(\frac{1}{r_\perp m_D}\right) + \text{const.}\right], \quad (9)$$

where we have assumed the quark momentum to be along the z axis for simplicity, i.e., $\boldsymbol{p}_\perp = p^- = 0$. Note that, the r.h.s of Eq. (9) is proportional to the forward dipole scattering amplitude for a dipole of size $r_\perp = \theta_{q\bar{q}} L$.

3. Angular structure of the antenna in the medium

Let us finally turn to the angular structure. In the small angle limit $\theta_{q\bar{q}} \ll 1$ and $\theta \ll 1$, we simplify $p^+/k^+ \approx E/\omega$, getting

$$\mathcal{P}^{\text{med}} \equiv -\frac{\boldsymbol{\kappa}_\perp \cdot \bar{\boldsymbol{\kappa}}_\perp}{(p\cdot k)(\bar{p}\cdot k)} = \frac{n_q\cdot n_{\bar{q}} - n_q \cdot n - n_{\bar{q}}\cdot n}{(n_q\cdot n)(n_{\bar{q}}\cdot n)}, \quad (10)$$

where $n_q \equiv p/E$, $n_{\bar{q}} \equiv \bar{p}/\bar{E}$ and $n \equiv k/\omega$. Yet, dividing this symmetrically into a quark and an antiquark part contribution, with $\mathcal{P}_q^{\text{med}} = \mathcal{P}^{\text{med}}/2$, and averaging over the azimuthal angle we obtain $2\langle \mathcal{P}_q^{\text{med}} \rangle = -\Theta(\cos\theta_{q\bar{q}} - \cos\theta)/(1-\cos\theta)$. Surprisingly, the medium-induced soft gluon radiation off the quark is suppressed inside the cone of opening angle $\theta_{q\bar{q}}$, as opposed to the standard angular structure obtained in vacuum, see Eq. (1). Furthermore, due to this feature the collinear pole present in Eq. (7) is automatically cut off. Thus, when $\omega \to 0$ the medium-induced gluon emission off the quark can be written as

$$dN_q^{\text{med}} = \frac{\alpha_s C_F}{2} A^{\text{med}} \frac{d\omega}{\omega} \frac{\sin\theta d\theta}{1-\cos\theta} \Theta(\cos\theta_{q\bar{q}} - \cos\theta), \quad (11)$$

with $A^{\text{med}} \simeq \alpha_s\, n_0\, m_D^2 L^+ r_\perp^2 \left[\ln\left(1/r_\perp m_D\right) + \text{const.}\right]/6\pi^2$. This demonstrates that there is a strict geometrical separation between radiation in vacuum and in medium. In Fig. (1), the gluon production cross-section off a $q\bar{q}$ pair with opening angle $\theta_{q\bar{q}} = 0.1$ traversing a medium with thickness $L = 4$ fm ($m_D = 0.5$ GeV, $\alpha_s = 1/3$ and $n_0 L = 1$), is plotted as a function of the angle θ for the energies, $\omega = 0.5$ GeV and 2 GeV. The dotted(red) curves correspond to the dominant contribution Eq. (7). Note that spectrum is indeed dominated by the bremsstrahlung term Eq. (7). Inside the cone delimited by the pair, of opening angle $\theta_{q\bar{q}}$, no radiation is allowed. The distribution jumps from zero inside the cone to a maximum value at $\theta = \theta_{q\bar{q}}$, it drops then as $\sim 1/\theta$ for $\theta > \theta_{q\bar{q}}$. Indeed, this remarkable feature survives even at larger values of ω due to some striking cancellations between different parts of the spectrum [10].

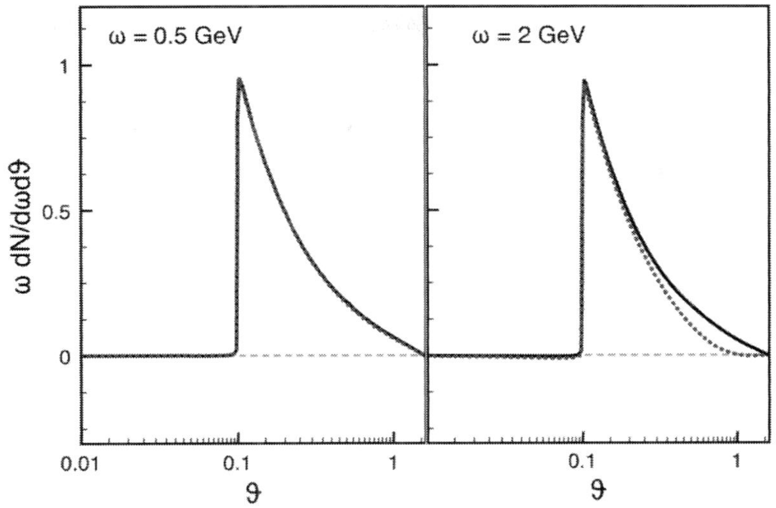

Figure 1. Angular distribution of the gluon spectrum for $\omega = 0.5$ GeV and $\omega = 2$ GeV for a jet with opening angle $\theta_{q\bar{q}} = 0.1$, see the text for details. The dotted (red) line corresponds to the dominant contribution in the soft-limit, $\mathcal{I}_{\rm brems}$, see Eq. (7), and the solid line corresponds to the full spectrum.

4. Summary

To summarize we have shown that medium induced gluon radiation off a quark-antiquark pair exhibits a logarithmic soft divergency, whereas the collinear singularity present in the vacuum case, is cut off, leading to anti-angular ordering of gluon emissions. Important implications on the phenomenology of jets in heavy-ion collisions are expected.

References

[1] Bassetto A, Ciafaloni M and Marchesini G 1983 *Phys. Rept.* **100** 201; Bassetto A, Ciafaloni M, Marchesini G and Mueller A H 1982 *Nucl. Phys.* B **207** 189
[2] Mueller A H 1981 *Phys. Lett.* B **104** 161
[3] Ermolaev B I and Fadin V S 1981 *JETP Lett.* **33** 269 [1981 *Pisma Zh. Eksp. Teor. Fiz.* **33** 285]
[4] Baier R, Dokshitzer Yu L Mueller A H, Peigné S and Schiff D 1997 *Nucl. Phys.* B **483** 291 ; 1997 **484** 265
[5] Zakharov B G 1996 *JETP Lett.* **63** 952 ; 1997 **65** 615
[6] Gyulassy M, Levai P and Vitev I 2000 *Phys. Rev. Lett.* **85** 5535 ; 2001 *Nucl. Phys.* B **594** 371
[7] Wiedemann U A 2000 *Nucl. Phys.* B **588** 303
[8] Mehtar-Tani Y 2007 *Phys. Rev.* C **75** 034908
[9] Salgado C A and Wiedemann U A 2004 *Phys. Rev. Lett.* **93** 042301
[10] Mehtar-Tani Y, Salgado C and Tywoniuk K arXiv:1009.2965 [hep-ph]

Journal of Physics: Conference Series **270** (2011) 012010 doi:10.1088/1742-6596/270/1/012010

Azimuthal anisotropy (v_2) of high-p_T π^0 and direct γ in Au+Au collisions at $\sqrt{s_{NN}} = 200$ GeV

Ahmed Hamed (STAR Collaboration)
Texas A&M University, College Station, USA
E-mail: ahamed@tamu.edu

Abstract. Preliminary results from the STAR collaboration of the azimuthal anisotropy (v_2) of π^0 and direct photon (γ_{dir}) at high transverse momentum (p_T) from Au+Au collisions at center-of-mass energy $\sqrt{s_{NN}} = 200$ GeV are presented. A shower-shape analysis is used to select a sample free of direct photons (π^0) and a sample rich in direct photons γ_{rich}. The relative contribution of background in the γ_{rich} sample is determined assuming no associated charged particles nearby γ_{dir}. The v_2 of direct photons ($v_2^{\gamma_{dir}}$) at mid-rapidity ($|\eta^{\gamma_{dir}}| < 1$) and high p_T ($8 < p_T^{\gamma_{dir}} < 16$ GeV/c) is extracted from those of π^0 and neutral particles measured in the same kinematic range. In mid-central Au+Au collisions (10-40%), the v_2 of π^0 ($v_2^{\pi^0}(p_T)$) and charged particles ($v_2^{ch}(p_T)$) are found to be ~ 0.12 and nearly independent of p_T. The measured $v_2^{\gamma_{dir}}(p_T)$ is positive finite and systematically smaller than that of π^0 and charged particles by a factor of ~ 3. Although the large $v_2^{\pi^0}$ at such high p_T might be partially due to the path-length dependence of energy loss, the non-zero value of $v_2^{\gamma_{dir}}$ indicates a bias of the reaction plane determination due to the presence of jets in the events. Systematic studies are currently in progress.

1. Introduction

The azimuthal distribution of the produced particles in heavy-ion collisions is expected to be sensitive to the initial geometric overlap of the colliding nuclei, and would result in anisotropic azimuthal distributions with respect to the reaction plane. The standard method to quantify the azimuthal anisotropy is to expand the particle azimuthal distributions in a Fourier series $\frac{dN}{d\phi} = \frac{1}{2\pi}[1 + \sum_n 2v_n \cos(n(\phi_{p_T} - \phi_{RP}))]$, where ϕ_{p_T} is the azimuthal angle of the produced particle with certain value of p_T, ϕ_{RP} is the azimuthal angle of the reaction plane, and v_n is the coefficient of the nth harmonic. The 2nd Fourier moment ($n = 2$) is referred to as the "elliptic flow" parameter v_2 and its differential form is given by

$$v_2(p_T) = \langle e^{2i(\phi_{p_T} - \phi_{RP})} \rangle = \langle \cos 2(\phi_{p_T} - \phi_{RP}) \rangle \tag{1}$$

where the brackets denote statistical averaging over different events.

While RHIC data show large amount of elliptic flow as predicted by the hydrodynamic models at low p_T, the results at high p_T deviate strongly from the hydrodynamic predictions as is expected [1]. The medium-induced radiative energy loss of partons (jet-quenching) has been proposed as the source for the large observed azimuthal anisotropy at high p_T, due to the path-length dependence of the parton energy loss [2]. The STAR results [3] show the amount of v_2 at high p_T is larger than the predicted values by pure jet-quenching models. Although recent measurements by PHENIX [4] show the produced π^0's in-plane outnumber those produced out-of-plane which may be consistent with the path-length dependence of energy loss, the reaction plane determination might have remaining bias toward the direction of the produced jets. STAR results show insensitivity of the path-length-dependence of energy loss [5] at high p_T through a comparison between γ_{dir}-charged particles and π^0-charged particles azimuthal correlations.

The v_2 measurement of direct photons would help to assess any remaining bias in the reaction plane determination leading to a positive v_2 signal. Furthermore the $v_2^{\gamma_{dir}}$ would give additional information to help disentangle the various scenarios of direct photon production through the expected opposite contributions to the v_2 [6, 7, 8, 9], and therefore could help to confirm the observed binary scaling of the direct photon [10].

2. Analysis and Results

2.1. *Electromagnetic neutral clusters*

The STAR detector is well suited for measuring azimuthal angular correlations due to the large coverage in pseudorapidity ($|\eta| < 1$) and full coverage in azimuth (ϕ). While the Barrel Electromagnetic Calorimeter (BEMC) [11] measures the electromagnetic energy with high resolution, the Barrel Shower Maximum Detector (BSMD) provides fine spatial resolution and enhances the rejection power for the hadrons. The Time Projection Chamber (TPC) [12] identifies charged-particles, measures its momenta, and allows for a charged-particle veto cut with the BEMC matching. Using the BEMC to select events (*i.e.* "trigger") with high-p_T γ, the STAR experiment collected an integrated luminosity of 535 μb^{-1} of Au+Au collisions in 2007. In this analysis, events having primary vertex within ± 55 cm of the center of TPC along the beamline, and with at least one electromagnetic cluster with $E_T > 8$ GeV are selected. More than 97% of these clusters have deposited energy greater than 0.5 GeV in each layer of the BSMD. A trigger tower is rejected if it has a track with $p > 3.0$ GeV/c pointing to it, which reduces the number of the electromagnetic clusters by only $\sim 7\%$.

2.2. v_2 *of neutral and charged particles*

The v_2 is determined using the standard method (Eq. 1), which correlates each particle with the event plane determined from the full event minus the particle of interest. The event plane is determined by

$$\phi_{RP} = \frac{1}{2}\tan^{-1}\left(\frac{\sum_i \sin(2\phi_i)}{\sum_i \cos(2\phi_i)}\right) \qquad (2)$$

where ϕ_i are the azimuthal angles of all the particles used to define the event plane. In this analysis, the charged-track quality criteria are similar to those used in previous STAR analyses [13]. The event plane is measured twice: 1) using all the selected tracks inside the TPC (full-TPC), and 2) using the selected tracks in the opposite pseudorapidity side to the particle of interest (off-η) in order to reduce the "non-flow" contributions (azimuthal correlations not related to the reaction plane). Since the event plane is only an approximation to the true reaction plane, the observed correlation is divided by the event plane resolution. The event plane resolution is estimated using the sub-event method in which the full event is divided up randomly into two sub-events (for full-TPC and off-η separately) as described in [14]. Biases due to the finite acceptance of the detector, which cause the particles to be azimuthally anisotropic in the laboratory system are removed according to the method in [15].

Figure 1 (left panel) shows the v_2 of charged particles (v_2^{ch}) at low p_T using the event plane method (off-η) compared to previous STAR measurements [13], and the v_2 of the charged and neutral particles using the full-TPC and off-η event plane methods at high p_T. At low p_T the v_2^{ch}(off-η) is smaller than the v_2 using the full TPC and agrees well with the $v_2\{4\}$ (4-particle cumulant) method, in which the contribution of the non-flow is expected to be small. At high

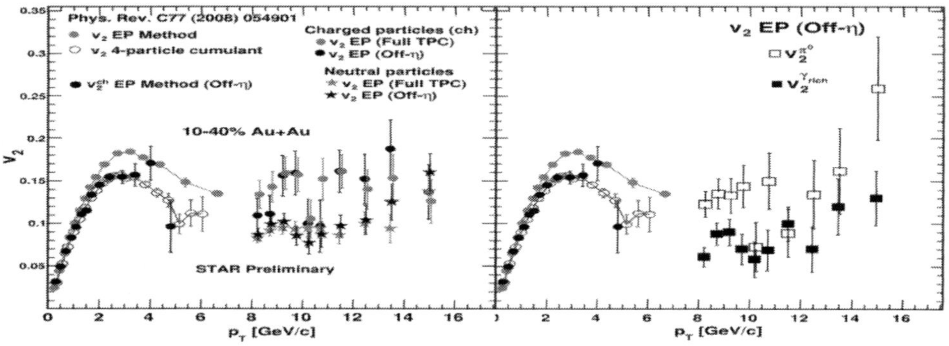

Figure 1. (Color online) For $p_T < 6$ GeV/c, both panels show previous STAR measurements [13] of v_2 as a function of p_T for charged particles with $|\eta| < 1$ in 10-40% $Au+Au$ collisions at $\sqrt{s_{NN}} = 200$ GeV using the Event-Plane method (closed red circles), and the 4-particle cumulant method (open circles). Also v_2 for charged particles ($|\eta| < 1$) using off-η event plane method is shown in closed black circles (this analysis). For $p_T > 6$ GeV/c: (left panel) v_2 of charged particles (red and black circles) and v_2 of neutral particles (red and black stars) using the full TPC and off-η event plane methods respectively; (right panel) v_2 of π^0 and γ_{rich} (open and closed squares, respectively) using the off-η method.

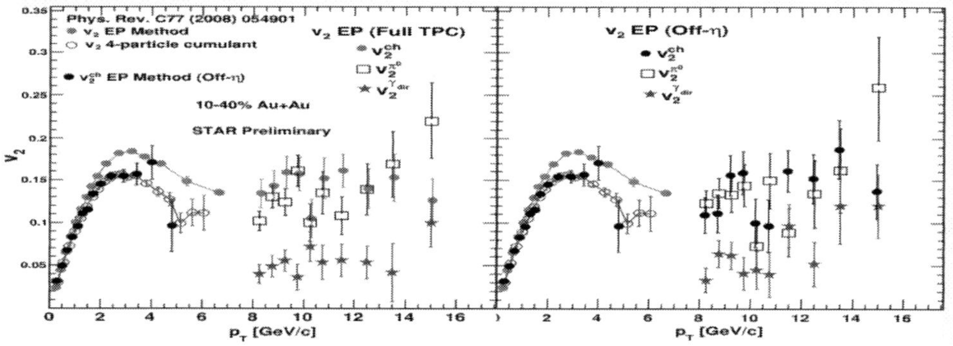

Figure 2. (Color online) For $p_T < 6$ GeV/c, both panels show measurements as in Fig. 1. For $p_T > 6$ GeV/c, both panels show v_2 of charged particles, π^0, and γ_{dir} (circles, squares, stars respectively) using the full TPC (left panel) and using off-η methods (right panel).

p_T the two different methods (full TPC and off-η) for the reaction plane measurements give similar results (both for the charged and neutral particles separately), which might indicate the dominance of non-flow contributions in v_2 measurements. The $v_2(p_T)$ of the neutral particles show systematically smaller values than those of charged particles due to the γ_{dir} contributions.

2.3. Transverse shower profile analysis

A crucial part of the analysis is to discriminate between showers from γ_{dir} and two close γ's from high-p_T π^0 symmetric decays. At $p_T^{\pi^0} \sim 8$ GeV/c, the angular separation between the two γ's resulting from a π^0 decay is small, but a π^0 shower is generally broader than a single γ shower. The BSMD is capable of $(2\gamma)/(1\gamma)$ separation up to $p_T^{\pi^0} \sim 20$ GeV/c due to its high granularity ($\Delta\eta \sim 0.007$, $\Delta\phi \sim 0.007$). The shower shape is quantified as the cluster energy, measured by the BEMC, normalized by the position-weighted energy moment, measured by the BSMD strips [5]. The shower profile cuts were tuned to obtain a nearly γ_{dir}-free (π^0_{rich}) sample and a sample rich in γ_{dir} (γ_{rich}). Since the shower-shape analysis is only effective for rejecting two close γ showers, the γ_{rich} sample contains a mixture of direct photons and contamination from fragmentation photons (γ_{frag}) and photons from asymmetric hadron (π^0 and η) decays.

Figure 1 (right panel) shows the v_2 of γ_{rich} sample and v_2 of the sample free of direct photons (π^0) using the off-η event plane method. As expected, the v_2 of γ_{rich} sample is smaller than that of π^0, while the $v_2^{\pi^0}$ is similar to v_2^{ch} which indicates the π^0 sample identified by the transverse shower-shape analysis, to be free of γ_{dir}.

2.4. v2 of direct photons

The $v_2^{\gamma_{dir}}$ is given by:

$$v_2^{\gamma_{dir}} = \frac{v_2^{\gamma_{rich}} - \mathcal{R} v_2^{\pi^0}}{1 - \mathcal{R}} \tag{3}$$

where $\mathcal{R} = \frac{N^{\pi^0}}{N^{\gamma_{rich}}}$, and the numbers of π^0 and γ_{rich} triggers are represented by N^{π^0} and $N^{\gamma_{rich}}$ respectively. The value of $\mathcal{R}$ is measured in [5] and found to be $\sim 30\%$ in central Au+Au. In Eq. 3 all background sources for γ_{dir} are assumed to have the same v_2 as π^0. Thus, excepting those γ_{frag} that have no near-side yield, all other sources of γ_{dir}'s background are removed.

Figure 2 (left and right panels) shows the $v_2^{\gamma_{dir}}$ and $v_2^{\pi^0}$ compared to v_2^{ch} at high p_T using the two different event-plane methods. While the $v_2^{\pi^0}$ and v_2^{ch} are similar ($\sim 12\%$), the $v_2^{\gamma_{dir}}$ is systematically lower than that of hadrons. The similarity of the v_2 results using the full-TPC and off-η at high p_T, along with the non-zero value of $v_2^{\gamma_{dir}}$, indicate the non-flow contributions to the measured v_2.

3. Conclusions

The STAR experiment has reported the first $v_2^{\gamma_{dir}}$ at high-p_T at RHIC. The non-zero value of $v_2^{\gamma_{dir}}$ is probably due to contributions from non-flow to the standard method of v_2 measurements, where the pseudorapidity gap between the particle of interest and the particles used for the reaction plane determination is smaller than two units in pseudorapidity. The positive value of $v_2^{\gamma_{dir}}$ demonstrates the negligible contribution of jet-medium photons [7] and possible contributions of γ_{frag} [6] to the direct photon productions over the covered kinematics range.

References

[1] STAR Collaboration, J. Adams *et al* , Nucl. Phys. A**757**, 102 (2005).
[2] E. V. Shuryak, Phys. Rev. C **66** 027902 (2002).
[3] STAR Collaboration, J. Adams *et al* , Phys. Rev. Lett. **93**, 252301 (2004).
[4] PHENIX Collaboration, A. Adare *et al* , Phys. Rev. Lett. **105**, 142301 (2010).
[5] STAR Collaboration, B. I. Abelev *et al* , Phys. Rev. C **82**, 034909 (2010).
[6] B. G. Zakharov, JETP Lett. **80**, 1 (2004).
[7] R. J. Fries, B. Müller, and D. K. Srivastava, Phys. Rev. Lett. **90**, 132301 (2003).
[8] S. Turbide, C. Gale, and R. J. Fries, Phys. Rev. Lett. **96**, 032303 (2006).
[9] G. Qin *et al* , Phys. Rev. C **80** 054909 (2009).
[10] PHENIX Collaboration, S. S. Adler *et al* , Phys. Rev. Lett. **94**, 232301 (2005).
[11] M. Beddo *et al* , Nucl. Instrum. Meth. A**499**, 725 (2003).
[12] M. Anderson *et al* , Nucl. Instrum. Meth. A**499**, 659 (2003).
[13] STAR Collaboration, B. I. Abelev *et al* , Phys. Rev. C **77** 054901 (2008).
[14] A. M. Poskanzer and S. A. Voloshin , Phys. Rev. C 58, **1671** (1998).
[15] E877 Collaboration, J. Barrette *et al* , Phys. Rev. C **56** 3254 (1997).

Journal of Physics: Conference Series **270** (2011) 012011 doi:10.1088/1742-6596/270/1/012011

Jet Fragmentation in Vacuum and Medium with γ-hadron Correlations in PHENIX

Matthew Nguyen

Research Fellow, CERN, Geneva, Switzerland

E-mail: Matthew.Nguyen@cern.ch

Abstract.
Jet fragmentation in p+p and Au+Au collisions is studied via back-to-back correlations of direct photons and charged hadrons. The direct photon correlations are obtained by statical subtraction of the background from decay photons. Results on the nuclear modification to the associated charged hadron yields are reviewed. Further studies of jet fragmentation in p+p using isolated direct photons are also presented. A k_T-smeared LO pQCD calculation is used to interpret the data. The sensitivity of the data to the underlying fragmentation function is tested and the results are found to be compatible with expectations of a sample dominated by quark jet fragmentation.

1. Introduction

The suppression of high p_T hadrons and the modification of back-to-back di-hadron correlations in nuclear collisions are generally understood to be manifestations of in-medium parton energy loss [1, 2]. The quantitive interpretation of such measurements is, however, complicated by the fact that the initial parton energy is unknown. The dominant mechanism for direct photon production at LO in pQCD is the quark-gluon Compton scattering process, $q + g \rightarrow q + \gamma$. Compared to partons, photons have a small cross section to interact with the medium. Modulo higher order effects, the direct photon can hence be used to determine the initial p_T of the recoil parton before energy loss, since they exactly balance. Using the photon as a proxy for the parton, the fragmentation function of the recoil jet, which may be effectively modified by the medium, can be studied in detail [3].

2. Photon-Hadron Correlations

Experimentally, the challenge of direct photon measurements is to separate them from the large background of photons produced from hadron decays. Direct photon correlations with charged hadrons may be obtained by a statistical subtraction of the decay component as follows. First the per-trigger yield (Y) of all photons is measured, which is the number of photon-hadron pairs divided by the number of photons, i.e., $Y_{\text{total}} \equiv N_{\gamma-h}/N_{\gamma}$. This quantity can be decomposed into contributions from direct and decay sources:

$$Y_{\text{total}} = \frac{N_{\text{direct}}}{N_{\text{total}}} Y_{\text{direct}} + \frac{N_{\text{decay}}}{N_{\text{total}}} Y_{\text{decay}}, \qquad (1)$$

where "direct" is taken to signify any photon not from hadron decay. We may solve for Y_{direct},

$$Y_{\text{direct}} = \frac{R_\gamma}{R_\gamma - 1} Y_{\text{total}} + \frac{1}{R_\gamma - 1} Y_{\text{decay}}, \tag{2}$$

where the direct photon excess, $R_\gamma \equiv N_{\text{total}}/N_{\text{decay}}$, is known from measurements of the direct photon cross section [4, 5]. Y_{decay} is estimated from π^0 and η-hadron correlations which compose more than 95% of the decay background. The daughter photon correlations are determined from those of their parent mesons by a simulation which takes into account decay kinematics as well as detector resolution, acceptance and efficiency [6].

Figure 1 shows examples of azimuthal correlations, $Y(\Delta\phi)$, for direct photons in p+p and central Au+Au collisions, as well as the corresponding total and decay photon correlations used in the statistical subtraction [6]. On the near-side Y_{direct} is consistent with zero, as would be expected at LO. The disappearance of back-to-back correlations in central Au+Au, a phenomenon well-known from di-hadron correlations, is also evident in Y_{direct}. Figure 2 shows the nuclear modification factor I_{AA}, which is the ratio of Y in Au+Au to Y in p+p, as a function of collision centrality as quantified by the number of participants. Also shown is I_{AA} for π^0-hadron correlations [7] as well as R_{AA} for single π^0's [8] which both demonstrate a level of modification consistent with the direct photon correlations within sizable uncertainties. The parton path-length through the medium is expected to vary amongst the three observables such that for a medium with an extended region of partial transmission, one should observe a different level of modification amongst these observables [9]. In contrast, the results suggest that the medium is sufficiently absorbent that the path-length difference is beyond the sensitivity of the current measurements. Although initial measurements of the fragmentation function using direct photon correlations have been made (not shown), no modification is apparent within the sizable statistical and systematic uncertainties [10].

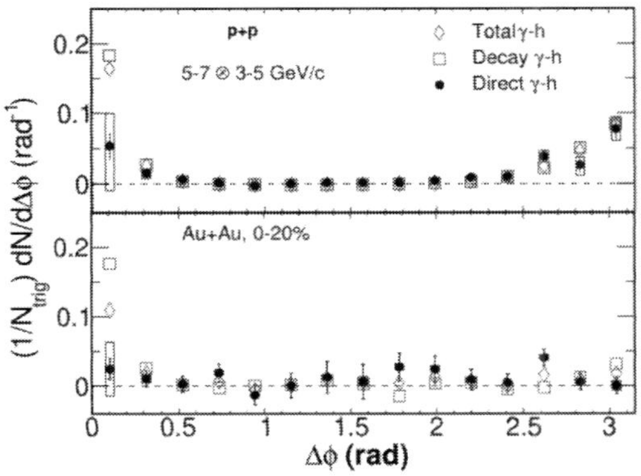

Figure 1. Azimuthal correlations of charged hadrons with inclusive, decay and direct photons in p+p and central Au+Au [6].

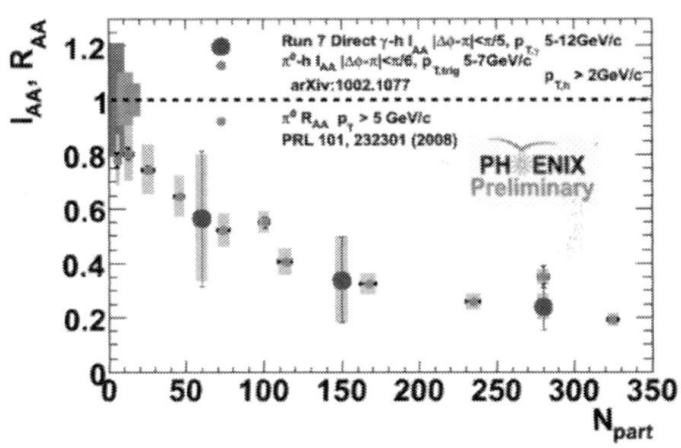

Figure 2. Nuclear modification factors for π^0 (R_{AA}) and direct photon-associated and π^0-associated charged hadrons (I_{AA}).

The uncertainties in the direct photon correlations are predominantly due to the statistical subtraction of the large decay background. Higher precision may be achieved by removing decay photons event-by-event. This is accomplished by tagging decay photon pairs from π^0 and η based on their invariant mass and by applying an isolation cut. Given a finite detector acceptance and efficiency, a residual decay photon component survives these cuts which is subtracted at the statistical level [11]. These methods are difficult to apply in high multiplicity nuclear collisions, but are relatively straight-forward to apply in p+p collisions. Figure 3 compares azimuthal correlations for direct photons using tagging and isolation to those from the purely statistical subtraction. The uncertainties are notably reduced. To study the fragmentation of the away-side jet, $x_{\mathrm{E}}\ (\equiv \vec{p}_{\mathrm{T},\gamma} \cdot \vec{p}_{\mathrm{T,h}}/|p_{\mathrm{T},\gamma}|^2)$ is used as a proxy z, a good approximation for $p_{T,\gamma} \approx p_{jet}$. Figure 4 shows the away-side per-trigger yield vs. x_{E} for isolated direct photon triggers in p+p.

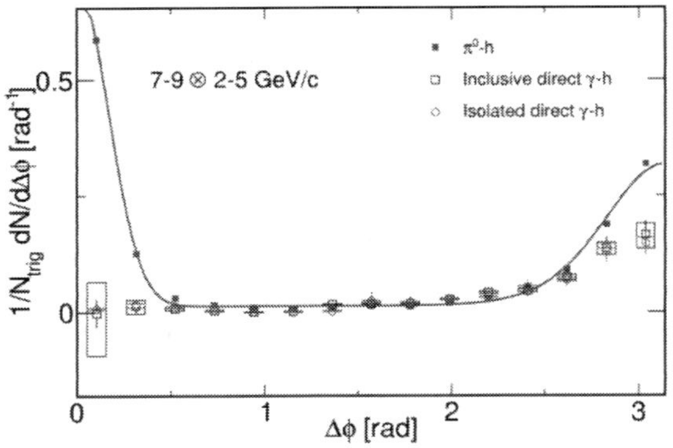

Figure 3. Azimuthal correlations of charged hadrons with π^0 's, inclusive and isolated direct photons [11].

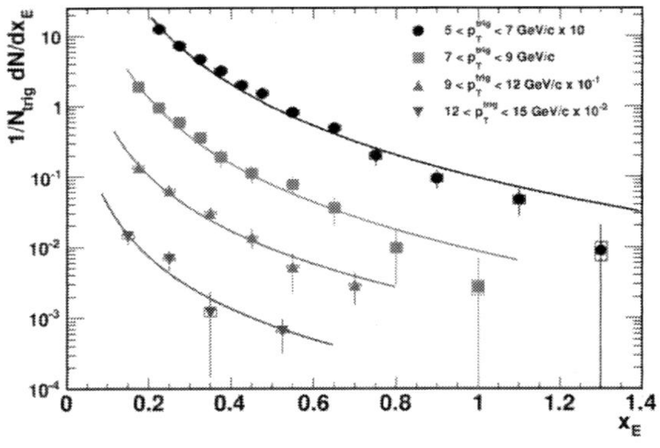

Figure 4. Charged hadron yield per isolated direct photon as a function of x_{E} (scaled for visibility) with power-law fits [11].

The values of the parameter n of power-law fits to the x_{E} distributions (dN/d$x_{\mathrm{E}} \propto x_{\mathrm{E}}{}^{-n}$) are shown in Fig. 5. To test the sensitivity of the data to the underlying fragmentation function, a LO pQCD calculation was performed adding a phenomenological Gaussian k_{T} smearing whose width was determined from data [11]. The calculation was performed for both the Compton scattering and annihilation sub-processes, which correspond to quark and gluon fragmentation, respectively. The data lie closer to the calculation for the Compton sub-process which is expected to dominate. Photons are more likely to oppose an up than a down quark due to the valence quark content of the proton and the dependence of the Compton scattering amplitude on the electric charge of the quark. Hence one should observe an excess of positive charge for away-side hadrons which is confirmed by data in Fig 6. Also shown is the LO+k_{T} smearing calculation.

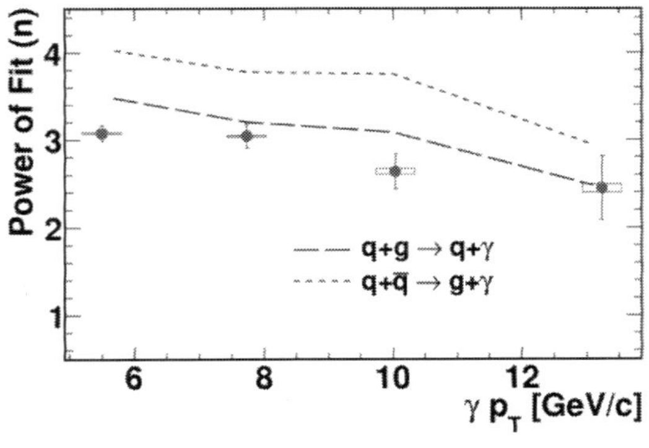

Figure 5. Power parameter n of fits to the x_{E} distributions of direct photon triggers [11].

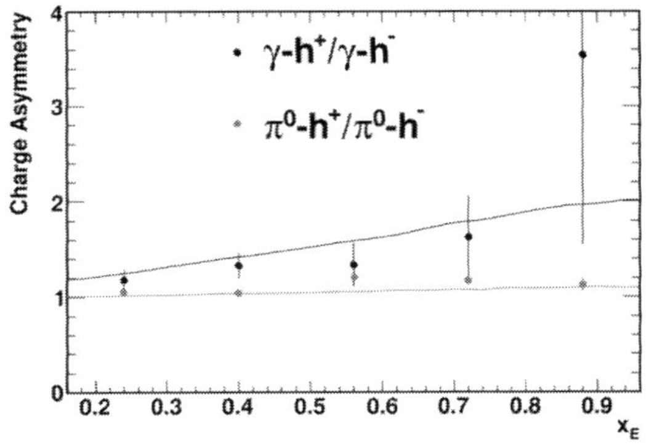

Figure 6. Charge asymmetry of partner hadrons for direct photon and π^0 triggers [11].

Journal of Physics: Conference Series **270** (2011) 012011 doi:10.1088/1742-6596/270/1/012011

3. Conclusions

We have presented results on two-particle correlations using direct photon triggers in p+p and Au+Au collisions. Using a purely statistical method, we observe the disappearance of back-to-back correlations in central Au+Au collisions similar to previous observations in the di-jet channel. The level of suppression is consistent to that observed for single and di-hadrons, within the current statistical and systematic errors. In p+p collisions, more precise results were obtained by the event-by-event identification of a large fraction of the decay photon background contribution. This was achieved using invariant mass tagging of decay pairs as well as isolation cuts. The data were interpreted in terms of a model in which a Gaussian k_T-smearing was applied to LO pQCD. The recoil jet was found to be compatible with quark fragmentation rather gluon fragmentation, as expect for sample dominated by LO quark-gluon Compton scattering. A charge asymmetry of the away-side jet fragments was observed, further confirming the dominance of the quark-like nature of the away-side jet fragmentation. The application of event-by-event methods points the way to future studies in which the uncertainties on the heavy-ion measurement may be reduced. Furthermore, the results demonstrate that direct photons are an effective quark jet tag, a feature which may be exploited in heavy-ion collisions to study the color factor dependence of parton energy loss.

References

[1] Adcox K *et al.* (PHENIX), *Phys. Rev. Lett.* **88**, 02230 (2002).
[2] Adcox K *et al.* (PHENIX), *Phys. Rev. Lett.* **97**, 052301 (2003).
[3] For a review of medium modifications to fragmentation functions see
Arleo F, *Eur. Phys. J* **C61** 603, (2009).
[4] Adler S S *et al.* (PHENIX), *Phys. Rev. Lett.* **94**, 23230 (2005).
[5] Adler S S *et al.* (PHENIX), *Phys. Rev. Lett.* **98**, 012002 (2006).
[6] The statistical subtraction method and estimate of decay photon correlations are described in detail in
Adare A *et al.* (PHENIX), *Phys. Rev.* **C80**, 024908 (2010).
[7] Adare A *et al.* (PHENIX), *Phys. Rev. Lett.* **104**, 252301 (2010).
[8] Adare A *et al.* (PHENIX), *Phys. Rev. Lett.* **101**, 232301 (2008).
[9] Zhang H, Owens J F, Wang E and Wang X-N, *Phys. Rev. Lett.* **98**, 12301 (2007).
Zhang H, Owens J F, Wang E and Wang X-N, *Phys. Rev. Lett.* **103**, 032302 (2009).
[10] Connors M, *Nucl. Phys. A* **830**, 447C-450C (2009).
[11] Adare A *et al.* (PHENIX), *Phys. Rev.* **D82**, 072001 (2010).

Journal of Physics: Conference Series **270** (2011) 012012 doi:10.1088/1742-6596/270/1/012012

Mach Cones in Viscous Matter

I. Bouras[1], A. El[1], O. Fochler[1], F. Lauciello[1], F. Reining[1], J. Uphoff[1], C. Wesp[1], E. Molnár[2,3], H. Niemi[2], Z. Xu[1,2] and C. Greiner[1]

[1] Institut für Theoretische Physik, Johann Wolfgang Goethe-Universität, Max-von-Laue-Str. 1, D-60438 Frankfurt am Main, Germany

[2] Frankfurt Institute for Advanced Studies, Ruth-Moufang-Str. 1, D-60438 Frankfurt am Main, Germany

[3] KFKI, Research Institute of Particle and Nuclear Physics, H-1525 Budapest, P.O.Box 49, Hungary

E-mail: bouras@th.physik.uni-frankfurt.de

Abstract. Employing a microscopic transport model we investigate the evolution of high energetic jets moving through a viscous medium. For the scenario of an unstoppable jet we observe a clearly strong collective behavior for a low dissipative system $\eta/s \approx 0.005$, leading to the observation of cone-like structures. Increasing the dissipation of the system to $\eta/s \approx 0.32$ the Mach Cone structure vanishes. Furthermore, we investigate jet-associated particle correlations. A double-peak structure, as observed in experimental data, is even for low-dissipative systems not supported, because of the large influence of the head shock.

1. Introduction

The elliptic flow coefficient v_2 measured in heavy-ion collisions at the Relativistic Heavy Ion Collider has a large value, leading to the indication that the created quark-gluon plasma (QGP) behaves like an almost perfect fluid [1, 2]. This is confirmed by recent calculations of viscous hydrodynamics [3] and microscopic transport calculations [4, 5]. Jet-quenching [6] has been discovered and very exciting jet-associated particle correlations have been observed [7]-[10]. They indicate the formation of shocks in form of Mach Cones induced by highly-energetic partons traversing the QGP [11]. Measurements of the Mach Cone angle could give us the possibility to extract the equation of state of the QGP.

In this work we address the question, under what conditions Mach Cones can develop in viscous gluon matter for given constant η/s values. Furthermore, we discuss a possible double-peak structure in jet-associated two-particle correlations.

2. The parton cascade BAMPS

The Boltzmann Approach of Multi Parton Scatterings (BAMPS) employed in this study is a microscopic transport model solving the Boltzmann equation

$$p^\mu \partial_\mu f(x,p) = C(x,p) \quad (1)$$

for on-shell particles with the collision integral $C(x,p)$. The algorithm for collisions is based on the stochastic interpretation of the transition rate [12, 13]. In this study, we consider only binary gluon scattering processes with an isotropic constant cross section, which is related to the η/s ratio via a simple relation [14, 15].

3. Shocks Waves and Mach Cones

Mach Cones, which are special phenomena of shock waves, have their origin in ideal hydrodynamics [16]. A very weak perturbation in a perfect fluid induces sound waves which propagate with the speed of sound $c_s = \sqrt{dp/de}$, where p is the pressure and e is the energy density. In the case where the perturbation with velocity $v_{\rm jet}$ moves faster than the sound waves, the sound waves lie on a cone with an emission angle $\alpha_w = \arccos(c_s/v_{\rm jet})$. Considering a massless gluon gas, with $e = 3p$ and $c_s = 1/\sqrt{3}$, and a perturbation with $v_{\rm jet} = 1$, the emission angle is

$$\alpha_w = 54,73^\circ\,. \tag{2}$$

In the case of stronger perturbations, the sound waves move faster than the speed of sound through the medium and therefore are called shock waves [17]. Then we can approximate the emission angle by

$$\alpha \approx \arccos\frac{v_{\rm shock}}{v_{\rm jet}}\,. \tag{3}$$

The velocity of the shock front $v_{\rm shock}$ depends on the pressure (energy density) in the shock front region p_0 (e_0) and in the stationary medium itself p_1 (e_1):

$$v_{\rm shock} = \left[\frac{(p_1 - p_0)(e_0 + p_1)}{(e_1 - e_0)(e_1 + p_0)}\right]^{\frac{1}{2}} \tag{4}$$

Eq.(4) has the following limits: For $p_0 >> p_1$ we obtain $v_{\rm shock} = 1$. If $p_0 \approx p_1$, that is, a very weak perturbation, we get the expected limit of the speed of sound $v_{\rm shock} \approx c_s$. In the latter case Eq.(3) reduces to the one given by α_w, as expected.

4. Transition from ideal to viscous Mach Cones in BAMPS

In this section we employ the microscopic transport model BAMPS to investigate Mach Cones with different strength of dissipations in the medium. All simulations are realized within a static and uniform medium of massless Boltzmann particles and $T = 400$ MeV. To save computational runtime we reduce our problem to two dimensions. The physical results compared to full three-dimensional simulations are similar except for the fact that the energy density decreases slower in our reduced geometry. Here we choose the xz-plane and apply a periodic boundary condition in y-direction.

A jet moving in positive z-direction is initialised at $t = 0$ fm/c at the position $z = -0.8$ fm. The jet is treated as a massless particle with zero spatial volume and zero transverse momentum, that is, $p_z = E_{\rm jet} = 200$ GeV and $v_{\rm jet} = 1$. The jet deposits energy and momentum to the medium via collisions with medium particles. In this scenario we neglect the deflection of the jet; its energy and momentum is set to its initial value after every collision.

In Fig.1 we demonstrate the transition from ideal Mach Cone to a highly viscous one by adjusting the shear viscosity over entropy density ratio in the medium from $\eta/s = 1/64\pi \approx 0.005$ to $1/\pi \approx 0.32$. The energy deposition of the jet is approximately $dE/dx = 11 - 14$ GeV/fm. The simulations are shown at $t = 2.5$ fm/c.

Using a non-physical small viscosity of $\eta/s = 1/64\pi$ we observe a strong collective behavior in form of a Mach Cone, as shown in the left panel of Fig.1. For comparison, the ideal Mach Cone caused by a very weak perturbation is given by a solid line. Its emission angle is given by (2). The shock wave, characterized by a higher energy density compared to the medium at rest, propagates with the emission angle α smaller than the ideal one, α_w. Due to the fact that the energy deposition is strong, the shock propagates through the medium faster than the speed of sound, given approximately by Eq. (3). A simulation where the energy deposition to the medium is much smaller reproduces the ideal Mach Cone with its emission angle α_w (2).

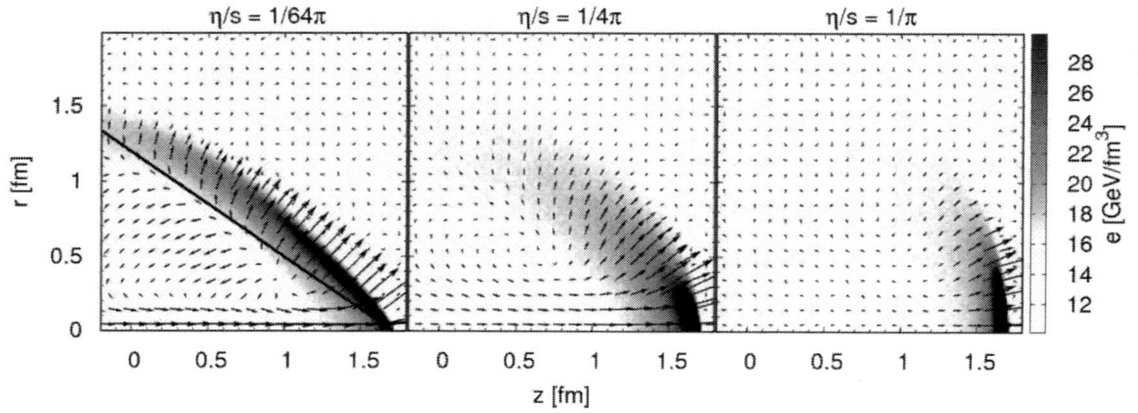

Figure 1. (Color online) The shape of a Mach Cone shown for different viscosities of the medium, $\eta/s = 1/64\pi$ (left), $\eta/s = 1/4\pi$ (middle), $\eta/s = 1/\pi$ (right). We show here the energy density plotted together with the velocity profile. Additionally, in the left panel the linear ideal Mach Cone for a very weak perturbation is shown by a solid line; its emission angle is (2).

Furthermore, a strong diffusion wake in direction of the jet, characterized by decreased energy density, and a head shock in the front are clearly visible. The results agree qualitatively with those found in [18]-[20].

If we increase the viscosity of the medium to larger values, shown in the middle and left panel of Fig.1, the typical Mach Cone structure smears out and vanishes completely. Due to stronger dissipation, the collective behavior gets weaker because of less particle interactions in the medium with a larger η/s. The results agree qualitatively with the analysis of one-dimensional shock waves realized in earlier studies using kinetic theory and viscous hydrodynamics [21]-[23], where a smearing-out of the shock profile is observed with higher viscosity.

5. Two-particle Correlation

The existence of a conical structure in Fig.1 for low viscosity are very exciting and naively lead to the assumption of a double-peak structure in jet-associated particle correlations. In Fig.2 we show the particle distribution $dN/(d\omega N)$ plotted vs. ω as extracted from BAMPS, where $\cos\omega = p_z/\sqrt{p_x^2 + p_z^2}$. We remark at this point that one has to mirror the results for 180° - 360°.

In the left panel we show the particle distribution for all particles. A peak in direction of the jet at 180° is observed. A big contribution of the diffusion wake and head shock is present and hinders therefore the appearance of a double peak structure. In the same figure the particle distribution without the diffusion wake is plotted. However, the contribution of the head shock is still too large to get a double peak structure. A systematic cut in space is shown in the right panel, where we try to cut simultaneously a part of the diffusion wake and head shock. As one can see, a double peak structure appears and is enhanced with larger cut in space. However, such a cut in space does not relate to any experimental observation in heavy ion collisions and has to be treated with caution.

We conclude that a double peak structure exists, but the contribution of the diffusion wake and head shock might be stronger. This phenomena will be discussed in future studies.

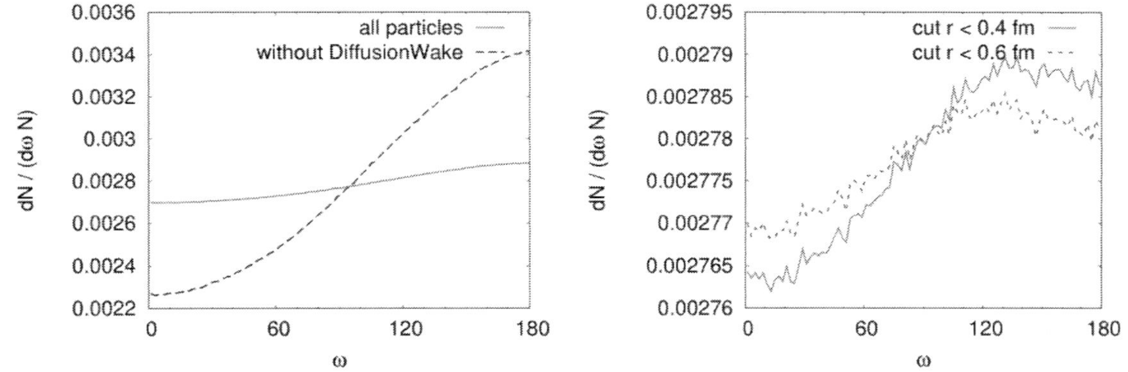

Figure 2. (Color online) Jet-associated particle correlations extracted from the simulation with $\eta/s = 1/64\pi$ as discussed in Fig.1. The left panel shows the correlations including all particles and one without the part of the diffusion wake. The right panel shows the correlations by systematically cuts in space to subtract the contribution of the diffusion wake and head shock. r is the transverse direction to the z-axis.

Acknowledgements
The authors are grateful to the Center for Scientific Computing (CSC) at Frankfurt University for the computing resources. I. B. is grateful to HGS-Hire. E. M. acknowledges support by OTKA/NKTH 81655 and the Alexander von Humboldt foundation. The work of H. N. was supported by the Extreme Matter Institute (EMMI).

This work was supported by the Helmholtz International Center for FAIR within the framework of the LOEWE program launched by the State of Hesse.

References

[1] S. S. Adler *et al.* [PHENIX Collaboration], Phys. Rev. Lett. **91**, 182301 (2003)
[2] J. Adams *et al.* [STAR Collaboration], *ibid.* **92**, 052302 (2004)
[3] M. Luzum and P. Romatschke, Phys. Rev. C **78**, 034915 (2008)
[4] Z. Xu, C. Greiner and H. Stöcker, Phys. Rev. Lett. **101**, 082302 (2008)
[5] Z. Xu and C. Greiner, Phys. Rev. C **79**, 014904 (2009)
[6] J. Adams *et al.* [STAR Collaboration], Phys. Rev. Lett. **91**, 172302 (2003)
[7] F. Wang [STAR Collaboration], J. Phys. G **30**, S1299 (2004)
[8] J. Adams *et al.* [STAR Collaboration], Phys. Rev. Lett. **95**, 152301 (2005)
[9] S. S. Adler *et al.* [PHENIX Collaboration], *ibid.* **97**, 052301 (2006)
[10] J. G. Ulery [STAR Collaboration], Nucl. Phys. A **774**, 581 (2006)
[11] H. Stöcker, Nucl. Phys. A **750**, 121 (2005)
[12] Z. Xu and C. Greiner, Phys. Rev. C **71**, 064901 (2005)
[13] Z. Xu and C. Greiner, Phys. Rev. C **76**, 024911 (2007)
[14] Z. Xu and C. Greiner, Phys. Rev. Lett. **100**, 172301 (2008)
[15] A. El, A. Muronga, Z. Xu and C. Greiner, Phys. Rev. C **79**, 044914 (2009)
[16] L. D. Landau and E. M. Lifshitz, *Fluid Dynamics*, Second Edition, Butterworth-Heinemann (1987)
[17] D. H. Rischke, H. Stoecker and W. Greiner, Phys. Rev. D **42**, 2283 (1990)
[18] I. Bouras *et al.*, J. Phys. Conf. Ser. **230**, 012045 (2010)
[19] B. Betz *et al.*, Phys. Rev. C **79**, 034902 (2009)
[20] D. Molnar, AIP Conf. Proc. **1182**, 791 (2009)
[21] I. Bouras *et al.*, Phys. Rev. Lett. 103, 032301 (2009)
[22] I. Bouras *et al.*, Nucl. Phys. A **830**, 741C (2009)
[23] I. Bouras *et al.*, Phys. Rev. C **82**, 024910 (2010)

Journal of Physics: Conference Series **270** (2011) 012013 doi:10.1088/1742-6596/270/1/012013

The Influence of the Underlying Event on Jet Measurements in Heavy Ion Collisions

Martin Spousta for the ATLAS Collaboration
Charles University in Prague, Faculty of Mathematics and Physics
E-mail: spousta@ipnp.troja.mff.cuni.cz

Abstract. We present a study of the performance of jet reconstruction algorithms for Pb+Pb collisions at $\sqrt{s_{NN}} = 5.52$ TeV based on a full GEANT simulation of the ATLAS detector. It is shown that ATLAS can provide measurements that are sensitive to different in-medium energy loss scenarios leading to the jet quenching. The capability of measurements of fragmentation function and j_T distribution are presented. We concentrate on some biases in these measurements that might be due to the detector effects or to the choice of a jet finding algorithm.

1. Introduction

Measurements of jets and their properties in heavy ion collisions provide information about the jet quenching phenomenon that is predicted for heavy ion collisions and already measured at RHIC. The jet quenching means that a hard parton, before fragmenting into a jet of hadrons, deposits a significant fraction of its energy in the medium, leading to a change of jet energy distribution and particle multiplicity inside a jet. At RHIC experiments the jet quenching phenomenon has been clearly established in the measurement of single- and di-hadron spectra [1]. Due to the limited p_T reach of these measurements and the large soft background, they suffer from biases towards particles that escape the medium having lost little energy. In contrast, the high center-of-mass energy and large rates for high-p_T jets at the LHC will make full jet measurements possible over a wide range of jet energies. In these proceedings we present capabilities of the ATLAS detector [2] to measure jets and their properties in heavy ion collisions, focusing on the problems related to the choice of the jet finding algorithm, the impact of the underlying event on the jet energy resolution, and the feasibility of measuring the jet internal structure.

2. Choice of the jet finding algorithm

The jet finding in Pb+Pb collisions at LHC will have to deal with a substantial contribution from the underlying event. The underlying event multiplicity in Pb+Pb collisions at the LHC energies is not precisely know. The existing models predict very different values of $\mathrm{d}N_{ch}/\mathrm{d}\eta = 1000-6000$ in the most central Pb+Pb collisions. For the jet studies, not only the underlying event multiplicity is important, but also the event-by-event fluctuations in the calorimeter energy deposits. Left plot of Fig. 1 illustrates the difference in energy deposition per one calorimeter

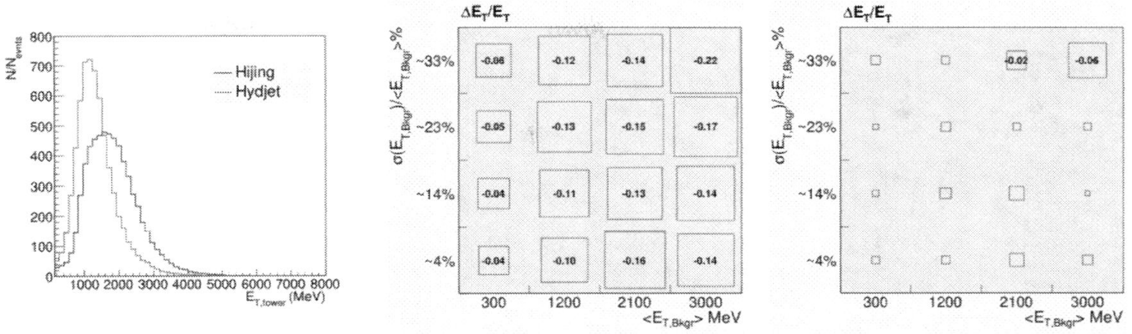

Figure 1. *Left:* E_{T} spectra of ATLAS calorimeter towers in simulation of HIJING and HYDJET central events. *Middle:* Mean difference in the energy of PYTHIA clean jets and jets reconstructed in a presence of calorimeter background of different characteristics. Jets are reconstructed using k_{T} algorithm. *Right:* The same as in the middle plot, but for jets reconstructed using anti-k_{T} algorithm.

tower[1] between two event generators, HIJING [3] and HYDJET [4] simulated with the same impact parameter range of $0-3$ fm. It is visible that the difference in the mean and in the width of distributions is quite sizable.

Bearing in mind these facts we decided to test a general performance of the jet finding algorithms with respect to a presence of the calorimeter background using a simple toy-model based on a random generation of the calorimeter background energy with Poisson-like distributions. This allows us to easily vary the size of the fluctuations in the background as well as the mean deposited energy. The background was then merged with the fully reconstructed PYTHIA [5] jets at the level of calorimeter towers. PYTHIA di-jet events with energy in the range of $30-80$ GeV have been used. The extreme case of the largest mean of the deposited energy ($\langle E_{\mathrm{T,bkgr}}\rangle \approx 3$ GeV per a calorimeter tower) and the largest size of fluctuations ($\sigma(E_{\mathrm{T,bkgr}})/\langle E_{\mathrm{T,bkgr}}\rangle \approx 33\%$) resembles the properties of the underlying event simulated by unquenched HIJING in the most central Pb+Pb collisions ($\mathrm{d}N_{ch}/\mathrm{d}\eta = 2700$, $b = 2$ fm) at $\sqrt{s_{NN}} = 5.52$ TeV. Events with and without the background were reconstructed using the jet finding algorithms implemented in FastJet package [6]. The mean value of the background has been subtracted after the jet reconstruction from each tower belonging to a given jet. The middle plot of Fig. 1 illustrates the relative difference in the transverse energy of clean PYTHIA jets and jets reconstructed in the presence of the background, both reconstructed using the k_{T} algorithm with the distance parameter $R = 0.4$. The x-axis shows the mean transverse energy of the calorimeter tower background, the y-axis shows the relative size of fluctuations in the background. One can see that k_{T} algorithm underestimates the jet energy in the presence of a sizeable background. This underestimation becomes larger for higher backgrounds. Jet energy scale, $\Delta E_{\mathrm{T}}/E_{\mathrm{T}}$, is shifted by as much as 20% for the background with the highest mean and the highest size of fluctuations. The underestimation of the jet energy is due to the fact that a softer part of a jet may preferably be clustered to the background rather than to a jet. This is reflected by the fact that with increasing calorimeter tower background the number of towers that are assigned to a jet by the k_{T} algorithm is decreasing. This is not the case for the anti-k_{T} algorithm or iterative cone algorithm where the number of towers remains almost constant across the phase space of different calorimeter tower backgrounds (see the left plot of Fig. 1). It was found that similarly to k_{T} also the Cambridge/Aachen algorithm suffers from the underestimation of the jet

[1] Calorimeter tower of the ATLAS detector has a size of $\Delta\eta \times \Delta\phi = 0.1 \times 0.1$.

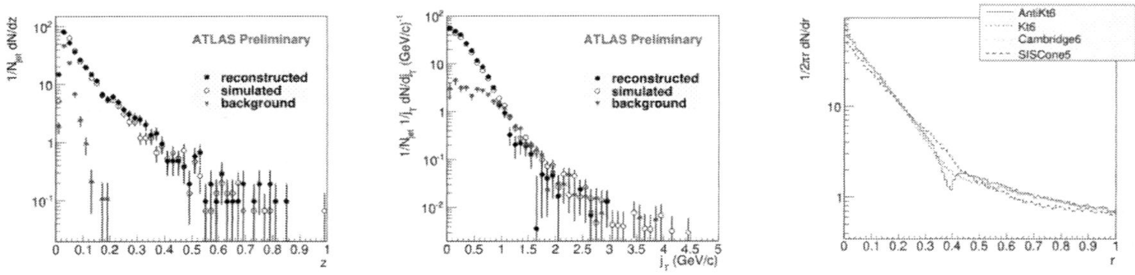

Figure 2. Comparison of the particle-level (*open circles*) fragmentation functions (*left*) and $j_{\rm T}$ distributions (*middle*) and the reconstructed distributions after the background subtraction and after the corrections on tracking efficiency and jet position resolution (*closed circles*) for the most central HIJING Pb+Pb collisions. The subtracted background distributions are shown separately (*triangles*). *Right:* Charged particle multiplicity measured as a function of the distance from the jet axis for different jet finding algorithms.

energy. In addition, it was also found that the jet position resolution and jet energy resolution is $\sim 2-6$ times better for the anti-$k_{\rm T}$ and iterative cone algorithm than in the case of $k_{\rm T}$ and Cambridge/Aachen algorithms.

We may conclude that the anti-$k_{\rm T}$ algorithm and iterative cone algorithm exhibit better performance in terms of the jet energy scale, jet energy resolution, and jet position resolution as compared to the $k_{\rm T}$ and Cambridge/Aachen algorithms. More details about the expected performance of the jet reconstruction in heavy-ion collisions can be found in Refs. [7, 8, 9].

3. Jet energy resolution

One of the basic variables used to evaluate the performance of the jet reconstruction is the jet energy resolution. Relative jet energy resolution of jets measured in the calorimeter can be written as $\frac{\sigma(E_{\rm T})}{E_{\rm T}} = \frac{p_0}{\sqrt{E_{\rm T}}} \oplus \frac{p_1}{E_{\rm T}} \oplus p_2$ [10]. The first term in the formula is the *stochastic term* which is an internal characteristic of a calorimeter and which describes the response to a particle showering that develops as a stochastic process. The second term is the *noise term* which reflects the calorimeter noise and/or underlying event fluctuations. The last term is the *constant term* which accounts for various instrumental effects and which can usually be neglected. In the Pb+Pb collisions the contribution of the underlying event to a jet is large. The energy deposition from the underlying event does not depend on the jet energy, thus it is reflected by the noise term which scales as $1/E_{\rm T}$ for the relative jet energy resolution $\sigma(E_{\rm T})/E_{\rm T}$. The size of the noise term is measured in units of energy by the constant p_1. The value of the constant p_1 is proportional to the size of a jet and its upper limit can be obviously estimated as $p_1 = \sqrt{A\sigma_b^2(E_{\rm T})}$, where $\sigma_b(E_{\rm T})$ is the root-mean-square of the underlying event energy deposition in a unit calorimeter area (i.e. in one calorimeter tower) and A is the area of a jet in these units (i.e. the number of towers that form a jet). In the case of HIJING background $p_1 \approx 15$ GeV at midrapidity, in the case of HYDJET p_1 is only 11 GeV. Thus, once we know a realistic size of the underlying event fluctuations we can estimate the jet energy resolution.

4. Jet internal structure

The high center-of-mass energy and large rates of high-$p_{\rm T}$ jets at the LHC will make full jet measurements possible even in the heavy ion environment. Particularly interesting are measurements of fragmentation function and $j_{\rm T}$ distribution. The fragmentation variable, z is the longitudinal fraction of the jet momentum carried by a particle, $z = (\hat{p}_{\rm jet} \cdot \vec{p}_{\rm particle})/|\vec{p}_{\rm jet}| \approx$

$p_{\mathrm{T,particle}} \cos R/p_{\mathrm{T,jet}}$, where $R = \sqrt{\Delta\eta^2 + \Delta\phi^2}$ is the opening angle between jet axis and position of a particle in $\eta \times \phi$ space. The fragmentation function is then defined as $D(z) = 1/N_{\mathrm{jet}} \mathrm{d}N/\mathrm{d}z$. The transverse momentum of particles with respect to the jet axis, j_{T}, is defined as $j_{\mathrm{T}} = |\hat{p}_{\mathrm{jet}} \times \vec{p}_{\mathrm{particle}}| \approx p_{\mathrm{T,particle}} \sin R$. The fragmentation functions and j_{T} distributions are expected to be modified in heavy ion collisions compared to p+p collisions and provide a direct information on the properties of the hot QCD medium created in heavy ion collisions [11].

Fig. 2 compares the truth fragmentation function and j_{T} distribution computed using charged particles and truth jets with the reconstructed distributions computed using tracks. The background distributions that are subtracted are also shown. For the reconstruction, the most central HIJING collisions ($b = 2$ fm, $\mathrm{d}N_{ch}/\mathrm{d}\eta = 2700$) were used. The transverse momentum cut used for particles and tracks is $p_{\mathrm{T}} \geq 2$ GeV/c. Jets were reconstructed using the iterative cone algorithm. The energy of generated jets is between 70 GeV and 140 GeV. The correction for tracking efficiency of 70% as well as the position resolution correction have been applied. As one can see there is a good agreement between the truth and reconstructed fragmentation function and j_{T} distribution.

One of the measurements sensitive to the in-medium modifications of jets is the measurement of particle multiplicity as a function of the distance from the jet axis. The right plot of Fig. 2 shows the particle multiplicity measured as a function of the distance from the jet axis (r) for p+p PYTHIA simulated minimum-bias events at $\sqrt{s} = 900$ GeV with jets reconstructed using different jet finding algorithms (anti-k_{T}, k_{T}, and Cambridge/Aachen algorithms with $R = 0.4$, and SISCone algorithm with $R = 0.4$ and split-merge fraction of 0.75). One can see that the multiplicity evolution with r depends on the choice of the jet finding algorithm. Particularly noticeable is a dip at around $r = R = 0.4$ present in the multiplicity of particles obtained from the anti-k_{T} algorithm. This effect shows that the anti-k_{T} optimizes the jet axis with respect to the jet periphery that is defined by the distance parameter R. This effect, if seen in the heavy ion data, could be miss-interpreted as a Mach cone. Clearly it is important to use different jet finding algorithms.

More details on the measurement of the jet internal structure can be found in Ref. [12].

5. Summary

ATLAS can provide measurements of jets and their properties that are sensitive to different energy loss mechanisms in the presence of the heavy ion environment. We are looking forward to collect first heavy ion data in the fall of 2010.

References

[1] A. Majumder, Nucl. Phys. **A830** (2009) 875c.
[2] The ATLAS Collaboration, G. Add *et al.*, Journal of Instrumentation **3** (2008) S08003.
[3] M. Gyulassy and X.-N. Wang, Comput. Phys. Commun. **83** (1984) 307.
[4] I.P. Lokhtin, A.M. Snigirev, Eur. Phys. J. **C46** (2006) 211.
[5] T. Sjostrand, S. Mrenna, P. Skands, Journal of High Energy Physics **0605** (2006) 026.
[6] Cacciari M., hep-ph/0607071.
[7] N. Grau for the ATLAS Collaboration, J. Phys. G: Nucl. Part. Phys. **35** (2008) 104040.
[8] N. Grau *et al.*, Eur. Phys. J. **C62** (2009) 191.
[9] M. Spousta for the ATLAS Collaboration, ATL-PHYS-PROC-2009-022.
[10] Fabjan W.C., Gianotti F., Rev. Mod. Phys. **75** (2003) 1243.
[11] Salgado C.A., Wiedemann U.A., Phys. Rev. Lett **93** (2004) 042301.
[12] M. Spousta for the ATLAS Collaboration, ATL-PHYS-PROC-2009-002.

Intrajet radiation study with the ALICE experiment at LHC

Swensy Jangal[1,2] for the ALICE Collaboration
[1] IPHC, 23 rue du Loess-BP 28, 67037 STRASBOURG cedex 2, France
[2] Subatech, 4 rue Alfred Kastler La Chantrerie BP 20722 44307 NANTES cedex 3, France
E-mail: swensy.jangal@ires.in2p3.fr

Abstract. The ALICE collaboration is interested in measuring the $\xi = ln(p_T^{jet}/p_T^{hadron})$ distribution of hadrons in jets known as the Hump-backed plateau (HBP). After an introduction to color coherence effects responsible for the shape of the HBP, we discuss the ALICE capabilities to perform such a study. Preliminary results from ALICE for non corrected charged particle and jet spectra measurement in p+p collisions at $\sqrt{s}$ = 900 GeV and 7 TeV are presented. Finally, we show the results of a MC based analysis of intrajet radiations. It supports the idea of jet collimation which increases with increasing jet energy as expected from QCD and show the feasability of this study with ALICE.

1. Motivations for intrajet radiation study

Hard processes in QCD produce high energy initial partons which give birth to a cascade of radiated partons: the jet. The way partons radiate in the parton shower is constrained by color coherence effects [1]. From one emission to the next one, the θ emission angle is always decreasing because of the color charge of the former emitted parton. The shower keeps evolving with emission of partons of momentum k as long as $\theta > (kR)^{-1}$ (where $1/R \propto$ hadron mass at the end of the shower). Consequently, the sooner the shower is stopped, the heavier might be the mass of the resulting hadron at the end of the shower and the larger is the last emission angle [2]. Both phenomena are responsible for the shape of the so-called hump-backed plateau (HBP) which summerizes how hadrons are distributed inside jets as a function of the $\xi = \ln(p_T^{jet}/p_T^{hadron})$ variable. First, the so-called splitting function P(z) favours low p_T emission since it is proportional to 1/z (z $= p_T^{hadron}/p_T^{jet}$) resulting in an increase of the HBP with increasing ξ. Then, the combination of angular ordering and hadronization scale leads to the early suppression of soft emissions in the parton shower resulting in the depletion of the HBP in its high ξ region.

One of the models predicting the shape of this HBP is the Modified Leading Logarithmic Approximation (MLLA) which contains two free parameters: a QCD scale and a scale stopping the evolution of the parton shower. Because partons eventually fragment in sprays of hadrons at the end of the shower, MLLA is usually supplemented by the Local Parton Hadron Duality (LPHD) hypothesis in order to describe the hadron p_T distribution inside jets. The LPHD hypothesis assumes a one-to-one correspondence between a last emitted parton and a hadron so that the parton distributions at the end of the shower are directly reflected in the hadron distribution. The shape of the HBP has already been measured for instance in e^+e^- collisions

in TASSO and OPAL experiments for $\sqrt{s}$ = 14 GeV and $\sqrt{s}$ = 192-209 GeV respectively as well as in hadronic p+$\bar{p}$ collisions by the CDF collaboration at $\sqrt{s}$ = 1.8 TeV [3]. All experimental results are in good agreement with the expected shape predicted by the MLLA+LPHD [4].
Figure 1 presents the non corrected HBP expected to be measured by the ALICE experiment in p+p collisions at $\sqrt{s}$ = 10 TeV. These results are based on PYTHIA jet simulations performed

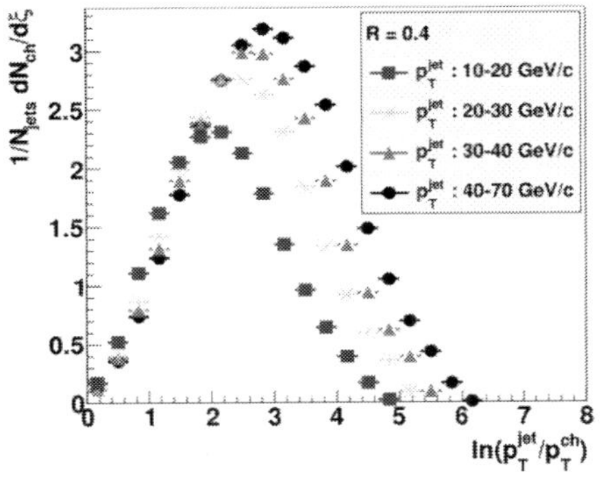

Figure 1. Hump-backed plateau expected in ALICE extracted from the fully reconstructed charged data coming from PYTHIA simulated p+p collisions at $\sqrt{s}$ = 10 TeV for four bins in jet transverse momentum: 10-20, 20-30, 30-40 and 40-70 GeV/c.

within the full simulation/reconstruction chain of ALICE using GEANT 3 to account for the transport of particles in the detectors. Jets have been reconstructed with a cone algorithm based on the UA1 jet finder running on charged particles only with a radius R = $\sqrt{\Delta\eta^2 + \Delta\phi^2}$ = 0.4. The same set of data is used for the intrajet radiation studies in section 3. The shapes of the HBP for 10-20 (squares), 20-30 (stars), 30-40 (triangles) and 40-70 (points) GeV/c jets are presented. It seems the expected trends from MLLA+LPHD predictions are observed: a) At higher jet p_T, the phase space available for particle emissions is larger which results in an increase of the number of entries with increasing ξ. b) At large ξ we observe a steeply falling shape. Nevertheless, it is not sufficient to prove the presence of coherence effects. Indeed, because of the small cone size used for jet reconstruction, fractions of intrajet radiations are missing, this could also explain the shape of the distributions. However, the moving position of the peak with increasing jet energy is a sign of the presence of color coherence effects [1]. The small cone size also induces a jet energy dependent bias and the Underlying Events lead to the presence of background in jets. These effects will have to be corrected in the future.
In the context of heavy ion collisions, the study of the HBP is of major interest to ALICE to investigate the properties of the dense partonic medium produced. The MLLA+LPHD predicts a distortion of the HBP in the presence of a dense partonic medium like the Quark-Gluon Plasma (QGP) where high energy initial partons are expected to loose their energy via gluon radiation leading to the suppression of the production of high p_T hadrons [4, 5]. A shift of the maximum position of the peak of the distribution towards high ξ values is expected. One should observe the enhancement of the amplitude of the distribution in the high ξ region due to the production of more soft hadrons. On the contrary, a decrease of the population in the low ξ region is expected as the result of the suppression of high p_T hadron production [4].
We first concentrate on the HBP measurement in p+p collisions in order to test ALICE capabilities to constrain pQCD but also to provide a baseline for future measurement in heavy ion collisions; our ultimate goal is to study and to extract properties of the QGP.

2. Track and jet reconstruction performances in the ALICE experiment

In ALICE, jets will be reconstructed at mid-rapidity with charged particle transverse momenta given by the tracking devices, the Inner Tracking System (ITS) and the Time Projection

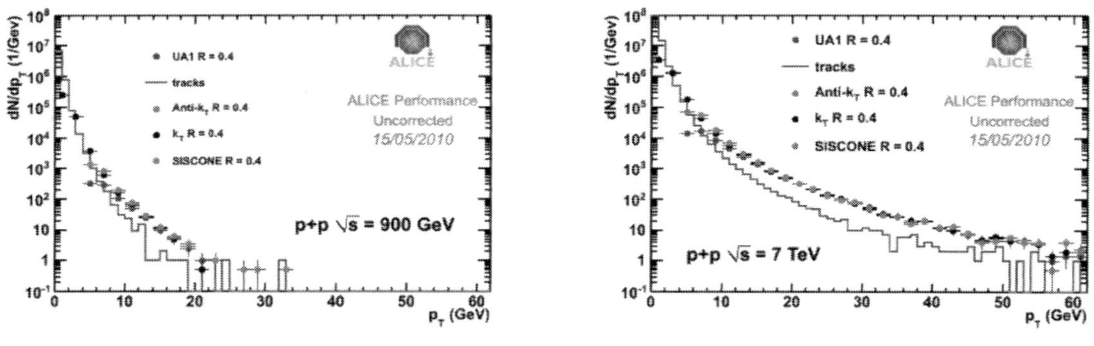

Figure 2. First charged hadrons (straight blue line) and jet raw spectra measurement of ALICE for p+p collisions at 900 GeV (left) and 7 TeV (right). Jets are reconstructed with a radius of 0.4, with SISCone (green), UA1 (blue), k_T (black) and anti-k_T (red) algorithms.

Chamber (TPC), and neutral particle transverse energies measured by the Electromagnetic Calorimeter (EMCal). As the EMCal is not yet fully installed in the ALICE cavern (4/12 SuperModules are installed), jets are currently reconstructed from charged particle p_T. ITS and TPC measurements endow ALICE with its excellent tracking capabilities from very low p_T ($\sim$ 200 MeV/c) to higher values ($\sim$ 100 GeV/c) with good efficiency (over 90%), good momentum resolution (still below 6% at 100 GeV/c in p+p and Pb+Pb collisions according to the design values) and excellent PID [6]. In figure 2 (full lines), are shown the non corrected p_T spectra of the charged tracks coming from $\sim$ 1.95 M p+p events at $\sqrt{s}$ = 900 GeV (left plot) and $\sim$ 16.7 M p+p events at 7 TeV (right plot) respectively. These plots show the large p_T range reachable from track measurement for the first data registered by ALICE. The figure 2 also presents the first measurements by the ALICE collaboration of the non corrected jet p_T spectra reconstructed from the same set of events collected in p+p collisions at 900 GeV (left) and 7 TeV (right). Jets have been reconstructed with cones and sequentials algorithms available in the ALICE analysis framework [7]: UA1 (blue) and SISCone (green) which are cone algorithms, k_T (black) and anti-k_T (red) which are sequential algorithms [7]. Both cone and sequential jet finders have their advantages and drawbacks, nevertheless they are not perfect and introduce biases in the reconstructed jet energy because of the use of parameters. Moreover, detector effects, background, charged to neutral fluctuations also bias the reconstructed energy. This last contribution dominates in our case as jets have been reconstructed from charged particles p_T only. The presented spectra are not yet corrected for the track and jet reconstruction efficiency. It can be done by correcting individually each bias or by using unfolding methods [8, 9]. Nevertheless, we observe a good agreement between the jet finders except at low p_T where UA1 reconstructs less jets than the others. This is due to its seed parameter as jet reconstruction is started only from particles which pass a threshold in p_T: the seed value. Figure 2, clearly shows that jet measurement can already be done over large p_T intervals which is promising for the study of the HBP over several bins of p_T.

3. Preliminary studies on intrajet radiation

In this section, we show the preliminary studies on intrajet radiation based on PYTHIA+GEANT simulations in which angular ordering features have been modelized. The angle and transverse momentum distributions of hadrons inside jets have been observed.
Figure 3 (left) shows the angle in three dimensions between the jet axis and the hadron transverse

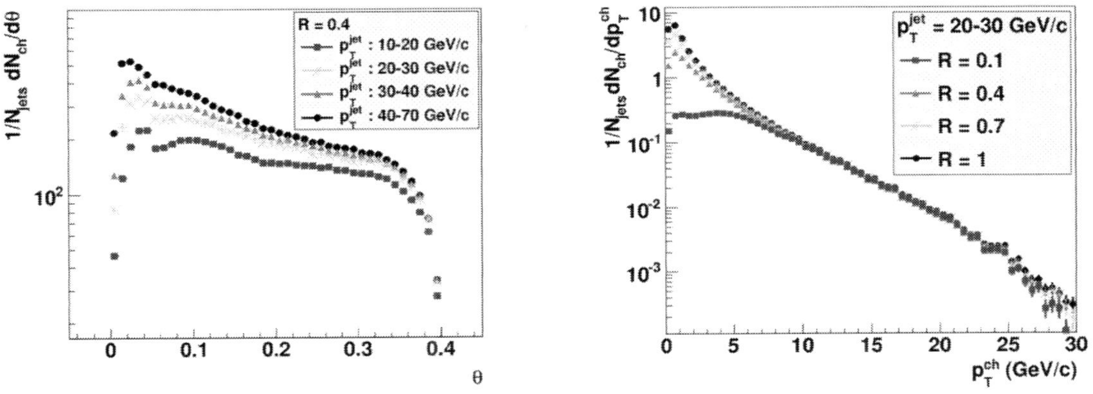

Figure 3. θ (left) and p_T (right) distributions of hadrons. It corresponds to fully reconstructed charged data coming from PYTHIA simulated p+p collisions at $\sqrt{s} = 10$ TeV.

momenta, θ, distribution. We observe that the number of hadrons increases with the jet energy and particularly near the jet axis.
In figure 3 (right) is plotted the p_T distribution of hadrons for jets of p_T taken between 20 and 30 GeV/c and reconstructed with different radii. We notice that all high p_T hadrons are recovered whatever the radius value. But more low p_T hadrons are recovered when the radius increases.
No corrections have been applied yet, nevertheless the features expected from QCD are observed: energetic jets contain more hadrons and particularly near their axis in agreement with the idea of collimation. Hard hadrons are emitted at small angles and when moving away from the jet axis toward larger angles, more soft hadrons can be found. But a lot of these soft hadrons come from the background which evolves in R^2. The background will have to be subtracted before going further in the interpretation of the data.

4. Summary

Color coherence effects constrain the way partons radiate inside a jet and are responsible for the shape of the HBP. The ALICE experiment is well-suited to perform such a study. Jets over a large p_T domain have been reconstructed which is a very promising basis for fragmentation function studies. A preliminary work on simulated data with PYTHIA+GEANT containing angular ordering features supports the idea of jet collimation: the latter is stronger for higher energy jets, more particles are expected close to the jet axis and low p_T hadrons are emitted with a larger probability at higher angles. This work shows that the intrajet measurement is feasible with ALICE.

References

[1] Dokshitzer Yu L, Khoze V A, Troian S I and Mueller A H, *Rev. Mod. Phys. 60:373*, 56 (1988).
[2] Dokshitzer Yu L, Khoze V A, Mueller A H and Troian S I, *Basics of perturbative QCD*, Tran Thanh Van J (Editions Frontières), 367 (1991).
[3] CDF collaboration, *Phys. Rev. D68 012003*, 20 (2003).
[4] Borghini N and Wiedemann U A, *arXiv:hep-ph/0506218v1*, 4 (2005).
[5] Sapeta S et Wiedemann U A, *Eur. Phys. J. C55:293-302*, 15 (2008).
[6] ALICE collaboration *et al*, *J. Phys. G: Nucl. Part. Phys.* ***32*** *1295*, 747 (2006).
[7] Soyez G, *arXiv:hep-ph/0807.0021*, 6 (2008).
[8] Cacciari M, Salam G and Soyez G, *JHEP 0804:005*, 42 (2008).
[9] Blobel V, *arXiv:hep-ex/0208022v1*, 10 (2002).

Jets in 200 GeV p+p and d+Au collisions from the STAR experiment at RHIC

Jan Kapitán (for the STAR Collaboration)
Nuclear Physics Institute ASCR, Na Truhlarce 39/64, 18086 Praha 8, Czech Republic
E-mail: kapitan@rcf.rhic.bnl.gov

Abstract. Full jet reconstruction in heavy-ion collisions is a promising tool for the quantitative study of properties of the dense medium produced at RHIC. Measurements of d+Au collisions are important to disentangle initial state nuclear effects from medium-induced k_T broadening and jet quenching. Study of jet production and properties in d+Au in combination with similar studies in p+p is an important baseline measurement needed to better understand heavy-ion results. We present mid-rapidity inclusive jet p_T spectra and di-jet correlations (k_T) in 200 GeV p+p and d+Au collisions from the 2007-2008 RHIC run. We discuss the methods used to correct the data for detector effects and for background in d+Au collisions.

1. Introduction

Jets are remnants of hard-scattered partons, which are the fundamental objects of pQCD. At RHIC, they can be used as a probe of the hot and dense matter created in heavy ion collisions [1, 2]. To quantify the signals observed in heavy ion collisions in comparison to p+p collisions, it is necessary to measure the cold nuclear matter effects in systems such as d+Au. There are several ways to describe cold nuclear matter effects. Partonic rescattering [3] leads to k_T broadening and to modification of jet p_T spectra. Nuclear parton distribution functions [4] influence jet p_T spectra, but the effects at mid-rapidity are not expected to be very strong.

2. Jet reconstruction

This analysis is based on $\sqrt{s_{NN}} = 200$ GeV p+p and d+Au data from the STAR experiment, recorded during RHIC run 8 (2007-2008). The Barrel Electromagnetic Calorimeter (BEMC) detector is used to measure the neutral component of jets, and the Time Projection Chamber (TPC) detector is used to measure the charged component of jets. In the case of a TPC track pointing to a BEMC tower, its momentum is subtracted from the tower energy to avoid double counting (electrons, MIP and possible hadronic showers in the BEMC). To reduce possible BEMC backgrounds, the jet neutral energy fraction is required to be within $(0.1, 0.9)$. An upper $p_T < 15$ GeV/c cut was applied to TPC tracks due to uncertainties in TPC tracking performance at high-p_T in run 8 (under further investigation).

The Beam Beam Counter detector, located in the Au nucleus fragmentation region, was used to select the 20% highest multiplicity events in d+Au collisions. The acceptance of TPC and BEMC together with experimental details (calibration, primary vertex position cuts) limit the jet fiducial acceptance to $|\eta| < 0.55$ ($R = 0.4$), $|\eta| < 0.4$ ($R = 0.5$), where R is the resolution parameter used in jet finding.

Journal of Physics: Conference Series **270** (2011) 012015 doi:10.1088/1742-6596/270/1/012015

Recombination jet algorithms kt and anti-kt from the FastJet package [5] are used for jet reconstruction. To subtract the background, a method based on active jet areas [6] is applied event-wise: $p_{\mathrm{T}}^{Rec} = p_{\mathrm{T}}^{Candidate} - \rho \cdot A$, with ρ estimating the background density per event and A being the jet active area. Due to the asymmetry of the colliding d+Au system, the background is asymmetric in η. This dependence was fit with a linear function in η and included in the background subtraction procedure.

To study detector effects, Pythia 6.410 and GEANT detector simulations were used. Jet reconstruction was run at MC hadron level (PyMC) and at detector level (PyGe). To study effects of the d+Au background, a sample with added background (PyBg) was created by mixing PyGe events with 0-20% highest multiplicity d+Au events (minimum bias online trigger).

3. Nuclear k_{T} broadening

Azimuthal correlations of jets in di-jet events in p+p and d+Au can provide information on nuclear k_{T} broadening. To increase di-jet yield, BEMC high tower (HT) online trigger was required (one tower with $E_{\mathrm{T}} > 4.3$ GeV), for both p+p and d+Au. A cut $p_{\mathrm{T}} > 0.5$ GeV/c applied for tracks and towers to reduce background and the jet finding was run with $R = 0.5$.

To select a clean di-jet sample two highest energy jets ($p_{\mathrm{T},1} > p_{\mathrm{T},2}$) in each event were used, with $p_{\mathrm{T},2} > 10$ GeV/c. Observed di-jet signal in d+Au collisions is shown in Figure 1. Distributions of $k_{\mathrm{T,raw}} = p_{\mathrm{T},1} \sin(\Delta\phi)$ were constructed for di-jets and Gaussian widths, $\sigma_{k_{\mathrm{T,raw}}}$, were obtained for the two jet algorithms and two ($10 - 20$ GeV/c, $20 - 30$ GeV/c) $p_{\mathrm{T},2}$ bins.

Figure 2 shows that the widths are similar for PyMC, PyGe and PyBg distributions, therefore detector and background effects are negligible, due to the interplay between jet p_{T} and $\Delta\phi$ resolutions. Figure 3 shows an example of the $k_{\mathrm{T,raw}}$ distributions for data. The Gaussian fit to p+p data is not ideal and the precise shape of the distribution is under study. RMS widths of these distributions have therefore been checked and they agree with the sigma widths of the fits. The values extracted from the Gaussian fits are $\sigma_{k_{\mathrm{T,raw}}}^{p+p} = 2.8 \pm 0.1$ (stat) GeV/c and $\sigma_{k_{\mathrm{T,raw}}}^{d+Au} = 3.0 \pm 0.1$ (stat) GeV/c. Possible nuclear k_{T} broadening therefore seems rather small.

The systematic errors on $\sigma_{k_{\mathrm{T,raw}}}$ coming from a weak dependence on the $|\Delta\phi - \pi|$ cut for back-to-back di-jet selection (varied between 0.5 and 1.0) and differences between the various selections ($p_{\mathrm{T},2}$ range, jet algorithm) are estimated to be 0.2 GeV/c. Additional uncertainty comes from Jet Energy Scale (JES), which will be discussed in more detail in section 5.

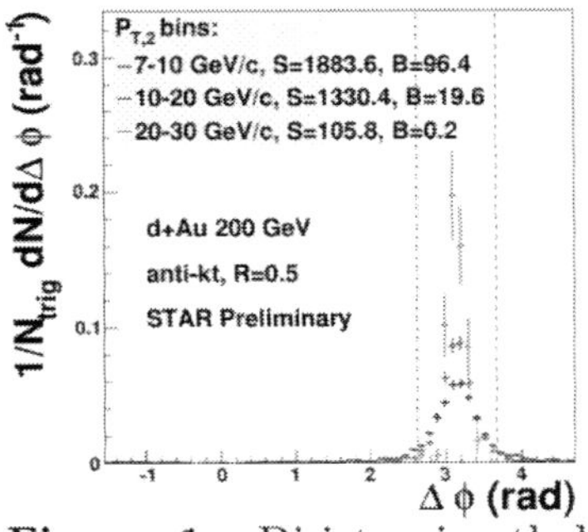

Figure 1. Di-jet azimuthal correlations in 3 $p_{\mathrm{T},2}$ bins.

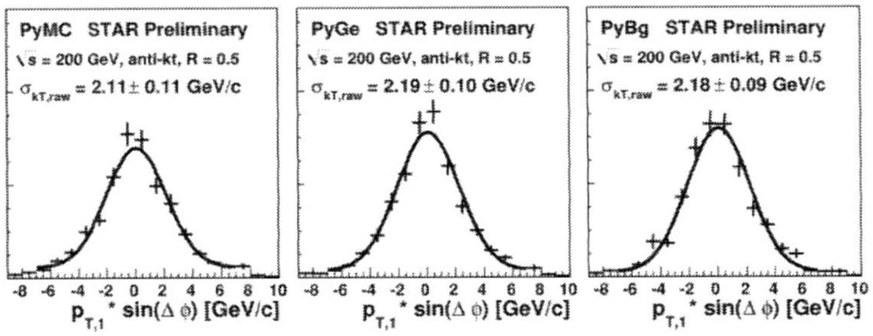

Figure 2. Distributions of $k_{\mathrm{T,raw}} = p_{\mathrm{T},1} \sin(\Delta\phi)$ for simulation ($10 < p_{\mathrm{T},2} < 20$ GeV/c).

4. Inclusive jet spectra

10M 0-20% highest multiplicity d+Au events with a minimum bias online trigger were used for this study. Anti-kt algorithm was used with $R = 0.4$, $p_{\mathrm{T}} > 0.2$ GeV/c cut was applied to tracks and towers. The jet p_T spectrum is normalized per event and the high multiplicity of d+Au events also guarantees the trigger efficiency is independent of the p_T of the hard scattering. Therefore, no correction related to trigger is needed for jet p_T spectra.

A bin-by-bin correction is used to correct for detector effects (tracking and tower efficiency, unobserved neutral energy, jet p_T resolution). It is based on the generalized efficiency, constructed as the ratio of PyMC to PyBg jet p_T spectra. These correction factors are then applied to measured jet p_T spectrum. As the detector effects on jet p_T spectrum differ substantially depending on jet p_T spectrum shape, the shapes have to be consistent between PyBg and measured jet p_T spectra. Figure 4 shows that this is indeed the case.

As the jet p_T spectrum is very sensitive to the jet energy scale, an additional correction was applied here to account for the lower TPC tracking efficiency in d+Au compared to that from the used p+p Pythia simulation. The d+Au efficiency was determined by simulating single pions and embedding them at the raw detector level into real d+Au minimum bias events. The tracking efficiency in the Pythia simulation was then artificially lowered, prior to jet finding at PyGe and PyBg level, so that it matches the one obtained from d+Au embedding.

To compare per event jet yield in d+Au to jet cross section in p+p collisions, an input from MC Glauber study is needed. In the following, we have used $\langle N_{\rm bin}\rangle = 14.6\pm1.7$ for 0-20% highest multiplicity d+Au collisions and $\sigma_{\rm inel,pp} = 42$ mb. These factors were used to scale p+p jet cross section measured previously by the STAR collaboration [7] using Mid Point Cone (MPC) jet algorithm with $R = 0.4$. The resulting d+Au jet p_T spectrum is shown in Figure 5 together with scaled p+p jet spectrum. Within systematic uncertainties, the d+Au jet spectrum shows no significant deviation from the scaled p+p spectrum. The discrepancy between simulation and embedding was only discovered after the d+Au spectrum was presented in the conference, and it was corrected only for these proceedings. The difference is however smaller than the combined systematic and statistical uncertainties, the conclusions therefore remain the same.

5. Systematic uncertainties

The JES uncertainty dominates the uncertainties of d+Au measurement and is marked by the black lines in Figure 5. It has two components: 5% on the neutral jet component (BEMC calibration uncertainty) and 10% on the charged jet component (uncertainty of TPC tracking efficiency in jets in d+Au collisions). Embedding of jets into real d+Au events at raw detector level will allow to decrease this uncertainty in the future. As JES uncertainty is expected to be largely correlated between run 8 p+p and d+Au data, we plan to measure jet p_T spectrum in run 8 p+p collisions to decrease these uncertainties and be able to construct jet $R_{\rm dAu}$.

Caution is needed due to the use of different jet algorithms and different pseudorapidity acceptances in Figure 5. Although neither of these effects can change our conclusion, the same algorithm and the same acceptance have to be used in p+p and d+Au to construct jet $R_{\rm dAu}$.

6. Summary

Di-jet k_T widths were measured in 200 GeV p+p and d+Au collisions: $\sigma^{p+p}_{k_{\rm T,raw}} = 2.8 \pm 0.1$ (stat) GeV/c, $\sigma^{d+Au}_{k_{\rm T,raw}} = 3.0 \pm 0.1$ (stat) GeV/c. No significant broadening due to Cold Nuclear Matter effects was observed. Jet p_T spectrum from minimum bias 200 GeV d+Au collisions is consistent with scaled p+p jet spectrum within systematic uncertainties. Precise tracking efficiency determination from jet embedding and cross section measurement in run 8 p+p data will allow to construct jet $R_{\rm dAu}$.

Acknowledgement

This work was supported in part by grants LC07048 and LA09013 of the Ministry of Education of the Czech Republic and by the grant SVV-2010-261 309.

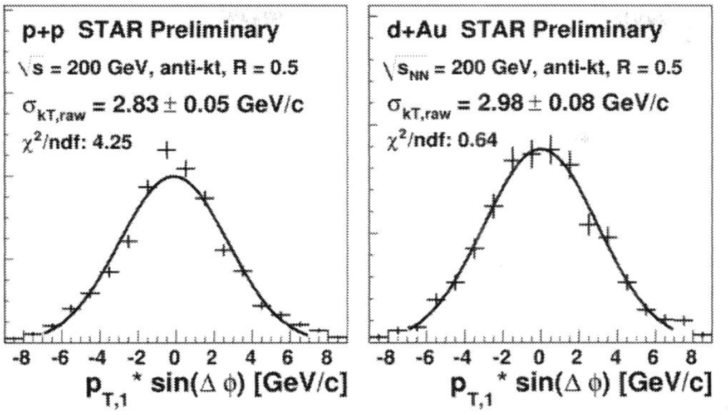

Figure 3. Distributions of $k_{\mathrm{T,raw}}$ for p+p, d+Au ($10 < p_{\mathrm{T},2} < 20$ GeV/c).

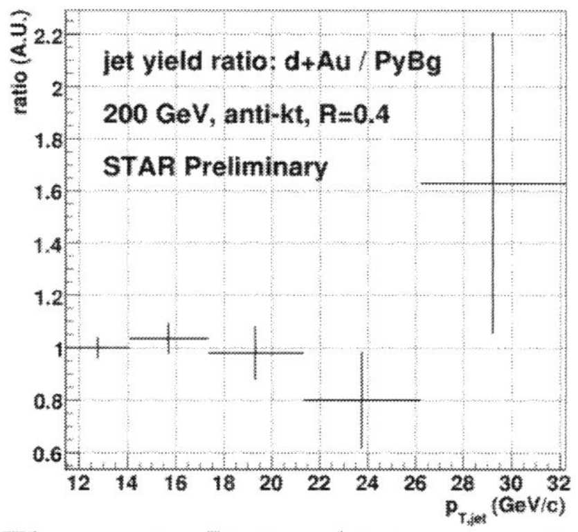

Figure 4. Ratio of jet p_{T} spectra between d+Au data and simulation.

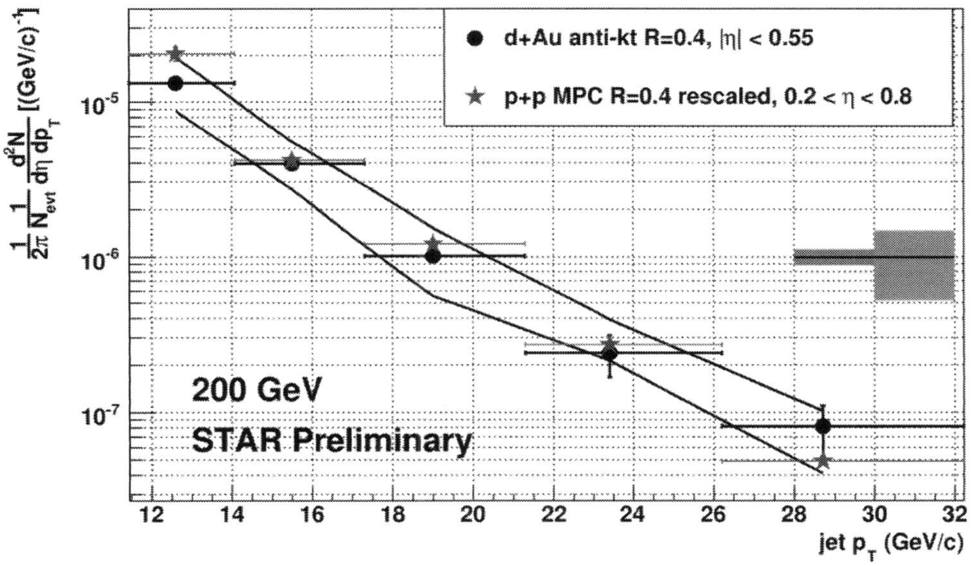

Figure 5. Jet p_{T} spectrum from d+Au collisions compared to scaled p+p spectrum [7]. Red box indicates uncertainty of $\langle N_{\mathrm{bin}} \rangle$, black lines indicate JES uncertainty in d+Au and the magenta box shows the total systematic uncertainty of p+p measurement (including JES uncertainty).

References

[1] Ploskon M 2009 *Nucl. Phys.* A **830** 255c-258c
[2] Bruna E 2009 *Nucl. Phys.* A **830** 267c-270c
[3] Vitev I 2003 *Phys. Lett.* B **562** 36-44
[4] Eskola K J, Paukkunen H, Salgado C A 2008 *JHEP* **0807** 102
[5] Cacciari M, Salam G 2006 *Phys. Lett.* B **641** 57
Cacciari M, Salam G, Soyez G 2008 *JHEP* **0804** 063
[6] Cacciari M, Salam G 2008 *Phys. Lett.* B **659** 119
Cacciari M, Salam G, Soyez G 2008 *JHEP* **0804** 005
[7] Abelev B *et al* 2006 *Phys.Rev.Lett.* **97** 252001

Study of particle production from quark and gluon jets in proton-proton collisions

Sona Pochybova[1,2]

[1] MTA KFKI RMKI, Konkoly-Thege Miklos 29-33, Budapest, Hungary
[2] Eötvös University, Pázmány Péter sétány 1/A, H-1117 Budapest, Hungary

E-mail: sona.pochybova@cern.ch

Abstract. We investigate whether and how different fragmentation properties of quarks and gluons affect identified particle spectra. We present a systematic study of π, K and p production in minimum bias (inelastic, non-diffractive), two- and three-jet events at RHIC, Tevatron and LHC energies. Through the study of two- and three-jet events and various jet-production channels we can directly access the fragmentation properties of quark and gluon jets. We present Monte-Carlo estimate for the contribution of quark and gluon jets to individual particle species spectra, that can be compared to experimental results and test our current knowledge of the physics behind particle production inside jets.

1. Introduction

Jets are produced in hard scatterings of colliding particles. Emerging from the early stages of collisions they are ideal tools to study final states, hadronisation processes and hadron production. Such questions can be addressed through investigation of fragmentation properties of quark and gluon jets in different event shapes (2- or 3- jet) and jet-production channels.

In the following, we show how gluon contribution to hadron spectra changes with collision energy, moreover, that this contribution is dominant for protons. Further, in the case of 2- and 3-jet events, we show that extra hard gluon radiation has an effect of 20% - 40% to the relative proton production. The presented Monte-Carlo analysis is an extension of multiple jet studies ([1] - [7]), with the perspective of implication in the future experiments at LHC, which offer high momentum particle identification capabilities.

The data were simulated using PYTHIA event generator [8] with the settings of P0 tune [9]. Three types of data-sets, were created for each collision energy - 200 GeV, 1800 GeV and 7 TeV including pure gluon (GG), pure quark (QQ) and mixed (QG) jet event production. Every sample contains 300 000 events. Separating the production channels, we were able to see the parton-type effect on single hadron spectra. Additionally, as a reference, three respective minimum bias (MB) samples were generated, each containing 1M events.

In order to study particle production in different event shapes, we selected 2- and 3-jet-like events based on the so called thrust variable, T [8]. Events with T smaller than 0.9 were treated as 3-jet-like, events with higher values as 2-jet-like. This separation was earlier proposed for multi-jet analysis at LEP, see e.g Ref. [10]. The effects of GG, QQ and QG production channels where studied via $p+\bar{p}/K^{+}+K^{-}$ (p/K) and $p+\bar{p}/\pi^{+}+\pi^{-}$ (p/π) ratios.

Journal of Physics: Conference Series **270** (2011) 012016 doi:10.1088/1742-6596/270/1/012016

Table 1. Integrated fraction of jet events in MB production (second column) and fraction of various production channels contained in a jet production.

$\sqrt{s}$ [TeV]	Jet/MB	GG/Jet	QQ/Jet	QG/Jet
0.2	0.1%	17.7%	27.3%	55%
1.8	34.2%	49.7%	7.6%	42.7%
7	95%	60%	5.3%	34.7%

2. Particle ratios in selected production channels

At first we examined the separate contributions of individual jet-production channels to the whole spectra (see Table 1) and to the p/π and p/K ratios (Figure 1).

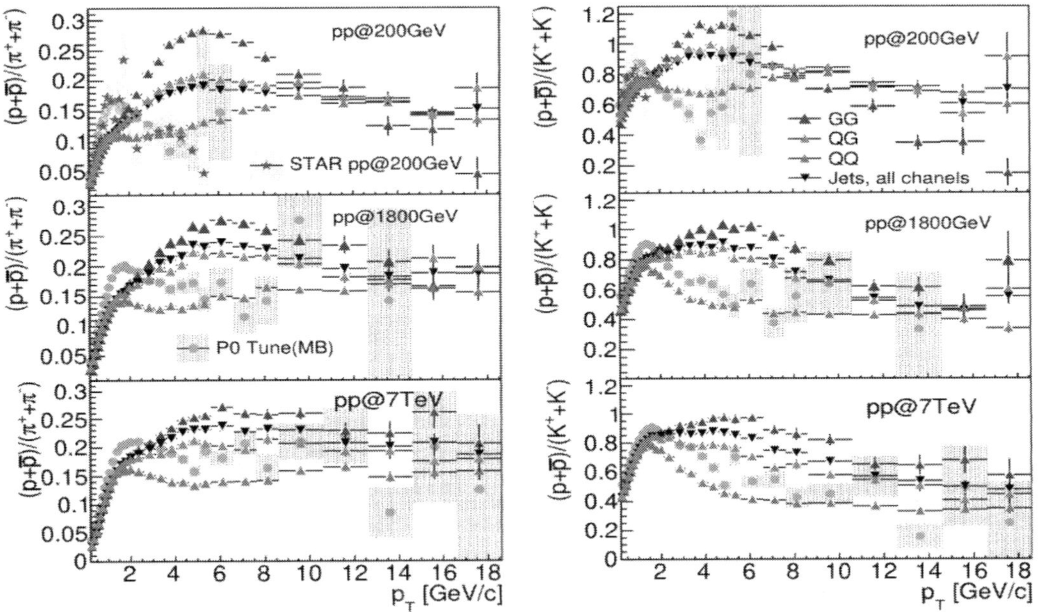

Figure 1. p/π (left panels) and p/K (right panels) for various production channels; GG (blue triangles), QG (magenta triangles), QQ (red triangles) compared to all production channels (black triangles) and MB (green circles). At $\sqrt{s} = 200$ GeV ratio is compared to STAR data [11].

In the case of jet sample generation, we have taken into account, that the cross section for jet production varies strongly with jet energy. For this reason, the jet sample was generated in three p_T^{hard} bins ($p_T^{hard} = \{15 - 50$ GeV/c$, 50 - 100$ GeV/c$, 100$ GeV/c $\leq\}$), where p_T^{hard} is the transverse momentum of the partons in the rest frame of the interaction. After proper cross section scaling, the partial p_T^{hard} samples where merged into one.

We can see, that the fraction of jet events contributing to the MB spectra rises with energy, furthermore, jet sample becomes gluon dominated.

For both p/π and p/K, in the $2 - 6$ GeV/c region, the highest value of the ratio is achieved for GG events and the ratio for all channels rises towards the GG value with collision energy

(Figure 1). Specifically, p/π ratio reaches values between $\approx 0.25 - 0.3$ and p/K ratio rises up to ≈ 1. This behaviour can be connected to how protons are formed within the popcorn fragmentation model in PYTHIA [8, 12]. Observing the ratios experimentally can contribute to further understanding of particle production mechanisms.

The MB points seem to prefer the region between QG and QQ values, especially in the region above 3 GeV/c, where hard scattering becomes important. In this sample we also observe a rising trend towards higher collision energies.

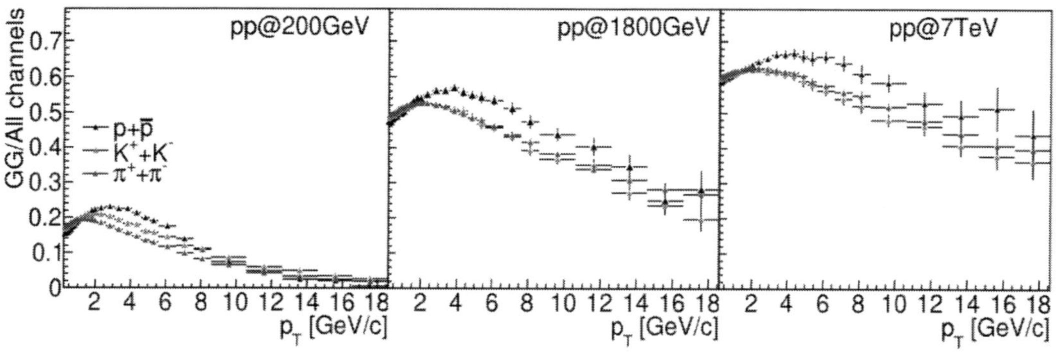

Figure 2. GG production channel contribution to all production channels for individual particle spectra.

In Figure 2 we plotted the GG channel contribution to the jet spectra for individual hadrons as it changes with hadron momentum and collision energy. The contribution weakens with momentum and rises with collision energy (also see Table 1). The highest contribution belongs to protons, which complements Figure 1.

3. Particle ratios in 2- and 3-jet like events

In the second part of our investigation, we focused on how the studied ratios are changing in the presence of extra hard gluon radiation, i.e in 3-jet events. In Figure 3 we plot the p/K and p/π ratios for 2- and 3-jet like events. The extra hard gluon radiation, present in 3-jet like events, causes an excess in proton spectrum w.r.t pion or kaon. The effect being stronger for p/π. This is true for regions below 6 GeV/c, where increase is up to $\approx 40\%$ for p/π and up to $\approx 20\%$ for p/K. The enhancement, caused by hard gluon radiation, in the individual spectra originates, as mentioned in section 2, in the fragmentation model used within PYTHIA.

Above 6 GeV/c, the differences between 2- and 3-jet like events vanish. This is consistent with Figure 2 in the high p_T region. The gluon contribution varies between protons as baryons, and pions and kaons as mesons.

4. Discussion

We presented a Monte-Carlo study of particles originating from hard scatterings. Through the investigation of p/K and p/π ratios in different jet-production channels we were able to see if and how the differences in the ratio values reflect the different fragmentation properties of quarks and gluons. Generally, we observed that gluon contribution to jet spectra rises with energy (see Table 1), and the gluon contribution to the spectra is highest for protons (Figure 1, Figure 2).

When comparing 2- and 3-jet like events, we saw that additional hard gluon radiation, which is present in the 3-jet case, causes an enhancement in proton spectrum w.r.t pion (up to $\approx 40\%$) as well as kaon (up to $\approx 20\%$) (Figure 3).

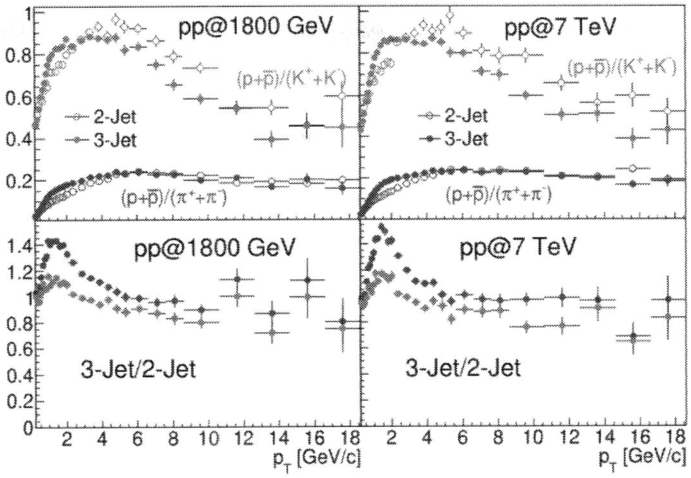

Figure 3. Top panels: p/π (blue) and p/K (red) ratios for all production channels spectrum. Bottom panels: (3-jet ratio)/(2-jet ratio).

To conclude, the differences in particle production in the individual production channels (GG, QG and QQ) as well as 2- and 3-jet events are present in the mid-p_T region ($2-6$ GeV/c) and are directly connected to the fragmentation model used in PYTHIA. Further they are related to the way how "soft" and "hard" parts of the event interplay. In this sense, the presented study suggests methods to investigate fragmentation and coalescence in an experimentally interesting momentum region for this kind of analysis.

Acknowledgements
I would like to thank Peter Lévai, Gergely G. Barnaföldi and Levente Molnár for fruitful discussions on the presented topic. The work has been funded by OTKA NK77816, PD73596 and Eötvös University, Budapest.

References
[1] P. Abreu *et al.* [DELPHI Collaboration], Z. Phys. C **70**, 179 (1996).
[2] R. Barate *et al.* [ALEPH Collaboration], Eur. Phys. J. C **17**, 1 (2000).
[3] G. Alexander *et al.* [OPAL Collaboration], Z. Phys. C **69**, 543 (1996).
[4] A. A. Affolder *et al.* [CDF Collaboration], Phys. Rev. Lett. **87**, 211804 (2001).
[5] V. M. Abazov *et al.* [D0 Collaboration], Phys. Rev. D **65**, 052008 (2002) [arXiv:hep-ex/0108054].
[6] S. Abachi *et al.* [D0 Collaboration], Phys. Rev. D **53**, 6000 (1996) [arXiv:hep-ex/9509005].
[7] D. E. Acosta *et al.* [CDF Collaboration], Phys. Rev. D **71**, 032002 (2005) [arXiv:hep-ex/0410018].
[8] T. Sjostrand, S. Mrenna and P. Z. Skands, JHEP **0605**, 026 (2006) [arXiv:hep-ph/0603175].
[9] Peter Z. Skands [Fermilab] FERMILAB-CONF-09-113-T, May 2009. 14pp [arXiv:hep-ph/0905.3418v1].
[10] F. Barreiro, S. Brandt and M. Gebser, Nucl. Instrum. Meth. A **240** (1985) 237.
[11] John Adams *et al* [STAR Collaboration] Phys. Lett. B **637** (2006) 161 [arXiv:nucl-ex/0601033].
[12] B. Andersson, G. Gustafson, G. Ingelman and T. Sjostrand, Phys. Rept. **97** (1983) 31.

Journal of Physics: Conference Series **270** (2011) 012017 doi:10.1088/1742-6596/270/1/012017

Jets and Underlying Events at LHC Energies

A. G. Agócs[1,2], G. G. Barnaföldi[1], P. Lévai[1]
[1]MTA KFKI RMKI, Konkoly-Thege Miklós út 29-33, H-1121 Budapest, Hungary
[2]Eötvös University, Pázmány Péter sétány 1/A, H-1117 Budapest, Hungary
E-mail: agocs@rmki.kfki.hu

Abstract. Jet-matter interaction remains a central question and a theoretical challenge in heavy-ion physics and might become important in high-multiplicity events in proton-proton collisions at LHC energies. Full jet measurement at LHC offer the proper tool to investigate energy loss process and fragmentation of hard parton in the medium. Since jet reconstruction will be constrained to small cone sizes, then study of the connection between jets and surrounding environment provides a further possibility to extend our exploration. We study jets at $\sqrt{s}$ = 14 TeV and *pp* collisions at $\sqrt{s}$ = 7 TeV. We analyze the flavor components in jet-like environments. We introduce a definition for surrounding cones/belts and investigate flavor dependence and correlation of different hadron species produced in jets. Here, we focus on proton-triggered correlations. Our analysis can be extended for heavy ion collisions.

1. Introduction

The state-of-the-art particle detectors and very-high energies reached at the Large Hadron Collider (LHC) opened a new window to create and investigate high momentum particle showers in hadron-hadron collisions. These showers originate from color partons and they are identified as *jets*. Jets were discovered in electron-positron (e^-e^+) collisions, where it was assumed: partons are travelling in QCD vacuum. However, in proton-proton and heavy-ion collisions partons are propagating in a color medium. The deconfined medium modifies the parton properties e.g. induces gluon radiation and parton energy loss [1]. This phenomena has been seen and investigated in details in heavy-ion collisions at RHIC energies [2, 3].

High-multiplicity proton-proton events may behave similarly displaying intense interaction between jets and the surrounding 'matter'. Thus, study of jet energy loss and jet-matter interaction can be accomplished with the proper determination of the jets and the remaining background, the so called *underlying events*.

Generally speaking, underlying events (UE) contain particles originating from partons outside identified jet(s). The first 'standard' definition of UE was given by the CDF Collaboration at Fermilab [4]. In an earlier work we introduced *surrounding belts* (SB) around the cones of identified jets [5]. Here, we focused on properties of near and away side jets, and their surrounding regimes, especially hadron contents and correlations.

We display our recent results on the generalized definition of the underlying event. We analyzed mean-p_T vs. multiplicity for the newly defined areas in case of $\sqrt{s}$ = 14 TeV proton-proton collisions. We performed PYTHIA simulations for $\sqrt{s}$ = 7 TeV *pp* collisions. We used PYTHIA 6.4 [6] with ATLAS-CSC tune [7] for both cases and generated ~ 20 M minimum bias events. Furthermore, we have studied proton-triggered correlations in surrounding belts characterized by various radii and thicknesses at $\sqrt{s}$ = 7 TeV.

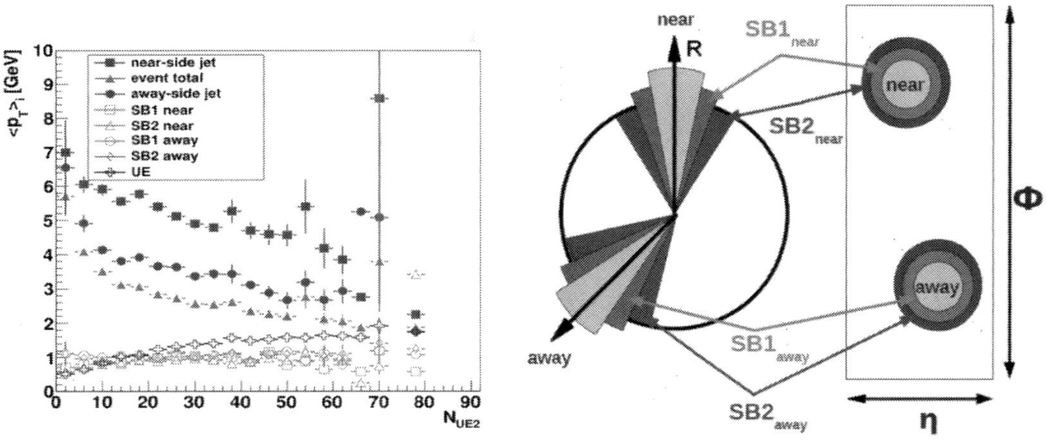

Figure 1. *Left panel:* mean-p_T vs. UE multiplicity outside surrounding cones in pp collisions at 14 TeV. *Right panel:* illustration of 'inner-belt' (SB1) and 'outer-belt' (SB2). (Color online.)

2. Mean-p_T vs. multiplicity for the generalized UE

Transition area between jet cones and underlying event carries information about jet-matter interaction. The investigation of this area demands proper definitions and solid basis for quantitative analysis.

First we considered the mean transverse momenta vs. multiplicity in 14 TeV proton-proton collisions. The newly defined areas based on *right panel* of Fig. 1. Here, the main areas are highlighted: 'near' and 'away' refers for the the near- and away-side jets respectively. We marked the inner surrounding belt with 'SB1' and the outer 'SB2' for both near and away side jets. Based on this definition, the underlying event is everything outside the outer belts.

Hadron multiplicities can be given in each region however, for reference we used N_{UE2}, which is the generalized underlying event area – outside both outer surrounding belts. We plotted the mean transverse ($\langle p_T \rangle$) momenta of above areas as the function of N_{UE2} on the *left panel* of Fig. 1. The ($\langle p_T \rangle$ for near- and away-side jet is drawn by *full squares* and *full disks* respectively. Mean-p_T for surrounding belts are plotted with *open squares* and *open triangles* for inner belts: '$SB1_{near}$' and '$SB1_{away}$'. Open circles and diamonds denote results on outer belts: '$SB2_{near}$' and '$SB2_{away}$', respectively. *Full triangles* display the $\langle p_T \rangle$ for the total event. Finally, we indicated $\langle p_T \rangle$ for the underlying event with *stars*.

As we plotted on Fig. 1 jet-like events has high-$\langle p_T \rangle$ values, with low multiplicity, due to the produced jets in the 14 TeV pp collisions. Mean p_Ts for these jet-like events are falling quickly as the N_{UE2} multiplicity is increasing. In parallel, the $\langle p_T \rangle$ for the total event has a similar structure. Surrounding belts have almost the flat value $\langle p_T \rangle = 1$ GeV/c, practically at every N_{UE2} value. Testing the underlying event, $\langle p_T \rangle_{UE}$ is increasing up to ~ 2 GeV/c, indicating a 'mini-jet' like structure at high multiplicities is starting to play the role, as we pointed out in Ref. [5].

Journal of Physics: Conference Series **270** (2011) 012017 doi:10.1088/1742-6596/270/1/012017

3. Geometrical setup for the analysis

The aim of our study is to map differences in surrounding belts, connected to near-side and away-side correlation peaks. Such differences carry information about hadronization processes and/or possible jet-matter interactions. The wanted differences can be amplified displaying the ratio of hadron yields in the different belts: $\mathrm{SB1}_{near}/\mathrm{SB1}_{away}$ and $\mathrm{SB2}_{near}/\mathrm{SB2}_{away}$.

Following the illustration of Fig. 1, we define 3 different 'geometrical' sets to be analyzed:

CDF-set : $\Delta\phi_{near} = 0° \pm 60°$; $\Delta\phi_{away} = 180° \pm 60°$; $\Delta\phi_{SB1} = \Delta\phi_{SB2} = \pm 6°$.
R-**set** : $\Delta\phi_{near} = 0° \pm 30°$; $\Delta\phi_{away} = 180° \pm 30°$; $\Delta\phi_{SB1} = \Delta\phi_{SB2} = \pm 6°$.
σ-**set** : $\Delta\phi_{near} = 0° \pm \sigma_{near}/2$; $\Delta\phi_{away} = 180° \pm \sigma_{away}/2$; $\Delta\phi_{SB1} = \Delta\phi_{SB2} = \pm 6°$.

The CDF-set is motivated by the original definition from CDF Collaboration [4]. The *R*-set is motivated by a smaller, jet-cone-like size. The σ-set is determined from PYTHIA simulations on correlations at 7 TeV at corresponding trigger and associated momentum regions. Here we will use $\sigma_{near} = 16.0°$ and $\sigma_{away} = 19.5°$ displayed on Fig. 1.

Comparing the above selections, the area of the underlying event is the smallest in the case of CDF-set, and it is the largest for σ-set. The widths of the surrounding belts (6°) is a result of an optimalization to have proper statistics for analysis and comparison.

4. Correlation studies with charged hadron triggers

We have calculated the wanted ratios of $\mathrm{SB1}_{near}/\mathrm{SB1}_{away}$ and $\mathrm{SB2}_{near}/\mathrm{SB2}_{away}$ for all possible charged hadron triggers, namely choosing one of the identified hadrons (π^+, π^-, K^+, K^-, p, $\bar{p}$) as trigger particle and extracting angular correlation with all other identified hadron species.

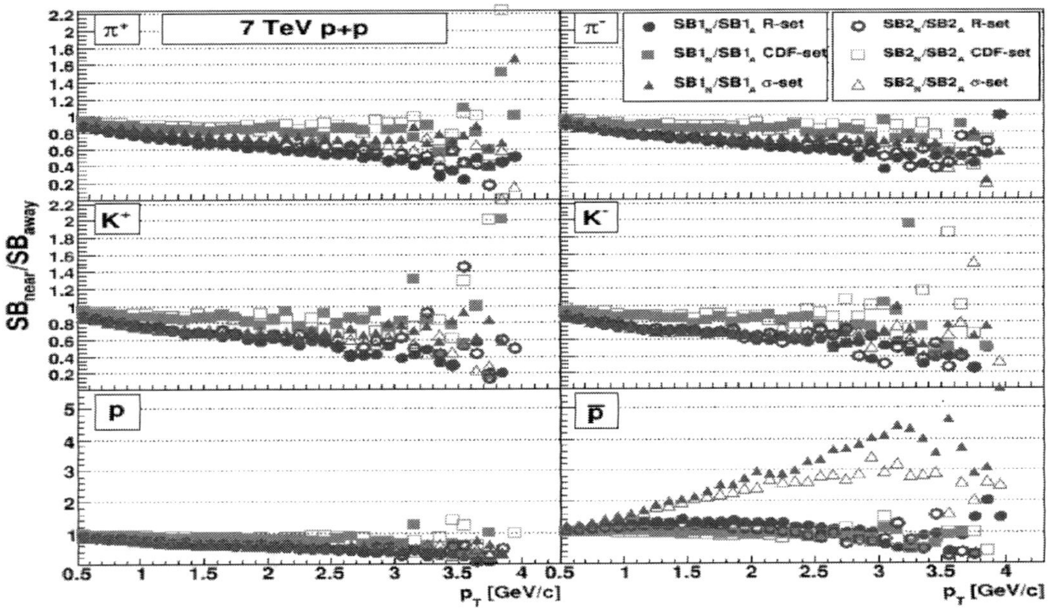

Figure 2. Differences in the proton-triggered correlation functions, enhanced by ratios of surrounding belts, $\mathrm{SB1}_{near}/\mathrm{SB1}_{away}$ and $\mathrm{SB2}_{near}/\mathrm{SB2}_{away}$ as a function of transverse momentum of the associated hadrons, $p_{T,assoc}$. See text for details on legend. (Color online.)

Fig. 2 displays our results extracted from PYTHIA simulations for *pp* at 7 TeV, using proton trigger in the momentum window defined earlier. Inner belt ratios, $SB1_{near}/SB1_{away}$ are marked by *full squares* for 'CDF-set', *full disks* for 'R-set', and *full triangles* for 'σ-set'. Outer belt ratios, $SB2_{near}/SB2_{away}$ are marked by *open squares* for 'CDF-set', *open circles* for 'R-set', and *open triangles* for 'σ-set'. Further results for other identified hadron triggers can be visualized similarly, but we want to focus on proton trigger.

Results from 'CDF-set', indicated by *full and open squares* are close to be unity (no effect) in all correlations. This means we are testing homogeneous UE region far from correlation peaks.

Results from 'R-set', indicated by *full and open circles* go below unity for most of the hadrons (except antiprotons), because of the expected momentum conservation for leading hadrons. Antiprotons behave differently due to a strong correlation in baryon-antibaryon production closer to the correlation peaks. This effect is based on baryon number conservation. A slight difference can be already seen between the ratios from inner belt and outer belt.

In the case of 'σ-set' (*full and open triangles*) we are closer to the correlational peaks, but the results are very similar to the 'R-set', except the deviation in proton-antiprotons correlation is even more strong. Thus, we are really in the fragmentation region with very strong baryon-antibaryon correlation. A slight difference between the inner belt and outer belt can be seen, also.

Taking K^+ trigger we will see strong $K^+ - K^-$ correlation driven by strangeness conservation. The K^+–proton and K^+–antiproton correlations are just similar to e.g. $K^+ - \pi^+$ ones.

5. Conclusions

We have investigated identified hadron correlations in *pp* collision at 7 TeV, by means of PYTHIA event generator. We introduced different geometrical setups. We have found strong proton-antiproton correlation, based on baryon number conservation and strong $K^+ - K^-$ correlation driven by strangeness conservation. Introducing different surrounding belts into the analysis further details were revealed. These correlations should be extracted from real data. The ALICE HMPID and the planned VHMPID detectors can play an important role in these explorations [9, 10].

Acknowledgments

This work was supported by Hungarian OTKA NK77816, PD73596 and Eötvös University. One of the authors (GGB) is also thanks to János Bolyai Research Scholarship of the HAS.

References

[1] M. Gyulassy, P. Levai and I. Vitev, Phys. Rev. Lett. **85**, 5535 (2000)
[2] P. Levai, *et al.* Nucl. Phys. A **698**, 631 (2002)
[3] I. Vitev, B. W. Zhang and S. Wicks, Eur. Phys. J. C **62**, 139 (2009)
[4] A. A. Affolder *et al.* [CDF Collaboration], Phys. Rev. D **65**, 092002 (2002).
[5] P. Levai and A. G. Agocs, PoS E **PS-HEP2009** (2009) 472.
[6] T. Sjostrand, S. Mrenna and P. Z. Skands, JHEP **0605**, 026 (2006)
[7] `https://twiki.cern.ch/twiki/bin/view/Atlas/UpgradeDefinitionPythiaTune`
[8] P. Levai, G. Fai and G. Papp, Phys. Lett. B **634**, 383 (2006)
[9] ALICE TDR for the HMPID, CERN/LHCC **9819** (1998)
[10] Submitted Letter of Intent for the proposed VHMPID detector for ALICE.

Journal of Physics: Conference Series **270** (2011) 012018 doi:10.1088/1742-6596/270/1/012018

Jet-triggered dihadron correlations

Andrew Adare for the STAR collaboration
Yale University, New Haven, CT
E-mail: andrew.adare@yale.edu

Abstract. Dihadron correlations in nuclear collisions continue to serve as an important tool for accessing energy loss and medium modification of jet shapes and yields, particularly in comparison to p+p reference data. However, there remain unsettled issues in the methodology and interpretation of dihadron correlations. In particular, the disentanglement and subtraction of event anisotropy from a correlated and (possibly highly quenched) jet signal in a large heavy-ion background is not a closed case. A new technique for reducing background in a triggered correlation analysis is to require trigger particles to reside within the cone radius of a reconstructed jet. A simulation study of the influence of background on dihadron correlations is presented together with improvements afforded by using jet reconstruction for purity enhancement of trigger particles. Additionally, progress using these techniques in Au+Au data from the STAR detector is reported along with discussion of their relation to existing results.

1. Introduction

Energy loss of fast partons escaping the hot matter produced in relativistic nuclear collisions is an important signature of the quark-gluon plasma. Angular distributions of correlated leading and associated charged hadrons (here denoted respectively as type "A" and "B") have exhibited significant modification in both shape and yield in central Au+Au collisions compared to a p+p reference [1, 2, 3]. In particular, a doubly-peaked structure has been reported on the recoil side. Three-particle correlations suggest the feature to be conical [4], and its explanation has attracted theoretical interest. Notably, at high trigger momenta ($p_T^A \gtrsim 7$ GeV/c), the doubly-peaked structure is not observed at any p_T^B [5]. If energy loss scales with parton energy such that a cone feature from medium response remains constant or rises relative to other correlation structures, such experimental results do not seem to support theoretical expectations. This talk explores how spurious "Mach cones" can arise from the presence of non-jet background and improper characterization of event shapes, and introduces $h^{\pm}_{\text{jet}}$–$h^{\pm}$ correlations as a tool to clarify understanding of energy loss and medium response.

2. The two source model and event geometry

Dihadron correlation analyses conventionally assume that anisotropic pairing in phase space arises from either jet fragmentation or event-wide mechanisms such as elliptic flow. Disentangling the two components is complicated by jet-medium interactions, initial and final-state radiation, and fluctuations at various scales. Despite such complications, the underlying event shape is often parametrized with Fourier coefficients v_n as

$$\frac{dN_{bkg}}{d\Delta\phi} = B_0 \left(1 + 2\sum_n v_n^{AB} \cos n\Delta\phi\right). \tag{1}$$

To date, most analyses have neglected all terms but $n = 2$, describing the background as a pure ellipse modulating a pedestal B_0, generated presumably by initial anisotropy and hydrodynamic flow. However, fluctuations have recently been shown to strongly affect the shape of individual events. In particular, Glauber Monte Carlo simulations indicate that some events bear a strong triangular shape, and a mapping of triangular anisotropy to v_3 in simulations such as AMPT has led to calculated v_3/v_2 ratios of 20-100% [6] .

3. Background in correlations: expected influences

In general, correlated signals S are very sensitive to the shape and normalization of the background B when S/B is very small. Because the low S/B region is precisely where 2-peak structures have been most prominent, it is important to understand how background influences correlations.

If all trigger particles are from jet-related sources, adding background to the population of associated particles results in a pedestal with magnitude $n_{bkg}^B/2\pi$ (where $n = N/N_{evt}$), since

$$\int d\Delta\phi \frac{1}{N_{jet}^A} \frac{dN_{jet-bkg}^{AB}}{d\Delta\phi} = \frac{n_{bkg}^B}{2\pi}. \tag{2}$$

Adding non-jet trigger particles does not change the total pair yield per trigger in each event, since the pair and trigger multiplicities rise at the same rate and thus cancel. However, the correlation is weakened. The jet peak yields are diluted by a factor $f \equiv N_{jet}^A/N^A$. But the total per-trigger yield is unchanged, again due to the cancellation. Thus the pedestal rises by $(1-f)n_{jet}^B$. These expectations are tested in a simulation described in the following section.

4. Jet-triggered dihadron correlations: simulations and real data

The trigger purity can be enhanced by requiring the trigger particle to be inside a reconstructed jet. In this analysis, the FastJet anti-k_T clustering algorithm is used due to its relatively low sensitivity to background fluctuations and its tendency to reconstruct conical jets. An angular distance is defined between each trigger particle and the nearest jet axis, $\Delta R \equiv \sqrt{(\phi_{jet}-\phi_A)^2 + (\eta_{jet}-\eta_A)^2}$. We distinguish $h_{jet}^{\pm} - h^{\pm}$ correlations from inclusive $h^{\pm} - h^{\pm}$ by requiring $\Delta R < R_c$, where $R_c = 0.4$.

In order to study how non-jet trigger particles influence correlations, a simple Monte Carlo study was conducted in which 20 GeV pythia jets were emebedded in an isotropic thermal background, where $dN_{bkg}/dp_T = N_0 \exp(-p_T/T)$. Fits to STAR 0-5% central $h^{\pm}$ spectra at 1-2 GeV/c motivated a choice of $T = 260$ MeV. N_0 was chosen to match central Au+Au collsions at RHIC: in the STAR acceptance, $N_0 = dN_{ch}/d\eta \cdot \Delta\eta \cdot N_{all}/N_{ch} = 650 \cdot 2 \cdot 3/2 = 1950$.

Figure 1 (top) shows $h_{jet}^{\pm} - h^{\pm}$ pair yields per trigger particle for pythia particles in the presence of background, where $2 < p_T^A < 3$ and $1 < p_T^B < 2$ GeV/c. To check the expectation of equation 2, the background pedestal level shown as a red line is calculated explicity as

$$\frac{n_{bkg}^B}{2\pi} = \frac{1}{2\pi} N_0 \frac{N_{bkg}(1-2\,\mathrm{GeV})}{N_{bkg}(\mathrm{all}\, p_T)} = \frac{1}{2\pi} \cdot 1950 \cdot 0.105 = 32.8. \tag{3}$$

When the background pedestal is subtracted, the original background-free yield is recovered

Journal of Physics: Conference Series **270** (2011) 012018 doi:10.1088/1742-6596/270/1/012018

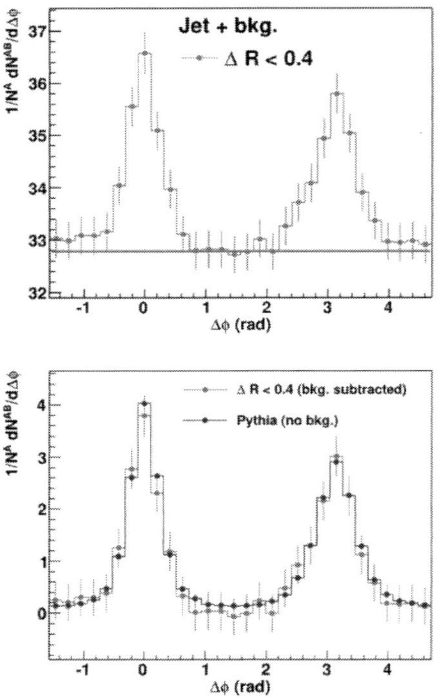

Figure 1. $h^{\pm}_{jet} - h^{\pm}$ for simulated jets in a thermal background.

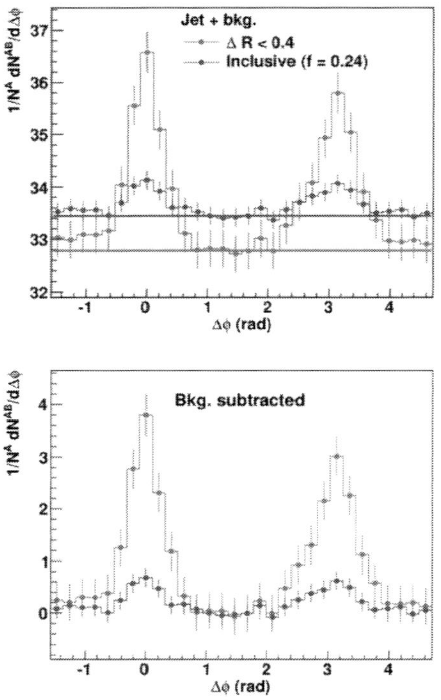

Figure 2. $h^{\pm}_{\rm jet} - h^{\pm}$ vs. $h^{\pm} - h^{\pm}$ for simulated jet-background study.

(figure 1, bottom). In contrast, if the $\Delta R < R_c$ cut is not applied, about 3/4 of the trigger particles are from the thermal background, leading to a pedestal rise of $(1 - f)n^B_j$ (figure 2, top). Pedestal subtraction leads to an integrated yield reduced by f compared to $h^{\pm}_{\rm jet} - h^{\pm}$ (figure 2, bottom), suggesting that the $\Delta R < R_c$ cut is highly effective in removing non-jet trigger particles.

A critical conclusion is that non-jet trigger particles diminish jet correlation signals, causing enhanced sensitivity to misrepresentation of the background shape. Thus, distortions in the background-subtracted result due to incorrect v_2 values, neglect of $v_{n\neq 2}$ components, and normalization errors are exacerbated by the diminished signal strength.

A first result from $h^{\pm}_{jet} - h^{\pm}$ correlations in real data was produced from a STAR Au+Au and p+p high tower (HT) triggered data sample rich in highly energetic jets. Moreover, only particles with $p_T > 2$ GeV/c are used in jet reconstruction, and jets are required to have a background-corrected $p_T > 10$ GeV/c.[1].

A comparison of $h^{\pm}_{jet} - h^{\pm}$ $(\Delta R < 0.4)$ and $h^{\pm} - h^{\pm}$ is shown for Au+Au and p+p for two different p^A_T bins in figures 3 and 4. Both measurements are from a common event selection; every event contains a jet with $p^{jet}_T > 10$ GeV/c. Thus, even the inclusive dihadron correlations (blue points) have enhanced correlations relative to those expected from minimum bias events. In the top panels, the background is included, and in the bottom panels, it has been subtracted. Apart from the event selection and ΔR cut, the analysis follows the procedure leading to the results in [1]. Thus, the background is shaped as in equation 1 with $v_n = 0$ for all $n \neq 2$. A

[1] Background subtraction is performed as $p^{jet}_T = p^{meas}_T - \rho A$, where ρ is the median p_T per unit area.

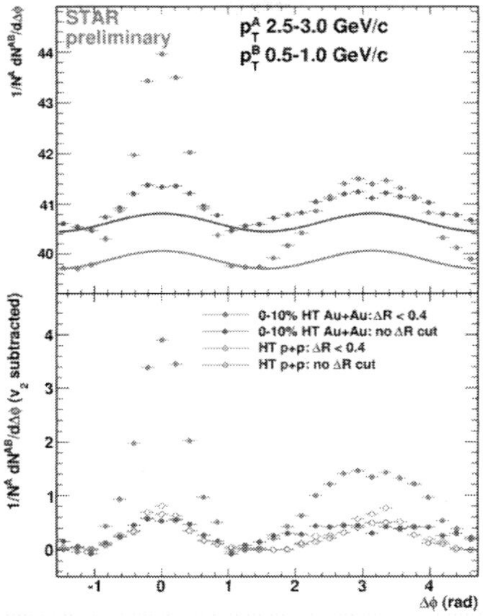

Figure 3. $h^{\pm}_{jet} - h^{\pm}$ and $h^{\pm} - h^{\pm}$ from STAR Au+Au and p+p at moderate p_T^A.

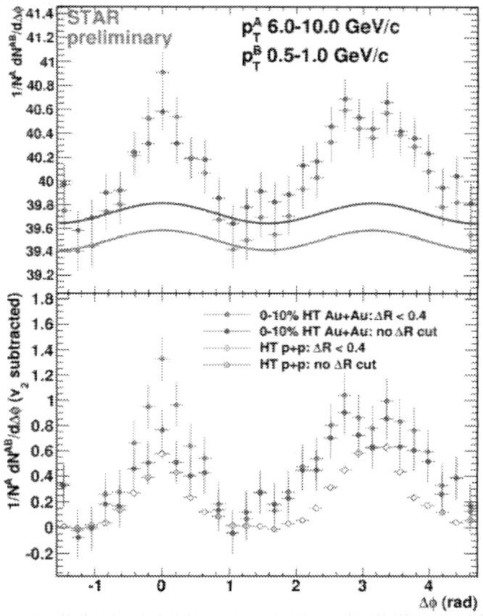

Figure 4. $h^{\pm}_{jet} - h^{\pm}$ and $h^{\pm} - h^{\pm}$ from STAR Au+Au and p+p at high p_T^A.

v_2 value from STAR measurements is used for both cases as an initial estimate, and the ZYAM assumption determines B_0.

5. Discussion and summary

The most striking feature of the results is the signal enhancement at low p_T^A. Apart from the fact that these results may depend sensitively on the jet definition, an explanation for the enhanced signal is not as straightforward as in the simulation for several reasons.

First, background sources in real data are more complicated. The trigger particle population may be contaminated by thermal fluctuations, recombination, and soft particles boosted by hydrodynamics or radial expansion. Moreover, many particles from hard scatterings unrelated to the primary di-jet are removed in $h^{\pm}_{\rm jet}$–$h^{\pm}$. Thus, $h^{\pm}_{\rm jet}$–$h^{\pm}$ does not sample the minimum-bias jet cross section. Additionally, the kinematics are not the same. While the simulation used essentially mono-energetic jets, the STAR results may reflect a selection of higher-Q^2 events in the $h^{\pm}_{\rm jet}$–$h^{\pm}$ case than for inclusive. At a given p_T^A, $h^{\pm}_{\rm jet}$–$h^{\pm}$ correlations sample harder collisions and lower-z hadrons.

In summary, triggering on more jet-like particles in Au+Au collisions enhances the correlation strength and diminishes the recoil-side double-peak feature precisely where it might be expected if energy loss which triggers medium response scales with parton energy. Although requiring careful interpretation, $h^{\pm}_{\rm jet} - h^{\pm}$ correlations add useful information to investigations of medium response and energy loss.

References

[1] Aggarwal M M *et al.* (STAR) 2010 (*Preprint* 1004.2377)
[2] Adare A *et al.* (PHENIX) 2008 *Phys. Rev.* **C78** 014901 (*Preprint* 0801.4545)
[3] Adams J *et al.* (STAR) 2005 *Phys. Rev. Lett.* **95** 152301 (*Preprint* nucl-ex/0501016)
[4] Abelev B I *et al.* (STAR) 2009 *Phys. Rev. Lett.* **102** 052302 (*Preprint* 0805.0622)
[5] Adare A *et al.* (The PHENIX) 2010 *Phys. Rev. Lett.* **104** 252301 (*Preprint* 1002.1077)
[6] Alver B and Roland G 2010 *Phys. Rev.* **C81** 054905 (*Preprint* 1003.0194)

Charm content in jets in proton-proton collisions with the ALICE experiment at CERN-LHC

Alessandro Grelli on behalf of the ALICE collaboration
ERC Starting Independent Research, group QGP, Institute of Subatomic Physics, Princetonplein 5, 3584 CC, Utrecht The Netherlands
E-mail: a.grelli@uu.nl

Abstract.
Charm and bottom quarks have been proposed as probes to study partonic matter produced in high-energy heavy-ion collisions. The detailed understanding of the production mechanisms in proton-proton collisions is of considerable interest as a QCD test tool and as reference calibration for heavy-ion studies. Measurements of the D meson yield in jets probe the production processes in which the observed D mesons are formed primarily from gluon splitting into $c\bar{c}$ or $b\bar{b}$ pairs. The charm content in jets is calculable in perturbative QCD, and the leading non-perturbative correction is expected to be significant at LHC energies. In this contribution we present a performance study of the reconstruction of charged $D^{*\pm}$ mesons in jets based on the first data collected by the ALICE experiment in minimum bias p+p collisions at $\sqrt{s}$ = 7 TeV . D^{*+} mesons are reconstructed through the decay sequence $D^{*+} \to D^0\pi^+$ and $D^0 \to K^-\pi^+$ (and its charge conjugate channel).

1. Introduction

The ratio of gluon splitting in $c\bar{c}$ ($g \to c\bar{c}$) pairs in gluon jets to that into light quarks pairs is calculable in perturbative QCD (pQCD). This offers not only the possibility to build a baseline for lead-lead collisions studies but even a clear possibility to test the pQCD predictions. At the energies available at Tevatron and RHIC the contribution to the total charm cross-section from this production mechanism is expected to be small [1, 2]. This production mechanism can be investigated studying the D^* content in jets and using the different fragmentation characteristic with respect to pair creation. Several attempts have been made to measure the D content in jets. In proton-proton collisions at $\sqrt{s} = 630$ GeV the UA1 collaboration has observed D^* in jets with transverse momenta over 40 GeV/c [3]. The D^* fractional momenta with respect to the jet energy are consistent with a production mechanism different from pair creation, in which the D^* originate from gluon splitting in $c\bar{c}$. The same result is found by the CDF collaboration at $\sqrt{s} = 1.8$ TeV [4]. The STAR collaboration at RHIC [5] shows that the contribution is about 6% and in agreement within the experimental errors with the pQCD predictions.

2. Experimental apparatus and data sample

The ALICE detector [6] is dedicated to the study of heavy ion collisions at LHC. It is divided into a barrel, that covers a $|\eta| < 0.9$ region with a solenoidal magnetic field of 0.5 T, and a forward muon arm that covers the region -2.5< $|\eta|$ <-4 with a 0.7 T dipolar magnetic field. The main tracking detectors used in the present study are placed in the barrel part of ALICE experiment and are the Inner tracking System (ITS) and the Time Projection Chamber (TPC). The ITS is made up of 6 layers of silicon detectors built using three different technologies [7] and it is the main vertex detector for heavy flavour physics. The TPC is the main tracking device of ALICE, it gives up to 160 spatial points and has excellent Particle IDentification (PID) capabilities. Besides the TPC, the PID is assured in a wide p_T range and for different particle species, by adding the informations of the Time Of Flight (TOF) and the ITS. The low material budget in the silicon tracker ($(X/X_0) \sim 7\%$) together with the excellent tracking performance of the TPC and the low magnetic field make the ALICE detector competitive even in p+p physics; in particular at low p_T. The results presented in this paper are obtained from a sample of 140×10^6 minimum bias collisions at $\sqrt{s} = 7$ TeV selected using an interaction trigger based on an "or" between two rings of scintillators (V0A and V0C) and the pixel tracker [8].

3. Jet reconstruction

In this analysis jets have been reconstructed using the TPC detector, based on the UA1[9] fixed cone algorithm with an energy threshold of 10 GeV and a cone radius of 0.4. Although ALICE offers the possibility to reconstruct jets with all the main algorithms available (see Fig.1, bottom), most of them are infrared safe, with the first data available we prefer to use the UA1 method with the reported specifications being a very robust and widely tested algorithm. On the top of figure 1 a di-jet event is shown using the ALICE event display. Based on Monte Carlo studies, the chosen cone size in the UA1 algorithm offers a good estimation of the jet axis .

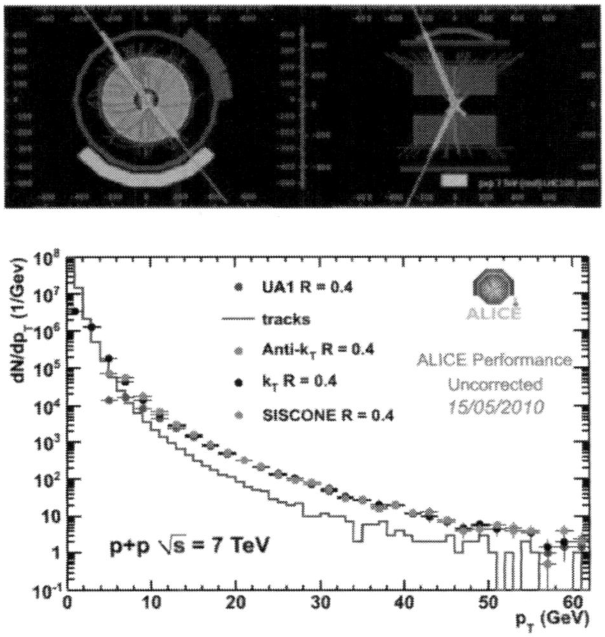

Figure 1. Top: Di-jet event captured by the ALICE event display. Bottom: Comparison of the uncorrected charged jet spectra for different reconstruction algorithms for $|\eta| < 0.5$ [10].

4. D* reconstruction

The reconstruction of D^* meson is performed in the channel $D^{*+} \to D^0\pi_s^+$ (BR = (67.7±0.5)%) where the D^0 is reconstructed in the hadronic channel $D^0 \to K^-\pi^+$ (BR = (3.91±0.05)%) and it is based on the invariant mass analysis and the D^0 displaced vertex reconstruction. We require as single track selection at least 70 spatial points in the TPC. In addition the tracks must have at least 4 out of six possible clusters in the ITS layers and at least one point in one of the pixel layers. In the case of the π_s (soft pion) we use in addition to the ITS+TPC reconstruction the ITS stand alone reconstruction. Once the single track selections are passed then the D^* candidates are built and a set of topological selections are applied to the candidate D^0 to suppress the combinatorial background. In the figure 2 a sketch of the D^0 decay topology is reported. The main cutting variables are the product of impact parameters ($d_0^K \times d_0^\pi$) and the cosinus of the pointing angle (θ_{point}). The impact parameter is defined as the distance of closest approach in the transverse plane between the back extrapolated trajectory of each of the D^0 daughters and the primary vertex. The θ_{point} is the angle between the D^0 momenta and its flight line.

To improve the background rejection factor a PID strategy has been implemented. In this analysis the TPC and TOF informations are combined. For the TPC, a 2σ cut around the kaon (pion) Bethe-Bloch lines ensures good kaon (pion) identification in a p_T region $300 \lesssim p_T \lesssim 800$ MeV/c. The additional information of the TOF allows to identify kaons up to 1.6 GeV/c. Finally the D^* peak is expected to be seen in the mass difference $\Delta M = M(K\pi\pi) - M(K\pi)$ on the edge of combinatorial background as a sharp peak centered at ΔM = 145.42 MeV/c^2. In figure 3 the D^* signal is shown, p_T integrated, as extracted using our selections from the first 140M events from the ALICE p+p minimum bias sample at $\sqrt{s} = 7$ TeV. The fit function is a convolution of a power law, used as threshold function, and a gaussian. The applied selections offer the possibility to extract the signal with a very good signal over background ratio (S/B ~ 2).

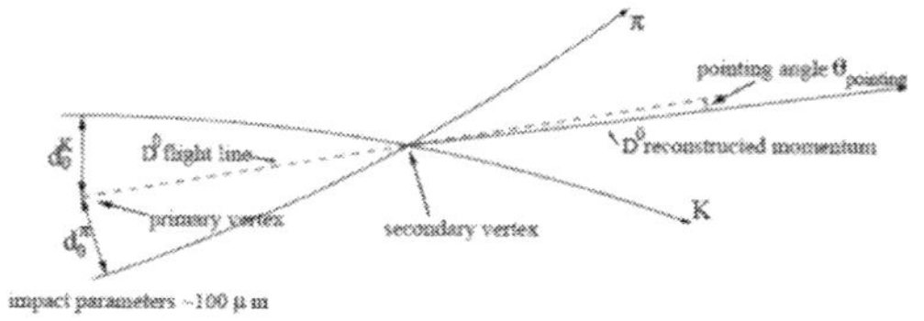

Figure 2. Sketch of the D^0 decay topology with the definition of the impact parameters (d) of the D^0 daughters and the pointing angle θ_{point} variables.

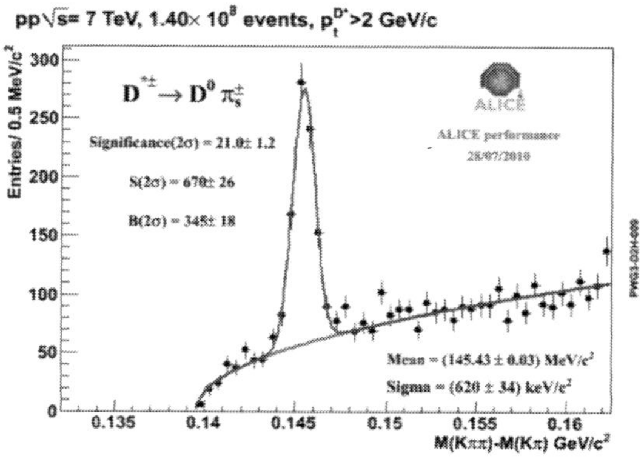

Figure 3. Invariant mass difference after all the selection cuts. The D* peak, p_T integrated, is clearly visible. A resolution of (620 ± 34) keV/c^2 is obtained

5. D* in jet reconstruction

In order to investigate the D^* content in jet we have analyzed 1M jet events extracted from the minimum bias sample requiring $E_{jet} \geq 10$ GeV and $|\eta| \leq 0.5$. The search is restricted to a region of radius 0.4 around the jet axis. Figure 4 shows the signal for the D^* reconstructed inside this sample of jets.

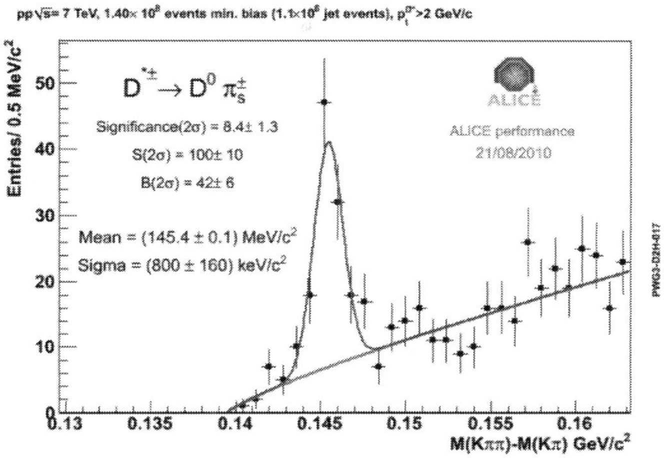

Figure 4. Invariant mass difference computed for candidates inside the jet cone. The D* peak is visible.

6. Conclusions

The first performance results on jet reconstruction and D^* meson reconstruction with the ALICE detector are reported. A set of selections to extract the D^* signal both in minimum-bias and in the jet sample is found. Since we expect to collect a sample of at least 10^9 min bias events in 2010 the measurement should not be statistics limited, therefore we tuned the cuts for high purity. The high purity will be an advantage for the correlation and fragmentation studies. In conclusion, we showed the feasibility of the reconstruction of the $D^{*\pm}$ mesons in a jet sample even with limited statistic and early data. This is the first step towards the D^* in jet analysis, the analysis of the correlations and fragmentation function will follow when additional statistics will be available.

References

[1] A.H Muller and P. Nason. *Phys. Lett.* **B 157**, 226 (1985).
[2] M.L. Mangano and P. Nason. *Phys. Lett.* **B 285**, 160 (1992).
[3] C. Albajar *et al.* (UA1 Collaboration), *Phys. Lett.* **B 244**, 566 (1990).
[4] F. Abe *et al.* (The CDF Collaboration), *Phys. Rev. Lett.* **D 64**, 348 (1990).
[5] B.I. Abelev *et al.* (STAR Collaboration), *Phys. Rev. Lett.* **D 79**, 112006 (2009).
[6] ALICE Collaboration: F Carminati *et al.* *J. Phys. G: Nucl. Part. Phys.* 30 1517 (2004).
[7] ALICE collaboration 2010 JINST 5 P03003.
[8] K. Aamodt *et al.*, *Eur. Phys. J.* **C65**, 111 (2010).
[9] S-L Blyth *et al.* *J. Phys. G*: Nucl. Part. Phys. **34** 271 (2007)
[10] Swensy Jangal, these proceedings.

Journal of Physics: Conference Series **270** (2011) 012020 doi:10.1088/1742-6596/270/1/012020

Multi-strange baryon measurements at LHC energies, with the ALICE experiment

Antonin Maire for the ALICE Collaboration
Institut pluridisciplinaire Hubert Curien (IPHC), 23 rue du loess - 67037 Strasbourg, France
E-mail: antonin.maire@cern.ch

Abstract. The status of the charged multi-strange baryon analysis (Ξ^-, $\overline{\Xi}^+$, Ω^-, $\overline{\Omega}^+$) at LHC energies is presented. This report is based on the results obtained with ALICE (A Large Ion Collider Experiment), profiting from the characteristic cascade-decay topology. A special attention is drawn to the early pp data-taking period (2009-2010) and subsequently, on the uncorrected p_T-spectra extracted at mid-rapidity for centre of mass energies of 0.9 TeV and 7 TeV.

1. Introduction and motivation

Measuring strange particle production in proton-proton (pp) collisions is a necessary benchmark for the physics of ultra relativistic heavy ions. This is important at the Large Hadron Collider (LHC), where the heavy-ion programme is scheduled to begin in November 2010 [1]. Moreover, strangeness in pp collisions is interesting in itself, as it may shed light on hadron production mechanisms. While the *hard* component of the event may be described by the perturbative Quantum Chromo-Dynamics (based on parton-parton scattering and fragmentation [2, 3]), the *soft* component must be treated in a more complex manner. Currently, the soft physics is described via thermal models [4, 5] or via QCD-inspired models (relying on multi-parton processes [6] or multiple scattering [7, 8], for instance). In either case, further improvements of such phenomenological approaches may be spurred by confrontation with experimental measurements.

In that respect, double- and triple-strange baryons (Ξ, Ω, ...), which are the focus of this publication, may provide relevant insights: due to identification via weak-decay topology reconstruction, they can be studied over a large momentum range. Starting from $p_T \approx 0.6$ GeV/c and up to ≈ 10 GeV/c, these spectra cover the region dominated by the soft processes and reach the energy scale where hard scattering mechanisms dominate.

Some measurements have already been performed at previous and current facilities. These include both p$\overline{\text{p}}$ collisions (Sp$\overline{\text{p}}$S, Tevatron) and pp collisions (RHIC), with centre of mass energies $\sqrt{s}$ ranging from 0.2 TeV up to 1.96 TeV [9–11]. The LHC, in operation since November 2009, can extend the existent 0.9-TeV measurements made by the UA5 collaborations in p$\overline{\text{p}}$, and perform new measurements at $\sqrt{s} = 7$ TeV, beyond the Tevatron energies.

The ALICE experiment [1] is well-suited for such spectrum measurements, due to a low p_T cut-off and excellent particle identification (PID) capabilities. The low p_T cut-off is made possible by the low magnetic field applied in the central barrel (≤ 0.5 T) and the low material budget in this mid-rapidity region (13% of radiation length [12]). The PID capabilities are supplied by a set of detectors utilizing diverse techniques (energy loss, time of flight, transition radiation, Cerenkov effect).

2. Data analysis and identification methods

2.1. Data collection and detector setup

The data presented here is from the minimum bias sample collected during the Nov-Dec 2009 LHC pp run at $\sqrt{s} = 0.9$ TeV [13] ($\sim 3 \times 10^5$ events), and from the 7 TeV pp run that started in March 2010 and is ongoing ($\sim 23.9 \times 10^6$ events were analysed here but more than 4×10^8 events have been collected since then). This study makes use of the ALICE central barrel [14], covering a pseudo-rapidity range $|\eta| < 0.9$ and the full azimuth, the whole being embedded in the large L3 solenoidal magnet which provides a nominal magnetic field of 0.5 T.

The cascade signals are obtained using essentially data collected by the two main tracking detectors: the Inner Tracking System (ITS), composed of 6 cylindrical layers of high-resolution silicon detectors [15], and the cylindrical Time Projection Chamber (TPC) [16].

2.2. Topological reconstruction

The multi-strange hadron identification is performed using a combination of displaced-vertex reconstruction, invariant mass analyses as well as single track PID methods, such as energy loss in the TPC. The reconstruction of the different multi-strange particles hinges on their respective charged weak decays, the so-called *cascade* structures. For each particle of interest, the main characteristics and utilized decay channels are listed in table 1. The anti-baryons are similarly reconstructed via the decay channel involving the charge conjugates.

Table 1. Main characteristics of the reconstructed particles [17].

Particles	mass (MeV/c^2)	$c\tau$	charged decay	B.R.
Λ^0 (uds)	1115.68	7.89 cm	$\Lambda^0 \rightarrow p + \pi^-$	63.9%
Ξ^- (dss)	1321.71	4.91 cm	$\Xi^- \rightarrow \Lambda^0 + \pi^-$	99.9%
Ω^- (sss)	1672.45	2.46 cm	$\Omega^- \rightarrow \Lambda^0 + \mathrm{K}^-$	67.8%

The guidelines of the reconstruction algorithm dedicated to cascades are sketched in Fig. 1. The Ξ^-, $\overline{\Xi}^+$, Ω^- and $\overline{\Omega}^+$ identification is based on the pairing of a Λ^0 or $\overline{\Lambda}^0$ baryon with an additional particle called *bachelor*. The Λ reconstruction is grounded in the secondary vertex finding, a *V0* structure built out of two *secondary* tracks of opposite charges, compatible with coming from the same vertex. It is checked that the V0 candidate thus obtained sits in the proper invariant mass window to validate the Λ hypothesis. The Λ candidate is then matched with a *secondary* track, to form a typical cascade structure. The reconstructed momentum of the multi-strange baryon must point in the direction of the primary vertex. The V0 vertex and the V0-bachelor matching are limited to a certain fiducial volume. A PID compatibility cut, based on the dE/dx measurement from the TPC, is required for each of the three decay products, known as *daughter* particles, to partially remove some combinatorial background.

2.3. Signal extraction

For each considered particle, we intend to extract a signal in successive p_{T} intervals (bins). For each p_{T} bin, the invariant mass signal sits on top of a given background. The signal extraction process, using "bin-counting" method, is illustrated in Fig. 2.

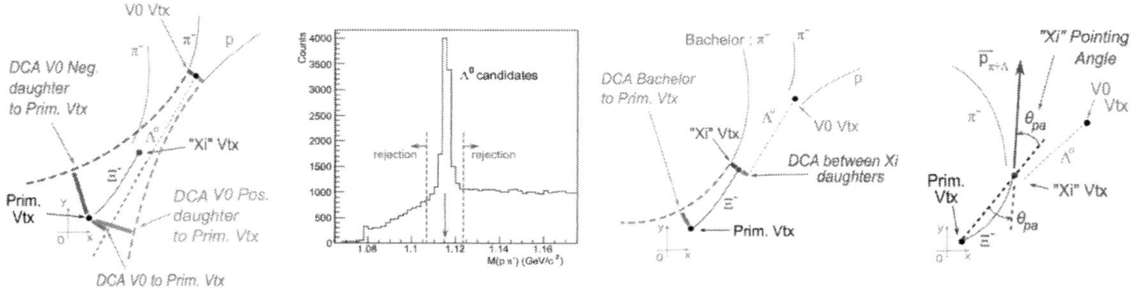

Figure 1. Cascade (Ξ^-, $\overline{\Xi}^+$, Ω^- and $\overline{\Omega}^+$) reconstruction principle. The acronym *DCA* stands for Distance of Closest Approach.

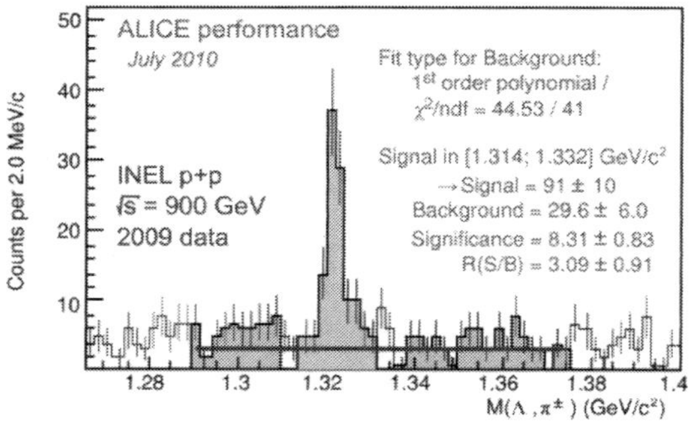

Figure 2. Signal extraction based on "bin counting" method, illustrated with the $\Xi^- + \overline{\Xi}^+$invariant mass distribution in $1.4 < p_T$ (GeV/c) < 2.0.

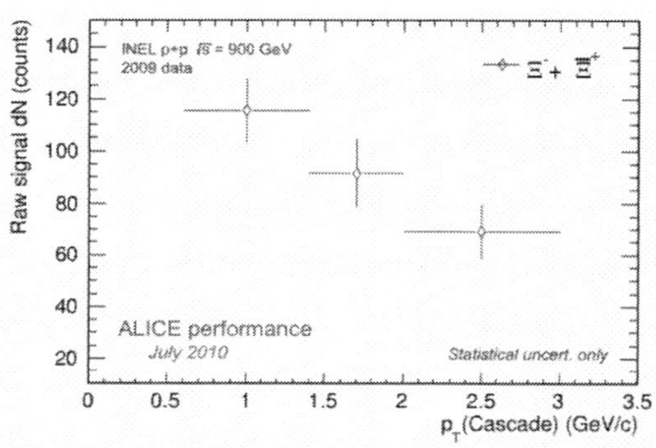

Figure 3. Raw signal by p_T bins for $\Xi^- + \overline{\Xi}^+$, in 2009 pp data at 0.9 TeV.

The signal is first approximated by a Gaussian lying on a polynomial background. Because of the non-Gaussian tails of the signal, this results in rough but sufficient estimates of the signal mean and width.

The background on each side of the signal (more than 6σ away from the Gaussian mean) is then sampled. The width of each background area is taken to be 7σ large.

The sum of signal and background (S+B) is sampled in the region defined by the Gaussian mean $\pm 4\sigma$. Consequently, we make use of the areas previously sampled on the side-signal bands to assess the background B under the signal S. The signal yield $S = (S + B) - B$ is thus computed without any assumption as to the signal shape.

3. Measurements at 0.9 and 7 TeV

The results for the 2009 pp sample at 0.9 TeV are shown in Fig. 3. The plot shows the signal counts (raw yields) for $\Xi^- + \overline{\Xi}^+$, as a function of p_T. The uncertainties correspond to both the statistical uncertainty related to the number of counts and the uncertainty from signal extraction.

The same particles are presently analysed for the pp data at 7 TeV. Due to higher statistics, baryons and anti-baryons can be studied separately and with larger counts: Ξ^- and $\overline{\Xi}^+$ or even Ω^- and $\overline{\Omega}^+$ can now be distinguished from one another, as illustrated in Fig. 4 and 5.

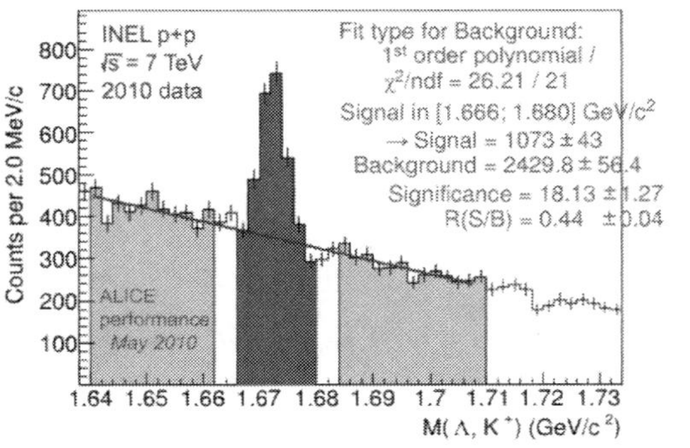

Figure 4. Invariant mass showing the $\overline{\Omega}^+$ signal in 2010 pp data at 7 TeV (23.9 M events).

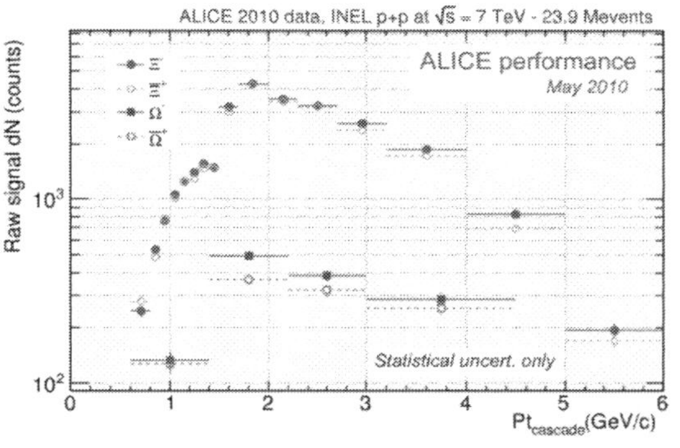

Figure 5. Raw signal by p_T bins, extracted for the different cascade species, in 2010 pp data at 7 TeV (23.9 M events).

4. Summary

A spectrum for $\Xi^- + \overline{\Xi}^+$ baryons is obtained for the first LHC pp run at 0.9 TeV, despite the limited statistics; the assessment of efficiency corrections and systematic uncertainties is under way. Due to the large statistics available for the 7-TeV pp data sample, Ξ^-, $\overline{\Xi}^+$, Ω^- and $\overline{\Omega}^+$ can be extracted distinctively. This bodes well for more accurate and differential analyses such as spectra as a function of p_T, rapidity or event multiplicity.

References

[1] B. Alessandro (ed.) *et al.* [ALICE Collaboration], *J. Phys.* G **32** (2006) 1295.
[2] T. Sjöstrand, S. Mrenna and P. Z. Skands, *J. High Energy Phys.* 0605 (2006) 026.
[3] R. Engel, J. Ranft and S. Roesler, *Phys. Rev.* D **52**, (1995) 1459.
[4] F. Becattini and U. W. Heinz, *Z. Phys.* C **76**, (1997) 269.
[5] P. Braun-Munzinger, K. Redlich and J. Stachel, arXiv:nucl-th/0304013.
[6] T. Sjöstrand and P. Z. Skands, *Eur. Phys. J.* C **39**, (2005) 129.
[7] K. Werner, *Nucl. Phys. Proc. Suppl.* **175-176**, 81 (2008).
[8] K. Werner, I. Karpenko, T. Pierog, M. Bleicher and K. Mikhailov, *Preprint* arXiv:1004.0805 [nucl-th].
[9] STAR Collaboration, B. I. Abelev *et al.*, *Phys. Rev.* C **75**, (2007) 064901.
[10] UA5 Collaboration, R. E. Ansorge *et al.*, *Nucl. Phys.* B **328**, (1989) 36.
[11] CDF Collaboration, *public note* CDF/PUB/QCD/PUBLIC/10084 (2010), www-cdf.fnal.gov.
[12] B. Hippolyte, *Eur. Phys. J.* C **62** (2009) 237 (*Preprint* arXiv:0901.3176 [hep-ex]).
[13] L. Evans and P. Bryant, (ed.), *JINST* **3**, (2008) S08001.
[14] K. Aamodt *et al.* [ALICE Collaboration], *JINST* **3** (2008) S08002.
[15] K. Aamodt *et al.* [ALICE Collaboration], JINST **5** (2010) P03003 [arXiv:1001.0502 [physics.ins-det].
[16] J. Alme *et al.* [ALICE Collaboration], arXiv:1001.1950 [physics.ins-det].
[17] C. Amsler *et al.* (Particle Data Group), *Phys. Lett.* B **667**, (2008) 1.

Probing dense nuclear matter with electromagnetic and rare hadronic probes

Manuel Lorenz for the HADES collaboration

E-mail: Lorenz@Physik.uni-frankfurt.de

Abstract. We present preliminary results on strangeness and dilepton production in Ar+KCl reactions at 1.76 AGeV. For the first time an ω signal could be extracted in heavy-ion collisions in this energy region. Furthermore we find a strong enhancement of yield in the low invariant lepton pair mass region over an elementary reference suggesting the onset of non trivial effects of the medium. This hint is supported by a nearly complete vanishing of the OZI suppression of the ϕ to ω ratio which is characteristic for elementary reactions.

1. Introduction

Heavy-ion collisions provide a unique tool to study strongly interacting matter under extreme conditions. While at SPS, RHIC and LHC energies it is common believe that a phase transition between hadronic and partonic degrees of freedom occurs, in the SIS/Bevalac energy regime, a partial restoration of the broken chiral symmetry is expected. The spontaneous and explicit breaking of chiral symmetry are usually related to the light quark and hadron masses. Therefore a (partial) restoration of this symmetry might result in a modification of hadron masses. Moreover, also the hadronic phase of the system differs in its characteristics compared to higher energies, where the produced pions outnumber the existing baryons resulting in a low baryochemical potential μ_B, whereas at SIS/Bevalac energies the pions are hidden in excited baryon states (Δ, N*) which act as energy accumulators. At a late stage of the reaction these resonances decay and release the pions.

Promising probes for such a strongly interacting system are particles containing newly produced strange quarks due to their steep excitation function, which makes them sensitive to the achieved compression and temperature, as well as dileptons as they do not suffer the strong interaction and therefore deliver undistorted information over the whole time evolution of the collision.

2. Overview of results

Following the pioneering experiments at the Bevalac [1] the production and propagation of strangeness has been investigated by the KaoS, FOPI and by the HADES collaborations. The results of the KaoS collaboration give evidence for a soft equation of state [2], a repulsive K-N potential and the importance of strangeness exchange reactions for subthreshold strangeness production [3]. FOPI data gave a first hint on a large ϕ contribution to subthreshold K^- production [4], which was quantified recently by HADES [5].

Both FOPI and HADES data confirm the evidence for a repulsive K-N potential [6, 7] and give evidence for the production of strange resonances [8, 9], as well as multistrange particles [10] already at such low energies.

Besides the unsettled production mechanism of multistrange particles, like the Ξ^-, the shape of the K^- spectra is not fully understood. The inverse slope parameter obtained from fits to transverse mass spectra are systematically (10 - 20 MeV) lower compared to the ones from K^+. Previous transport model calculations were able to reproduce the slope by introducing a strongly attractive K^--N potential resulting in dominant feeding of the K^- abundance from strangeness exchange reaction like $\pi^0 + \Lambda \rightarrow K^- + p$[1] and consequently resulting in a later emission and freeze-out time for K^- compared to K^+ [11]. But in [12] it was shown that the differences in the slope can be explained, within errors, by the superposition of the K^- that are coming from ϕ decays and a thermal source of K^- with the same initial slope as the measured one of the K^+. As input for the simulation the measured ϕ/K^- ratio of 0.37 ± 0.13 and inverse slopes of the charged kaons and the ϕ meson in the system Ar+KCl at 1.76 AGeV measured with HADES are used [5].

The emission of dileptons has been investigated by the DLS collaboration at the Bevalac [13]. DLS observed a large enhancement over a "hadronic cocktail" (provided by e^+e^- decay products of long living mesons π^0, η, ω), in the invariant mass region of $0.15 < M_{ee} < 0.55$ GeV/c^2 of e^+e^- pairs in the C+C system and the Ca+Ca system. Transport models were unable to explain this excess for a decade and this situation became known as the "DLS puzzle". The situation changed when HADES confirmed the DLS C+C data at 1 AGeV in 2008 [14]. Since then, various theoretical attempts have been proposed to explain the data, but lacking a final conclusion on the origin of the excess. In 2009, HADES data on proton on proton and on proton on quasifree neutron at 1.25 GeV showed that the excess is already present in elementary reactions [15]. Hence the so-called "DLS-puzzle", i.e. the unexplained enhancement in the low invariant mass region of e^+e^- pairs in the system C+C, results from an insufficient treatment of radiation from elementary N+N collisions. After first hints to a surprisingly strong pn channel in [16, 17], recent calculations have improved in reproducing these elementary channels but are still missing some yield in the pn channel [18].

In this contribution, we present preliminary results on dilepton production in Ar+KCl collisions at 1.76 AGeV, suggesting an additional excess over the elementary reference. The data set also allows, for the first time at this energy, for a determination of the ω multiplicity, which is at the moment still preliminary and in the process of being finalized.

3. Data analysis

The event selection was started with a first level trigger requiring a multiplicity > 16 in the Time-of-Flight wall, selecting the 35% most central events. It was followed by a second level trigger which selected at least one lepton candidate. Details on the apparatus, the lepton identification and of the experiment can be found in [5, 19].

For minimization of systematic errors, the lepton identification is performed using three independent methods. The dominant background source from γ conversion is suppressed using a cut on the lepton pair opening angle $\alpha_{ee} > 9^o$. The remaining background is modeled using same-event like-sign pairs in the invariant mass region $M_{ee} < 0.4$ GeV to handle the correlated background resulting from $\pi^0 \rightarrow \gamma\gamma$ decays followed by double conversion. In the higher invariant mass region mixed event pairs normalized to the like-sign background is used in order to reduce statistical fluctuations. The spectrum is normalized to the π^0 multiplicity estimated using the measured charged pions [20].

[1] Note that this reaction is endothermal in vacuum.

4. Low mass dilepton enhancement and up-lifted OZI suppression

As already mentioned, the yield in the low invariant mass region of e^+e^- pairs in the C+C system at 1 AGeV can be explained by a superposition of p+p and p+n data at 1.25 GeV. In order to learn more about the origin of the hadronic contribution and possible medium effects, we investigate how the situation changes as a function of energy and system size. Let us start by comparing the elementary reference to our C+C data set at 2 AGeV. First, we subtract the η contribution normalized to the neutral pion multiplicity using TAPS measurements [21]. As one sees in the left side of Fig. 1, where the C+C data at 2 AGeV are divided by the elementary reference provided by [(p+p)+(p+n)]/2 at 1.25 GeV (both scaled to the multiplicities of neutral pions), also these data are well described by the reference since the energy dependence is taken out in first order by the normalization to neutral pions. We conclude that the origin of the additional yield scales with energy like the pion production.

If we compare the elementary reference to the Ar+KCl data set at 1.76 AGeV [2] we find an additional excess factor of roughly three. This excess might be interpreted as the onset of non-trivial medium effects.

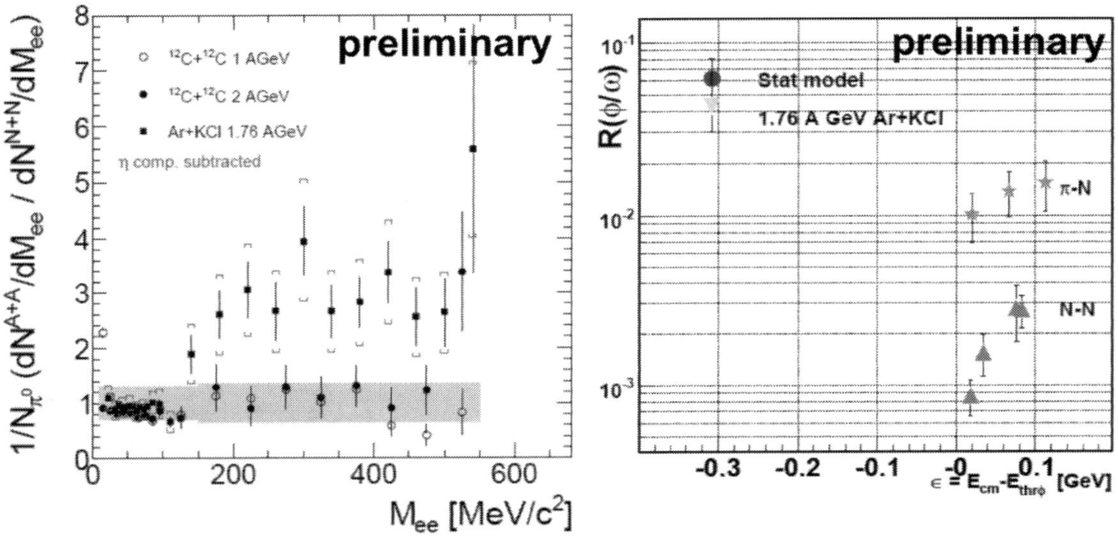

Figure 1. Left: Ratio of dielectron multiplicities in heavy-ion cllisions (C+C at 1AGeV: circles, C+C at "AGeV: bullets, Ar+KCl at 1.76 AGeV: squares) to dielectron multiplicity in nucleon-nucleon collisions as a function of invariant mass. The η contribution in heavy-ion data was subtracted using TAPS data [21]. Right: ϕ to ω ratio as a function of the ϕ excess energy for p+p (red triangles), π+N (magenta stars) and Ar+KCl (green square) data, as well as for a statistical model fit to the system Ar+KCl (blue circle) [26].

To reinforce this statement we look at the ϕ/ω ratio using a preliminary value of the ω multiplicity and the ϕ meson yield from the channel $\phi \rightarrow K^+K^-$ [5]. The ϕ meson production cross section in elementary reactions is strongly suppressed due to the Okubo-Zweig-Iizuka (OZI) rule [22, 23, 24] and, according to SU(3) flavor symmetry can proceed only because of a non ideal singlet and octet mixing in the ϕ wave and is commonly normalized to the cross section of

[2] The η contribution was averaged from TAPS measurements at 1.5 and 2.0 AGeV in the same system [21].

simultaneously measured ω mesons given as the ratio:

$$R_{(\phi/\omega)} = \frac{d\sigma(A+B\rightarrow\phi X)}{d\sigma(A+B\rightarrow\omega X)} = \tan^2(\delta\Theta_v)f = 4.2\cdot10^{-3}f, \quad (1)$$

where f is the ratio of the available phase space. Historically such ratios have been discussed in terms of strangeness content of the nucleon. While already in elementary reactions a deviation from Eq. 1 was observed [25], our data suggest that in heavy-ion reactions close to threshold the suppression seems to be fully neutralized and is, within errors, in agreement with a statistical model calculation (Fig. 1, right panel). More details on the statistical model fit can be found in [26]. The measured ϕ/ω ratio seems to be affected by the surrounding medium.

5. Summary and Outlook

We present preliminary results on dilepton production in Ar+KCl reactions at 1.76 AGeV. We find a strong enhancement in the low invariant lepton pair mass region over an elementary reference suggesting the onset of effects of the medium. This hint is supported by a nearly complete vanishing of the OZI suppression in the ϕ to ω ratio as well as the production of multiple strange objects like the Ξ^- more than 600 MeV below their free NN threshold. To obtain further systematics of these effects it is planned to take data using medium and heavy sized ion collision systems (AgAg, AuAu) from late 2010 on with HADES.

The collaboration gratefully acknowledges the support by BMBF grants 06DR9059, 06MT238 T5, and 06MT9156 TP5 (Germany), by the DFG EClust 153, by GSI TMKRUE, by grants MSMT LC07050 and GA ASCR IAA100480803 (Czech Rep.), by grants RII3-CT-2005-515876, FPA2006-09154, and CPAN:CSD2007-00042(Spain).

[1] S. Schnetzer et al., Phys. Rev. Lett. **49** (1982) 989; Phys. Rev. C. **40** (1989) 640.
[2] C. Sturm et al. (KaoS), Phys. Rev. Lett. **86** (2001) 39.
[3] A. Förster et al. (KaoS), Phys. Rev. C **75** (2007) 024906.
[4] A. Mangiarotti et al. (FOPI), Nucl. Phys. A **714** (2003) 89.
[5] G. Agakishiev *et al.* (HADES), Phys. Rev. C **80** (2009) 025209.
[6] M. L. Benabderrahmane *et al.* (FOPI), Phys. Rev. Lett. **102** (2009) 182501.
[7] G. Agakishiev *et al.* (HADES), arXiv:1004.3881 [nucl-ex].
[8] X. Lopez *et al.* (FOPI), Phys. Rev. C **76** (2007) 052203.
[9] X. Lopez *et al.* (FOPI), Phys. Rev. C **81**, 061902 (2010).
[10] G. Agakishiev *et al.* (HADES), Phys. Rev. Lett. **103** 132301 (2009).
[11] C. Hartnack, H. Oeschler and J. Aichelin, Phys. Rev. Lett. **90**, 102302 (2003).
[12] M. Lorenz *et al.* (HADES), Proceedings 48th International Winter Meeting on Nuclear Physics, Bormio, Italy, 25-29 Jan 2010.
[13] R. J. Porter *et al.* (DLS), Phys. Rev. Lett. **79** (1997) 1229.
[14] G. Agakichiev *et al.* (HADES), Phys. Rev. Lett. **98**, 052302 (2007).
[15] G. Agakishiev *et al.* (HADES), Phys. Lett. B **690**, 118 (2010).
[16] L. P. Kaptari and B. Kampfer, Nucl. Phys. A **764**, 338 (2006).
[17] L. P. Kaptari and B. Kampfer, Phys. Rev. C **80**, 064003 (2009).
[18] R. Shyam and U. Mosel, arXiv:1006.3873 [hep-ph].
[19] G. Agakishiev *et al.* (HADES), Eur. Phys. J. A **41**, 243 (2009).
[20] P. Tlusty *et al.* (HADES), arXiv:0906.2309 [nucl-ex].
[21] R. Averbeck, R. Holzmann, V. Metag and R. S. Simon, Phys. Rev. C **67**, 024903 (2003).
[22] S. Okubo, Phys. Lett. **5**, 165 (1963).
[23] G. Zweig, "An SU(3) Model For Strong Interaction Symmetry And Its Breaking. 2", Cern (1964).
[24] J. Iizuka, Prog. Theor. Phys. Suppl. **37**, 21 (1966) .
[25] A. Sibirtsev and W. Cassing, Eur. Phys. J. A **7**, 407 (2000).
[26] G. Agakishiev *et al.*, arXiv:1010.1675 [nucl-ex].

Femtoscopy and energy-momentum conservation effects in proton-proton collisions at 900 GeV in ALICE

Nicolas Bock

Department of Physics, The Ohio State University, 191 W Woodruff Ave, Columbus Ohio 43212, USA

E-mail: bock@mps.ohio-state.edu

Abstract. Two particle correlations are used to extract information about the characteristic size of the system for proton-proton collisions at $\sqrt{s}$=900 GeV measured by the ALICE (A Large Ion Collider Experiment) detector at CERN. The correlation functions obtained show the expected Bose-Einstein effect for identical particles, but there are also long range correlations present that shift the baseline from the expected flat behavior. A possible source of these correlations is the conservation of energy and momentum, especially for small systems, where the energy available for particle production is limited. A new technique, first introduced by the STAR collaboration, of quantifying these long range correlations using energy-momentum conservation considerations is presented here. It is shown that the baseline of the two particle correlation function can be described using this technique.

1. Introduction

Since November 2009 the Large Hadron Collider (LHC) has produced 14M proton-proton collisions at $\sqrt{s}$=900 GeV in ALICE (A Large Ion Collider Experiment) [1].The ALICE detector is designed to investigate matter at the highest energy densities, where it is supposed to be deconfined. This state of matter is composed of free quarks and gluons, therefore known as the Quark Gluon Plasma (QGP), and it is believed that right after the Big Bang matter was in this state. The QCD phase diagram (Fig. 1) shows the different phases of nuclear matter, and it is of great interest to study the properties of the QGP like the temperature of the phase transition, the critical point, the energy density, time evolution, lifetime of the system and flow-like effects. The present study is concerned with the measurement of the size of the particle emission region in proton proton collisions at $\sqrt{s}$=900 GeV using femtoscopy, and how it is affected by energy momentum conservation effects.

2. Femtoscopy in ALICE

Femtoscopy [2], or HBT [3], is a technique that measures the shape and size of the particle emission region by measuring pairs of identical particles simultaneously, see Fig. 2. Pions are most commonly used in this analysis, and because they are indistinguishable bosons the two-particle wave function has to be symmetrized as follows:

$$\Psi(p_1, r_1, p_2, r_2) \sim e^{i(p_1 \cdot r_1 + p_2 \cdot r_2)} + e^{i(p_1 \cdot r_2 + p_2 \cdot r_1)} \quad (1)$$

Journal of Physics: Conference Series **270** (2011) 012022 doi:10.1088/1742-6596/270/1/012022

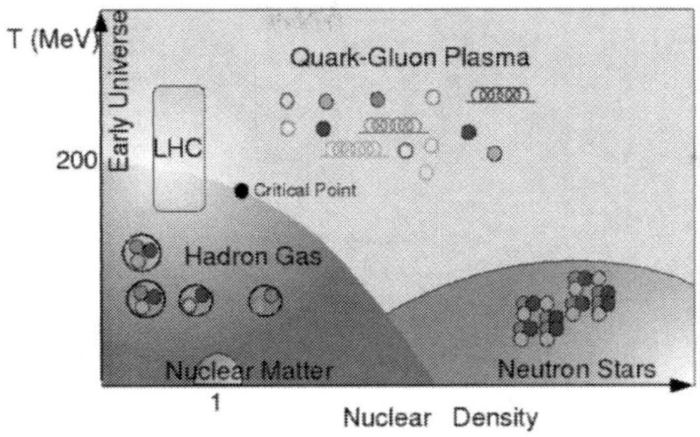

Figure 1. The QCD phase diagram shows the different phases of nuclear matter.

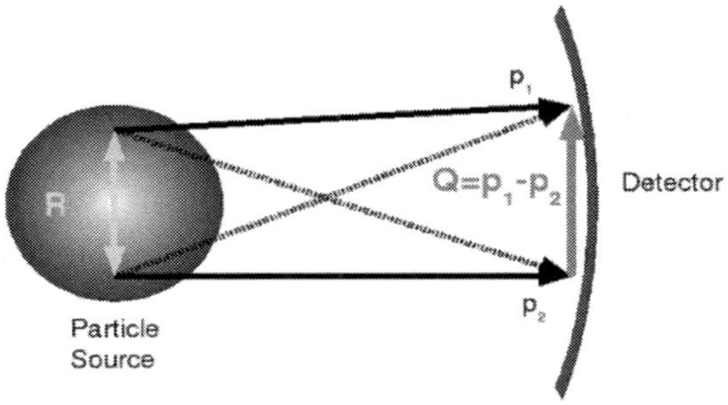

Figure 2. Two identical particles are emitted and detected simultaneously. There are two possible ways in which the particles could have been emitted.

The probability of detecting two such particles is :

$$|\Psi^2| \sim 1 + \cos((p_1 - p_2) \cdot (r_1 - r_2)) = 1 + \cos(Q \cdot \Delta r), \tag{2}$$

where $Q = p_1 - p_2$ is the momentum difference and $\Delta r = r_1 - r_2$ is the relative emission point of the particles (Fig. 2). Pairs with small relative momentum Q will be observed with higher rates because Bose-Einstein particles are likely to be found in the same state.

For a source of particle emission $\rho(r)$ a correlation function can be obtained by integrating over all possible pairs of pions resulting in the Fourier transform of the source function:

$$C_{th}(Q) = \int P(Q)\rho(r_1)\rho(r_2)dr_1dr_2 = 1 + \hat{\rho}(Q) \tag{3}$$

In particular for a Gaussian source $\rho(r) \sim \exp^{-r^2/2R^2}$ one obtains:

$$C_{th}(Q) = 1 + \lambda e^{-R^2Q^2}, \tag{4}$$

where λ is the correlation strength. This equation is used to fit the experimental correlation function, defined as the ratio of the two particle and single particle distributions:

$$C_{exp}(Q) = \frac{f(p_1, p_2)}{f(p_1)f(p_2)} \tag{5}$$

The two particle distribution is obtained by correlating pions from the same event, and the single particle distribution is obtained from mixed events. The data shown in Fig. 3 is from 250 k proton-proton events at $\sqrt{s} = 900$ GeV recorded by ALICE in December 2009. The higher counts at low Q are the Bose-Einstein enhancement, but there are also long range correlations present, which are caused by non-femtoscopic effects such as jets and energy momentum conservation induced correlations (EMCICs for short).

In Fig.(4) $C_{exp}(Q)$ is plotted together with a Gaussian fit. It is evident that the simple Gaussian fit neither describes the low Q enhancement, nor the long range correlations. This implies that the particle emission source for proton-proton collisions is non Gaussian and other functional forms such as exponential or Lorentzian can be used. In this paper we keep using a Gaussian as we are interested in describing the baseline.

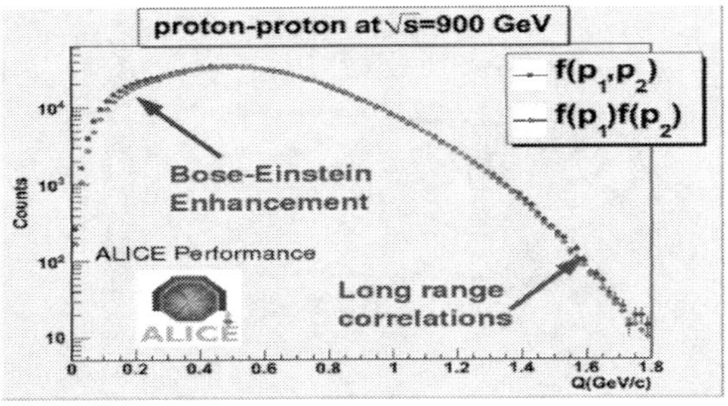

Figure 3. Two particle distributions for real and mixed events.

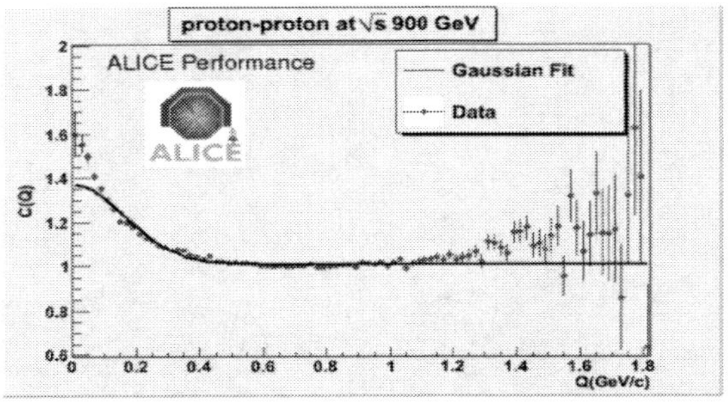

Figure 4. Correlation function with a Gaussian fit.

3. Energy-momentum conservation induced correlations

In order to determine the source size better it would be very useful to find a functional form for the sloped baseline observed in Fig.(4). A linear or quadratic slope can be used, but these are ad-hoc parameterizations, and it would be better if the baseline could be determined from first principles. It has been shown [4] that particle distributions are affected by EMCICs and if there are no other correlations present the two particle correlation function becomes:

$$C_{EMCIC}(p_1, p_2) = 1 - \frac{1}{N}\left(2\frac{\vec{p}_{1,T}\cdot\vec{p}_{2,T}}{\langle p_T^2\rangle} + \frac{p_{1,z}\cdot p_{2,z}}{\langle p_z^2\rangle} + \frac{(E_1 - \langle E\rangle)(E_2 - \langle E\rangle)}{\langle E^2\rangle - \langle E\rangle^2}\right) \tag{6}$$

where N is the event multiplicity. Equation 6 characterizes the EMCICs and it would be possible to remove their effect from the experimental correlation function. However, in experiment $N, \langle p_T^2\rangle, \langle p_z^2\rangle, \langle E\rangle$ and $\langle E^2\rangle$ cannot be measured, simply because not every single particle is detected. These unknown quantities can be parametrized instead, and the resulting equation can be used as a fit to the data:

$$C_{EMCIC}(p_1, p_2) = \left(1 - M_1\cdot\overline{\{\vec{p}_{1,T}\cdot\vec{p}_{2,T}\}} - M_2\cdot\overline{\{p_{1,z}\cdot p_{2,z}\}} - M_3\cdot\overline{\{E_1\cdot E_2\}} + M_4\cdot\overline{\{E_1 + E_2\}} - \frac{M_4^2}{M_3}\right) \tag{7}$$

where the M parameters are defined as: $M_1 = \frac{2}{N\langle p_T^2\rangle}$, $M_2 = \frac{1}{N\langle p_z^2\rangle}$, $M_3 = \frac{1}{N(\langle E^2\rangle - \langle E\rangle^2)}$, $M_4 = \frac{\langle E\rangle}{N(\langle E^2\rangle - \langle E\rangle^2)}$ and the notation $\overline{\{X\}}$ represents histograms of two-particle quantities that can be measured in experiment (see Fig 5).

Monte Carlo simulations, which do not have femtoscopic correlations but do guarantee energy and momentum conservation, have been used to test Eq. (7) as a fitting function for the baseline. Figure 6 shows a good agreement between data and the fit function.

When all correlations are present the total correlation function can be written as:

$$C(Q) = \Phi_{femto}(Q) \times C_{EMCIC}(Q) \tag{8}$$

where $\Phi_{femto} = 1 + \lambda e^{-R^2Q^2}$ is the pure femtoscopic effect. Equation (8) is then used as the fitting equation to the real data. The results are shown in Fig. 7 and it can be seen that the baseline is described quite well by this method. The oscillations in the fitting function at large Q arise from the limited statistics.

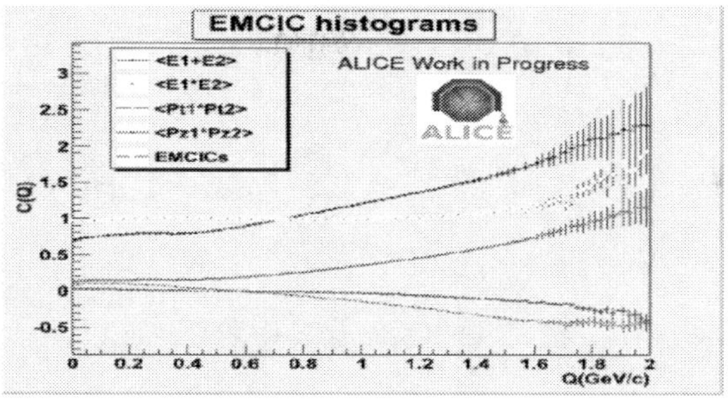

Figure 5. The different EMCIC histograms $\overline{\{X\}}$ are shown together with $C_{EMCIC}(Q)$.

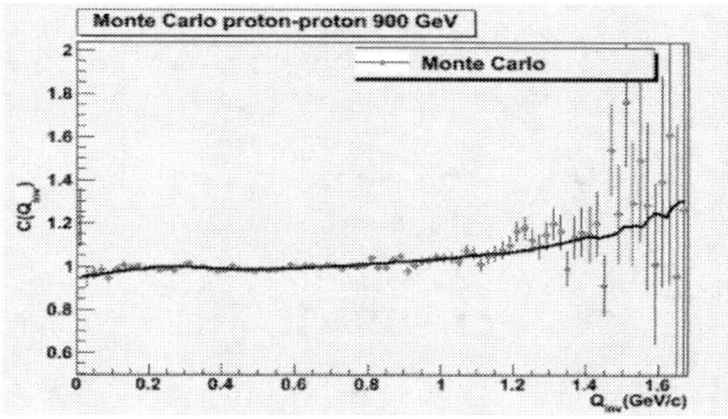

Figure 6. Eq. (7) is used to fit Monte Carlo Data.

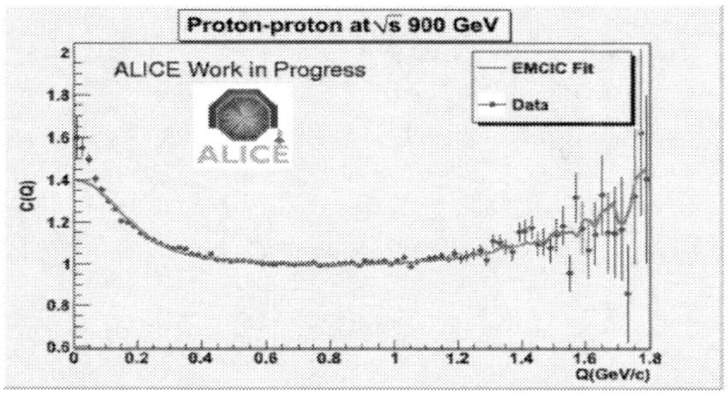

Figure 7. Fit to the data using Eq. (8).

4. Summary

Two particle correlations for proton-proton collisions at 900 GeV exhibit long range correlations that make the determination of the baseline difficult. These correlations are primarily from energy momentum conservation and it has been shown here that they can be quantified and used to determine the baseline correctly. The EMCICs introduce more parameters into the fitting function, but the parameterization is obtained from first principles and these preliminary results look promising. Further work will be done to separate the correlation function and EMCICs in pair average momentum bins $\left(k_T = \frac{p_1+p_2}{2}\right)$ and fit all bins simultaneously with one set of parameters M_1 through M_4, thereby constraining more the parameter space. Three dimensional correlation functions in $Q_{out}, Q_{side}, Q_{long}$ and its decomposition in spherical harmonics will also be studied.

Acknowledgments

The author would like to thank Thomas Humanic and Mike Lisa at The Ohio State University for all the insight in this field and their continued support and motivation, and the femtoscopy group at the ALICE experiment. This work was supported by the U.S. National Science Foundation under grant PHY - 0653432.

References

[1] The ALICE Collaboration, K. Aamodt et al, 2008 JINST 3 S08002
[2] M. A. Lisa, S. Pratt, arXiv:0811.1352 [nuclex]
[3] R. Hanbury-Brown and R. Q. Twiss, Nature 177 (1956) 27.
[4] Z. Chajecki, M. Lisa, Phys.Rev.C78:064903,2008.

Journal of Physics: Conference Series **270** (2011) 012023 doi:10.1088/1742-6596/270/1/012023

Fluctuations of the azimuthal particle distribution in NA49 at the CERN SPS

Tomasz Cetner and Katarzyna Grebieszkow for the NA49 Collaboration[1]

Warsaw University of Technology, Faculty of Physics, Koszykowa 75, 00-662 Warszawa, Poland

E-mail: `Tomasz.Cetner@cern.ch, kperl@if.pw.edu.pl`

Abstract. Event-by-event fluctuations and correlations in azimuthal angle are currently widely investigated in various experiments. In this paper the Φ measure (earlier used in experiments to evaluate fluctuations in p_T) is now applied to azimuthal angle ϕ. Properties of this Φ_ϕ function are investigated through fast generators and with complex models. Preliminary results of NA49 on Φ_ϕ are also presented.

1. Φ_ϕ measure

1.1. *Motivation*

The reasons to perform event-by-event azimuthal angle fluctuations were to investigate plasma instabilities [1], search for the Critical Point and Onset of Deconfinement, and measure fluctuations of elliptic flow [2, 3]. In this work the Φ measure [4] is chosen as it is a strongly intensive measure of fluctuations (does not depend on volume and on volume fluctuations). The Φ measure was already successfully used by NA49 to study average p_T fluctuations [5, 6] and charge fluctuations [7]. There are several other effects that influence the values of this measure, such as resonance decays, momentum conservation, flow, and quantum statistics. The Φ measure is not well suited to the analysis of these effects which represent a background for the present study.

1.2. *Φ_ϕ definition*

Let ϕ be a particle's azimuthal production angle. One can define a single-particle variable $z_\phi \equiv \phi - \overline{\phi}$, where $\overline{\phi}$ is an average over the single-particle inclusive azimuthal angle distribution. Let us also define an event variable $Z_\phi \equiv \sum_{i=1}^{N}(\phi_i - \overline{\phi})$, where the summation runs over particles in a given event. Then the Φ_ϕ measure is defined as: $\Phi_\phi \equiv \sqrt{\frac{\langle Z_\phi^2 \rangle}{\langle N \rangle}} - \sqrt{\overline{z_\phi^2}}$, where $\langle ... \rangle$ represents averaging over events. When particles are emitted by a number (N_s) of identical sources, which are independent of each other and $P(N_s)$ is the distribution of this number, then $\Phi_\phi(N_s)$ is independent of (N_s) (an intensive measure) and its distribution $P(N_s)$ (a strongly intensive measure). Additionally, if particles are produced independently (no inter-particle correlations), Φ_ϕ equals zero.

[1] for complete NA49 author list see http://na49info.web.cern.ch/na49info/na49/Collaboration/authors.html

2. Model studies

General properties of the Φ_ϕ measure were already studied using analytical calculations [8]. Fluctuations due to elliptic flow, quantum statistics and resonance decays in the hadron gas were evaluated. In this paper we present model studies that were conducted to investigate in detail the behavior of Φ_ϕ. Several toy models were employed to test separately the influence of various physics effects. Full event generators such as UrQMD [9], Pythia [10], and Hijing [11] were used as a direct reference to experimental results[2].

2.1. Toy models of elliptic flow and momentum conservation

To simulate elliptic flow events were generated with azimuthal angle distribution of particles following $\rho(\phi) = 1+2v_2cos(2(\phi-\phi_R))$. For each event the reaction plane angle ϕ_R was generated from a flat distribution in azimuthal angle, and the multiplicity N of particles in an event was taken from a Negative Binomial distribution with given $\langle N\rangle$ and dispersion $D_N \approx 0.5 \cdot \langle N\rangle$. The value of v_2 was a simulation parameter, constant for each simulation series (Fig. 1) or varying from event to event according to a Gaussian distribution with σ_{v_2} (Fig. 2). The elliptic flow effect results in positive Φ_ϕ values which are increasing with increasing v_2. Moreover, event-by-event fluctuations in v_2 cause an additional increase of Φ_ϕ values.

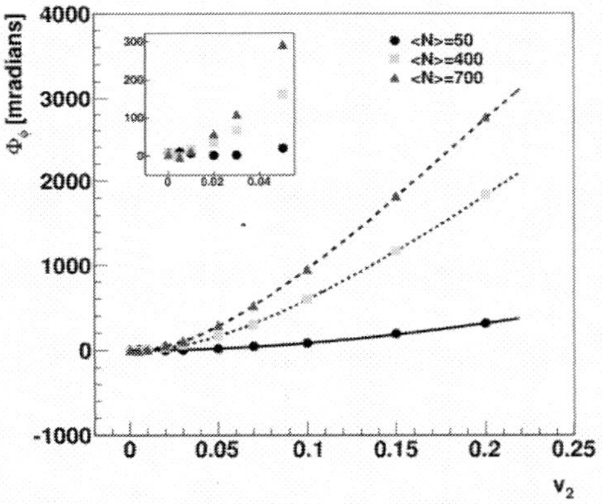

Figure 1. Φ_ϕ as a function of v_2 for elliptic flow simulation with constant v_2. Lines correspond to analytical formulas from [8].

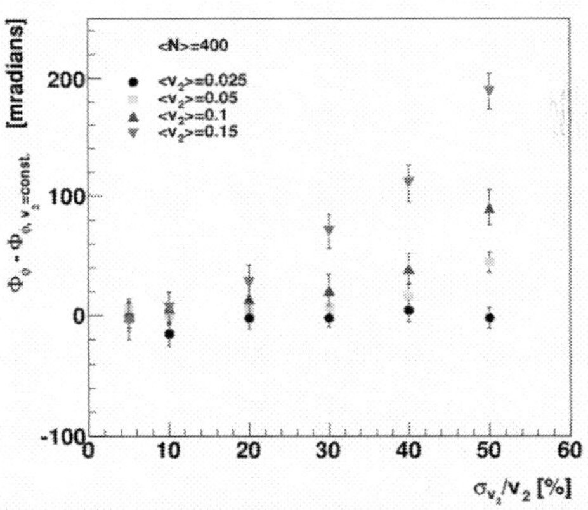

Figure 2. Difference between Φ_ϕ values for constant and fluctuating v_2 as a function of relative size of v_2 fluctuations.

To simulate global momentum conservation, each particle had its transverse momentum p_T drawn from the distribution $P(p_T) \sim p_T e^{\frac{-p_T}{T}}$, where $T = 200$ MeV, and its azimuthal angle from a flat distribution. Afterwards, for every particle p_x and p_y was modified: $p_x' = p_x - \frac{\sum_{i=1}^{N} p_{x_i}}{N}$, and $p_y' = p_y - \frac{\sum_{i=1}^{N} p_{y_i}}{N}$ to obey momentum conservation in the whole event (N is the event multiplicity generated from a Negative Binomial distribution). The simulation results in negative Φ_ϕ values (anti-correlation) which weakly depend on multiplicity (increase from about -370 miliradians for $\langle N\rangle = 10$ to around -300 miliradians for $\langle N\rangle = 1000$).

2.2. UrQMD – a full event generator

A simulation using UrQMD 3.3 was performed for p+p collisions in a wide energy range (SPS, RHIC, LHC) (Fig. 3). Default parameters of the generator were used, which for this version means simulation of hard processes (for energies $\sqrt{s} > 10$ GeV), but no hydrodynamics stage.

[2] see conference slides (http://hq2010.bnl.gov) for Pythia and Hijing results

For all energies and charge combinations Φ_ϕ values are negative (domination of anti-correlations). Their magnitude increases for negatives and decreases for positives and all charged. At LHC energies the values become charge independent.

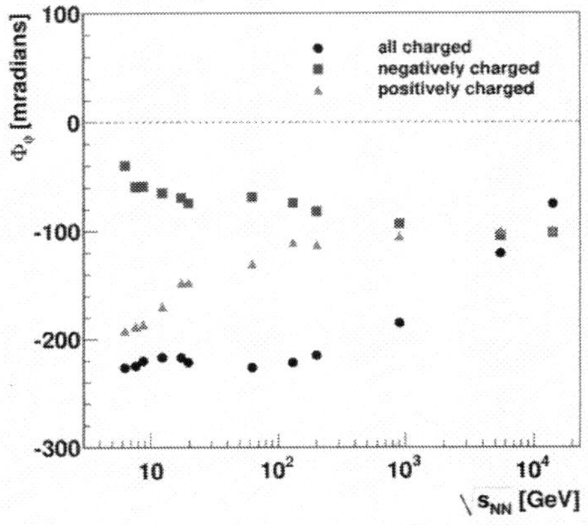

Figure 3. Φ_ϕ as a function of energy for p+p collisions generated using UrQMD 3.3.

3. NA49 experiment

The NA49 experiment [12] at the CERN SPS registered data on p+p, C+C, Si+Si, and Pb+Pb interactions at a center of mass energy of 6.3 - 17.3 GeV per N+N pair. This fixed target hadron spectrometer (four TPCs, two of them, VTPC-1 and VTPC-2, inside a magnetic field) allowed for a precise determination of the collision centrality by measuring energy of projectile spectators (VCAL).

3.1. Detector

The NA49 detector has a limited azimuthal angle acceptance; the acceptance losses are concentrated mainly around the up-down regions. Due to the magnetic field in VTPC-1/2 particles of opposite charge are almost completely separated in the MTPCs (Fig. 4); the MTPCs cover mainly the forward rapidity region. As the detector is left-right symmetric, acceptance for positively and negatively charged particles is the same, provided the azimuthal angle for one charge is reflected. To allow a quantitative comparison of Φ_ϕ for positively and negatively charged particles we rotate particles of one charge (Fig. 5).

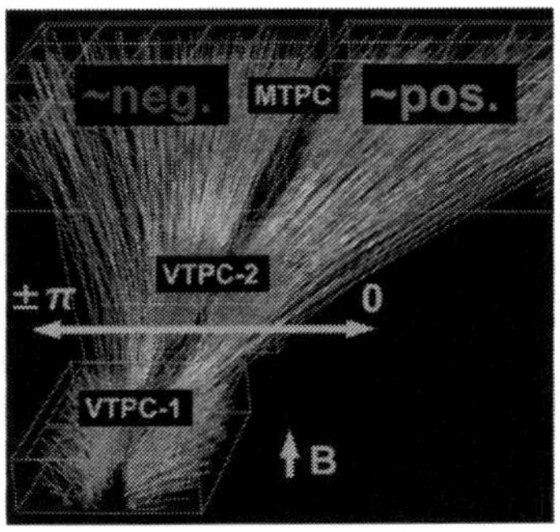

Figure 4. Tracks registered in NA49; four TPCs with marked regions where particles of given charge are dominant. VCAL (see the text) was located behind MTPCs.

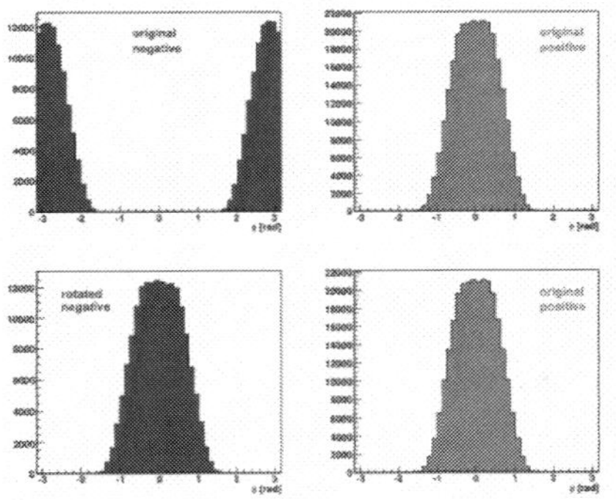

Figure 5. Examples of ϕ distributions at forward rapidity for positively and negatively charged particles before (upper plots) and after (lower plots) the shift.

3.2. Preliminary results

Figure 6 shows the energy dependence of Φ_ϕ for the 7.2% most central Pb+Pb interactions. There is no significant energy dependence, however negatively charged particles show positive Φ_ϕ values. The results from UrQMD, obtained with the same kinematic restrictions, show values consistent with zero. In the simulation the limited NA49 acceptance causes a major reduction in the absolute values of the Φ_ϕ measure (for details see conference slides and Fig. 3, showing UrQMD results for p+p data, where the complete rapidity and azimuthal angle range was used). The main effect is due to the rapidity cut. For the SPS energy range, Φ_ϕ values for a complete rapidity range are around -150 miliradians for positively charged and around -80 miliradians for negatively charged. The rapidity cut brings them to around -20 miliradians and -10 miliradians, respectively. The additional azimuthal angle restriction results in a Φ_ϕ value consistent with zero. Figure 7 presents the system size dependence of Φ_ϕ at 158*A* GeV (top SPS energy). Significant positive values of Φ_ϕ are observed with a maximum for peripheral Pb+Pb interactions. This result is qualitatively similar to the results of multiplicity [13] and average p_T [5] fluctuations in NA49. Further studies are planned to understand these intriguing effects.

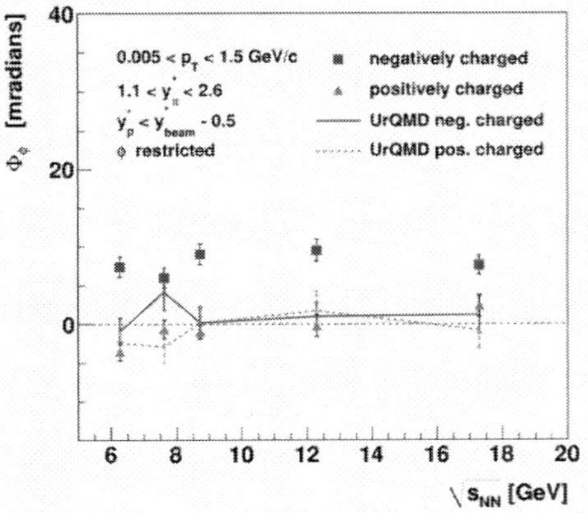

Figure 6. Φ_ϕ as a function of energy. NA49 data compared to UrQMD 1.3 with the same kinematic cuts. See [6] for details of kinematic cuts and precise azimuthal acceptance description.

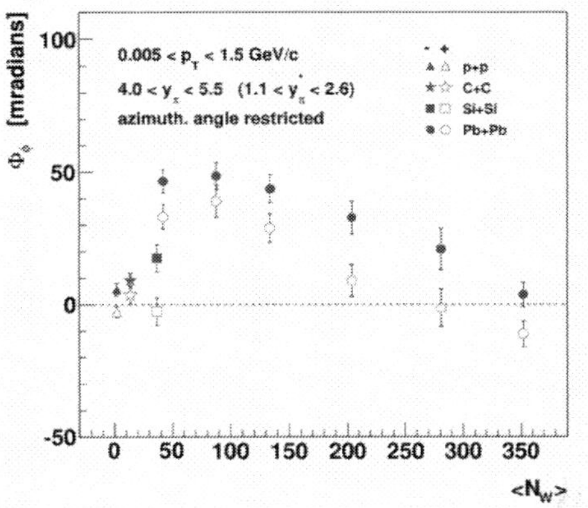

Figure 7. Φ_ϕ as a function of system size (mean number of wounded nucleons). See [5] for details of kinematic cuts and precise azimuthal acceptance description.

4. Summary

Event-by-event azimuthal fluctuations were analyzed using the Φ_ϕ measure, for models and for NA49 data. The properties of Φ_ϕ were studied using toy models and UrQMD. Elliptic flow results in positive values of Φ_ϕ which increase with the magnitude of flow and are sensitive to its fluctuations. Contrary to flow, momentum conservation produces negative Φ_ϕ values. The UrQMD model shows negative Φ_ϕ values with a weak energy dependence. The NA49 measurements for central Pb+Pb collisions showed weak energy dependence and positive Φ_ϕ values for negatively charged particles. On the other hand, a significant system size dependence is observed at the highest SPS energy. This result is qualitatively similar to the behavior found for multiplicity [13] and transverse momentum [5] fluctuations.

References

[1] Mrówczyński S 1993 *Phys. Lett.* B **314** 118-21
[2] Mrówczyński S and Shuryak E 2003 *Acta Phys. Polon.* B **34** 4241-56
[3] Miller M, Snellings R arXiv:nucl-ex/0312008
[4] Gaździcki M and Mrówczyński S 1992 *Z. Phys.* C **54** 127
[5] Anticic T et al. (NA49 Collab.) 2004 *Phys. Rev.* C **70** 034902
[6] Anticic T et al. (NA49 Collab.) 2009 *Phys. Rev.* C **79** 044904
[7] Alt C et al. (NA49 Collab.) 2004 *Phys. Rev.* C **70** 064903
[8] Mrówczyński S *Acta Phys. Polon.* 2000 B **31** 2065
[9] Bass S A et al. 1998 *Prog. Part. Nucl. Phys.* **41** 225-370; Bleicher M et al. 1999 *J. Phys.* G **25** 1859-1896
[10] Sjstrand T, Mrenna S, Skands P arXiv:0710.3820
[11] Wang X-N, Gyulassy M arXiv:nucl-th/9502021
[12] Afanasiev S et al. (NA49 Collab.) 1999 *Nucl. Instrum. Meth.* A **430** 210
[13] Alt C et al. (NA49 Collab.) 2007 *Phys. Rev.* C **75** 064904

Journal of Physics: Conference Series **270** (2011) 012024 doi:10.1088/1742-6596/270/1/012024

Transverse hydrodynamics and the early-thermalization problem at RHIC

Radoslaw Ryblewski
The H. Niewodniczański Institute of Nuclear Physics,
Polish Academy of Sciences,
PL-31-342 Kraków, Poland

E-mail: Radoslaw.Ryblewski@ifj.edu.pl

Abstract. The problem of early thermalization of matter produced in relativistic heavy-ion collisions at RHIC is discussed in the framework of a hybrid model that consists of the transverse-hydrodynamics stage followed by the standard perfect-fluid hydrodynamics stage and freeze-out. The two hydrodynamic regimes are connected with the help of Landau matching conditions. A satisfactory description of the soft hadronic observables is achieved. This indicates that the early-thermalization problem and also the HBT puzzle may be to large extent circumvented.

1. Introduction

The space-time evolution of matter produced in heavy-ion collisions at RHIC can be very well described by the relativistic hydrodynamics of an almost perfect fluid [1, 2, 3, 4, 5, 6, 7, 8]. However, the use of hydrodynamics encounters the problem of early thermalization. The fast equilibration of matter assumed in the hydrodynamic picture is naturally understood in the framework of sQGP [9] or wQGP with strong plasma instabilities [10]. On the other hand, the fast thermalization is difficult to explain in the framework of the microscopic models of early stages of collisions (e.g., in the string models). The detailed analysis of such models indicates that only transverse degrees of freedom undergo thermalization [11]. In a certain time interval after the collision, the system remains highly anisotropic with the transverse pressure much larger than the longitudinal one. The calculations using this idea (the so called transverse hydrodynamics) have turned out to be consistent with the data [12, 13, 14, 15].

Despite the initial pressure asymmetries, the RHIC data suggest that the full thermalization eventually takes place. This observation has inspired us to construct a hybrid hydrodynamic model [16] which includes the step-like transition from the transverse-hydrodynamics stage to the perfect-fluid hydrodynamics stage.

We assume that the system formed after the collision undergoes only transverse thermalization and in the proper time interval $\tau_\mathrm{i} \leq \tau \leq \tau_\mathrm{tr}$ can be described by the tranverse hydrodynamics. The time τ_i is the initialization time for transverse hydrodynamics, whereas τ_tr is the transition time to the perfect-fluid hydrodynamics. During this transition the energy and momentum are conserved. Additionally, we require that the entropy increases. After the transition, for $\tau > \tau_\mathrm{tr}$, the system is treated as the perfect fluid. The single freeze-out model is applied on the hypersurface of constant temperature T_{3f} to calculate physical observables. The hydrodynamic code is coupled to the statistical hadronization Monte-Carlo model THERMINATOR

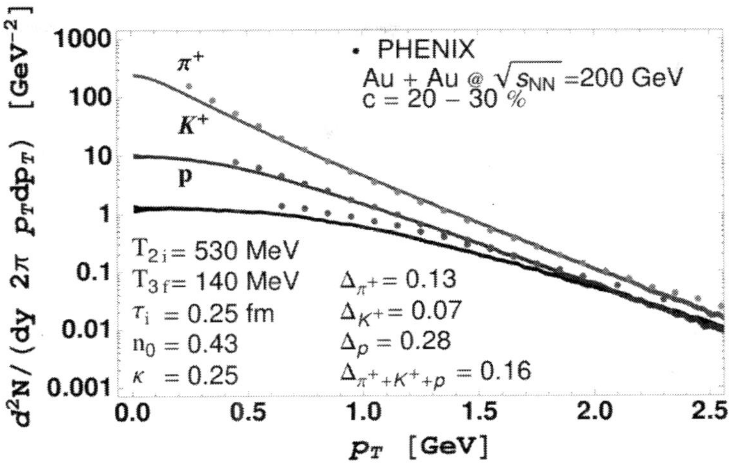

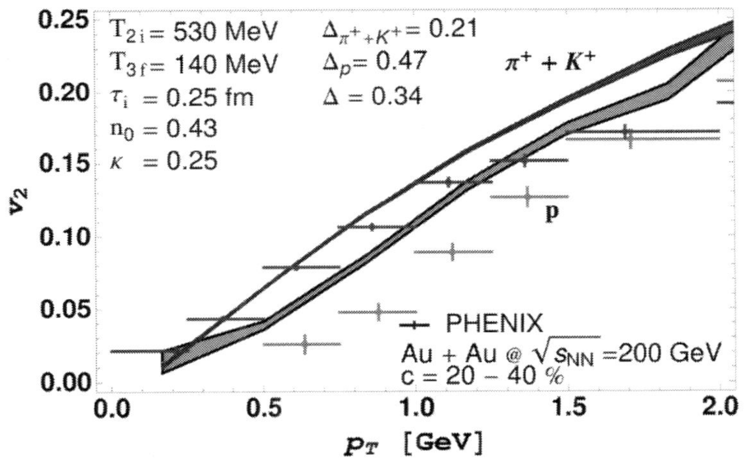

Figure 1. (Color online) Upper part: The comparison of the model (lines) and experimental (dots) transverse-momentum spectra of pions, kaons, and protons. The data are taken from Ref. [19]. Lower part: The comparison of the model and experimental elliptic flow coefficient v_2 for pions+kaons and protons. The bands indicate the error of the Monte-Carlo calculations. The data are taken from Ref. [20].

[17], which is used to calculate the soft hadronic observables: the transverse-momentum spectra of pions, kaons, and protons, the elliptic flow of pions+kaons and protons, and the HBT radii of pions. The radii are calculated with the method described in [18].

The idea of the initial transverse-hydrodynamics stage is used to circumvent the problem of very early thermalization. There exists, however, another problem connected with the hydrodynamic picture. The so called HBT puzzle refers to a difficulty of simultaneous correct description of the hadronic transverse-momentum spectra, the elliptic flow coefficient, and the Hanbury-Brown–Twiss interferometry data. Interestingly, the inclusion of the transverse-hydrodynamics stage improves the HBT results [16].

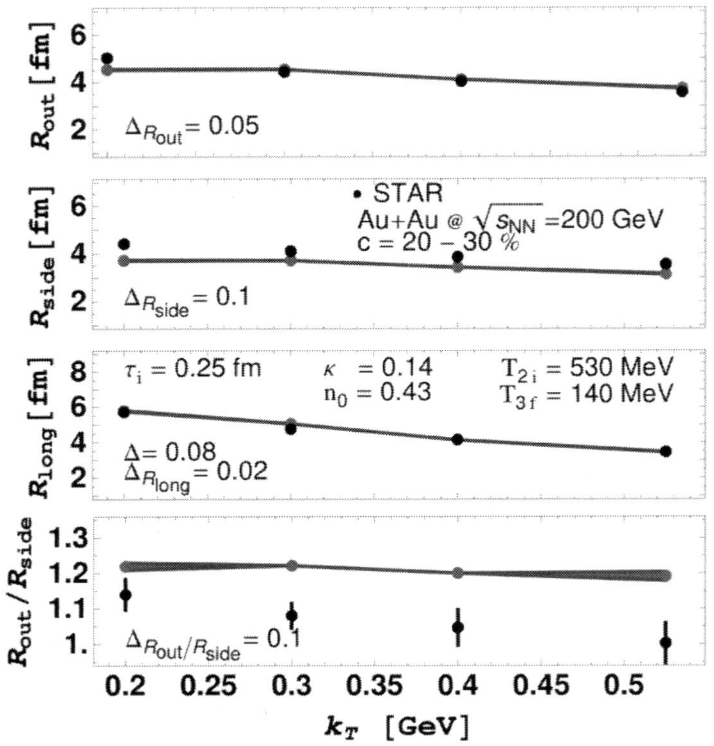

Figure 2. (Color online) The comparison of the model (red lines) and experimental (black dots) results for the pionic HBT radii. The data are taken from Ref. [21]. The model parameters the same as in Fig. 1.

2. Results

In the upper part of Fig. 1 we show our model results for the transverse-momentum spectra of pions, kaons and protons compared to the PHENIX data [19] for Au+Au collisions at the energy of $\sqrt{s_{NN}}$ = 200 GeV and for the centrality class 20-30%. One observes overall good agreement between the model predictions and the data, at the level of 16%. The model results have been obtained with the initial central temperature T_{2i} = 530 MeV and the freeze-out temperature T_{3f} = 140 MeV. The fraction of binary collisions in the initial profile of the density of sources of particle production is 0.25 and the density of transverse clusters in rapidity $n_0 = 0.43$ (our fitting procedure is described in more detail in [16]).

In the lower part of the Fig. 1 we show our results for the elliptic flow coefficient v_2 compared with the PHENIX [20] experimental data. The pion+kaon data are reproduced at the level of 21% while the model results for the protons overestimate the data by about 50 %. The reason for too large proton v_2 is the lack of hadron rescattering in the final state and/or the lack of viscous effects.

In Fig. 2 we present the model HBT radii compared to the STAR [21] data. We observe very good agreement between the experimental data and the results taken from the model analysis. All radii are reproduced with the accuracy better than 10%.

3. Conclusions

Our analysis shows that it is possible to obtain a satisfactory description of the soft hadronic RHIC data in a model that does not assume a very fast thermalization of the system. Initially, only transverse degrees of freedom are thermalized, as suggested by the particle production from the fluctuating strings. Our model reproduces well the HBT radii as well as other soft hadronic observables. This indicates that the two main RHIC problems may be to large extent circumvented. The proton elliptic flow that is too large may be corrected by the inclusion of the shear and bulk viscosity in the standard hydrodynamic stage or by the inclusion of the hadronic rescattering after freeze-out.

Recently we have also analyzed a physical scenario where the transverse hydrodynamics was followed by a sudden isotropization and freeze-out [14, 15]. The present model differs from [14, 15] by the inclusion of the extended phase described by the perfect-fluid hydrodynamics. The similarities between our present results and the results of Refs. [14, 15] indicate that the details of modeling the equilibration transition might be not very much important. However, the correct description of the HBT radii is achieved only if the full equilibration of the system is reached.

The present approach may be generalized in several ways. In particular, the perfect-fluid stage may be replaced by the viscous hydrodynamics. The viscous effects may be also included in the transverse hydrodynamics. Naturally, the next step should be a construction of a continuous change from transverse hydrodynamics to perfect-fluid hydrodynamics (first steps in this direction have been done very recently in [22]).

Finally we note that the large initial anisotropy of pressure may be explained only by introducing a very large shear viscosity, hence our model describes the initial processes which are out of range of application of the standard viscous hydrodynamics.

References

[1] P. F. Kolb, U. Heinz, in Quark-Gluon Plasma 3, edited by R.C. Hwa and X.-N. Wang (World Scientific, Singapore, 2004), p. 634, nucl-th/0305084.
[2] P. Huovinen, in Quark-Gluon Plasma 3, edited by R.C. Hwa and X.-N. Wang (World Scientific, Singapore, 2004), p. 600, nucl-th/0305064.
[3] E. V. Shuryak, Nucl. Phys., **A750** (2005) 64.
[4] D. Teaney, J. Lauret, E. V. Shuryak, nucl-th/0110037.
[5] Y. Hama, *et al.*, Nucl. Phys., **A774** (2006) 169.
[6] T. Hirano, U. W. Heinz, D. Kharzeev, R. Lacey, Y. Nara, J. Phys., **G34** (2007) S879.
[7] C. Nonaka, S. A. Bass, Phys. Rev., **C75** (2007) 014902.
[8] P. Bozek, I. Wyskiel, Phys. Rev., **C79** (2009) 044916.
[9] E. Shuryak, J. Phys., **G30** (2004) S1221.
[10] S. Mrowczynski, Acta Phys. Polon., **B37** (2006) 427.
[11] A. Bialas, Phys. Lett., **B466** (1999) 301.
[12] A. Bialas, M. Chojnacki, W. Florkowski, Phys. Lett., **B661** (2008) 325.
[13] M. Chojnacki, W. Florkowski, Acta Phys. Polon., **B39** (2008) 721.
[14] R. Ryblewski, W. Florkowski, arXiv:0910.0985.
[15] W. Florkowski, R. Ryblewski, arXiv:0912.3451.
[16] R. Ryblewski, W. Florkowski, arXiv:1004.1594.
[17] A. Kisiel et al., Comput. Phys. Commun. **174** (2006) 669.
[18] A. Kisiel et al., Phys. Rev. **C73** (2006) 064902.
[19] S. S. Adler, *et al.*, PHENIX, Phys. Rev., **C69** (2004) 034909.
[20] S. S. Adler, *et al.*, PHENIX, Phys. Rev. Lett., **91** (2003) 182301.
[21] J. Adams, *et al.*, STAR, Phys. Rev., **C71** (2005) 044906.
[22] W. Florkowski, R. Ryblewski, arXiv:1007.0130.

Journal of Physics: Conference Series **270** (2011) 012025 doi:10.1088/1742-6596/270/1/012025

Heavy flavor in heavy ion collisions

Lijuan Ruan

Physics Department, Brookhaven National Laboratory, Upton, NY 11973, USA

E-mail: ruan@bnl.gov,ruanlj@rcf.rhic.bnl.gov

Abstract. The recent results on heavy flavor at the Relativistic Heavy Ion Collider will be reviewed. The results on charm cross section, heavy flavor collectivity and energy loss, color screening effect and quarkonia production mechanism will be highlighted. Precise measurements with future detector upgrades will be discussed.

1. Introduction

Data taken in the last decade have demonstrated that the Relativistic Heavy Ion Collider (RHIC) has created a strongly interacting hot, dense medium with partonic degrees of freedom, the Quark Gluon Plasma (QGP) in central Au+Au collisions at $\sqrt{s_{NN}} = 200$ GeV [1, 2, 3, 4]. Understanding the properties of this matter, such as the colored degrees of freedom and the equation of state, is the physics goal of RHIC and of broad interest. For example, we would like to study the temperature of QGP, the chemical composition, whether the system is thermalized and the color screening effect. Heavy flavor measurements serve as an ideal probe for the QGP and will directly address those above properties. The mass of charm and bottom quark is much larger than the possible temperature in QGP thus $c\bar{c}$ and $b\bar{b}$ pairs are believed to be produced at initial impact from hard processes with high Q^2 transfer and therefore they can be calculated in perturbative Quantum Chromodynamic (pQCD) framework [5]. Any modification on the heavy flavor production would be sensitive to the system evolution, light flavor thermalization, medium modifications and properties. On the other hand, heavy quarkonia are thought to be an excellent probe for the color screeening effect in the medium. Different quarkonia states are predicted to dissolve at different temperatures due to different binding energies. Precise measurements of transverse momentum (p_T) distributions of quarkonia at different centralities, collision systems and energies will serve as a thermometer of the QGP [6, 7]. I will review recent heavy flavor results at RHIC. The results on charm cross section, heavy flavor collectivity and energy loss, color screening effect and quarkonia production mechanism will be highlighted. Precise measurements with future detector upgrades will be discussed. The results presented here were from collision energy at $\sqrt{s_{NN}} = 200$ GeV.

2. Open heavy flavor measurements at RHIC

At STAR, there are three measurements related to open heavy flavor production: D^0 reconstruction through $K\pi$ decay [8], single muons at low p_T (0.17-0.25 GeV/c) from charm decay [9], and single electrons from charm and bottom decay (non-photonic electrons) in the range of 1-10 GeV/c [8]. The D^0 and muon measurements at low p_T help constrain the cross section of $c\bar{c}$ production. The D^0 signal was obtained by mix-event technique. The muons

Journal of Physics: Conference Series **270** (2011) 012025 doi:10.1088/1742-6596/270/1/012025

from charm decay were derived by a two-component fit to the distribution of its distance of closest approach to the collisions vertex (DCA) [9]. The two components are muons from charm decay and from π, K decay. The muons at low p_T were found to be able to uniformly sample the charm cross section [10]. The non-photonic electrons are the inclusive electrons after subtraction of the photonic backgrounds, which are dominantly from gamma conversion and π^0, η Dalitz decays. The photonic electrons were reconstructed using the electron pair topology technique, in which the leading electrons, identified either through the Time of Flight detector or Electromagnetic Calorimeter, were combined with the other tracks from the Time Projection Chamber to reconstruct the opening angle or invariant mass distribution. By cutting on a small opening angle or a small invariant mass and correcting for the partner finding efficiency, the background was obtained and then subtracted from inclusive electrons [8]. The three measurements together cover 90% kinematical range of total charm cross section at mid-rapidity.

At PHENIX, there are two ways to measure the non-photonic electrons: cocktail method and converter method [11]. In the cocktail method, the photonic sources were simulated with PHENIX geometry and subtracted from inclusive electrons. In the converter method, the material with known radiation length were added in PHENIX experiment during the normal run and the non-photonic electrons were derived from the extrapolation. The electron measurements from these two are consistent. The $c\bar{c}$ cross section were derived from the electron measurements at $p_T > 0.4$ GeV/c [11]. In addition, there were also constraints from di-lepton measurements [12] and it was found that the $c\bar{c}$ cross section derived from non-photonic electrons at $p_T > 0.4$ GeV/c is consistent with that from di-lepton measurements in p+p collisions.

2.1. Charm cross section measurements at RHIC

There is a factor of two difference for the charm cross section measured by STAR and PHENIX. This is still under investigation. The low material run and the large Time of Flight detector coverage in year 2009 at STAR will help to solve this puzzle. However, the charm cross section from both experiments was found to follow the number of binary collision scaling from 200 GeV p+p, d+Au, peripheral Au+Au to central Au+Au collisions, which indicates that charm production is exclusively at the initial impact.

There was also a factor of two difference for high p_T electrons measured by STAR and PHENIX between 3 and 10 GeV/c without a significant p_T dependence in 200 GeV p+p and Au+Au collisions. Recently the high p_T electrons at 3-10 GeV/c were updated at STAR in p+p collisions and it was found that the measurement is consistent with that from PHENIX in the overlapping p_T region [13]. The Au+Au results are under investigation and the low material run of 200 GeV Au+Au collisions during 2010 will be helpful to solve the puzzle.

2.2. heavy flavor modification and collectivity

Non-photonic electrons, which come from heavy flavor charm and bottom decay, show a similar magnitude of suppression as light hadrons [8, 11]. The pQCD calculations including charm and bottom decay with collisional and radiative energy loss show a systematically higher R_{AA} value than experimental data [14, 15]. Further calculations indicate that with the charm contribution only, non-photonic electrons are expected to reproduce the data [15]. Using the azimuthal angle correlations between non-photonic electrons and charged hadrons (e-h) and between non-photonic electrons and D^0 ($e - D^0$), the bottom contribution factor to non-photonic electrons were measured [16]. It was found that at $p_T > 5$ GeV/c, the bottom contribution is very significant. This together with non-photonic electron R_{AA} measurements challenge the pQCD energy loss model calculations; they may indicate collisional dissociation of heavy mesons [17], in-medium heavy resonance diffusion [18], and multi-body mechanisms [19] might play an important role for heavy quark interactions with the medium.

In addition, it was found that charm freezeout earlier with moderate or smaller radial flow compared to stable particles from the non-photonic electron spectrum measurements in Au+Au collisions [9]. The elliptic flow of non-photonic electrons was found to be significant at low p_T, indicating the QGP created at RHIC is strongly-interacting [11].

3. Color screening effect on high p_T J/ψ?

The dissociation of quarkonia due to color screening in a QGP is a classic signature of deconfinement in relativistic heavy-ion collisions [20]. Results at RHIC show that the suppression of the J/ψ as a function of centrality (the number of participants) is similar to that observed at the SPS, even though the energy density reached in collisions at RHIC is significantly higher [21, 22]. Possible production mechanisms such as sequential suppression [23], $c\bar{c}$ recombination [24, 25, 26, 27] were proposed to explain this. Recent Lattice QCD calculations indicate that direct J/ψ is not dissociated in the medium created at RHIC while the suppression observed for J/ψ comes from the dissociation of χ_c and ψ' [28]. However, the direct J/ψ might be dissociated at RHIC at high p_T, which was predicted in the hot wind dissociation picture, in which the AdS/CFT approach was used and the dissociation temperature for J/ψ was predicted to decrease as a function of p_T [29]. The AdS/CFT approach was applied to hydro framework and predicted that J/ψ R_{AA} decreases versus p_T [30].

Figure 1 shows J/ψ R_{AA} as a function of p_T in 0-20% and 0-60% Cu+Cu collisions from STAR [31] and 0-20% Cu+Cu collisions from PHENIX [32]. The average of two STAR 0-20% data points at high p_T is $R_{AA} = 1.4\pm0.4(stat.)\pm0.2(syst.)$. Compared to low p_T PHENIX measurements, the results indicate that R_{AA} of J/ψ increases from low p_T to high p_T with a 97% confidence level (C.L.). The R_{AA} of high p_T J/ψ is in contrast to strong suppression for open charm [14, 17, 33], indicating that J/ψ might be dominantly produced through color singlet configuration. However, even though there is significant improvement from the next-next-to-leading order (NNLO) pQCD calculations with the color singlet model, the calculation still fails to reproduce the high p_T part [34]. The R_{AA} trend of J/ψ is contradictory to AdS/CFT+hydrodynamic calculations at the 99% C.L.. This might indicate two things: 1) Cu+Cu system is not big enough so that the calculation is not applicable. The larger system produced in Au+Au collisions may be necessary to observe or exclude the effect predicted by AdS/CFT; 2) the formation time effect for high p_T J/ψ is important since the AdS/CFT+hydrodynamic calculation shown in Fig. 1 requires that the J/ψ is produced as an on-shell J/ψ fermion pair, almost instantaneously, at the initial impact with no formation time. A calculation combining effects of J/ψ formation time, color screening, hadronic phase dissociation, statistical $c\bar{c}$ coalescence and B meson feed-down contribution can describe the data [35]. The calculation suggests a slight increase in the R_{AA} at higher p_T.

In additional to nuclear modification factors for heavy quarkonia, which can help further constrain their production mechanisms, it is important to understand the higher states and B decay contribution to J/ψ. For example, from $J/\psi - h$ correlations, it was found that B decay contributes to $13 \pm 5\%$ of the inclusive J/ψ at $p_T > 5$ GeV/c [31]. Through the di-electron decay, it was found that ψ' decay contributes to $8.5 \pm 2.6\%$ of the inclusive J/ψ [36]. Besides, the recent spin alignment result for inclusive J/ψ [37] is consistent with both CSM [38] and COM [39] calculations at low p_T, as shown in Fig. 2. The future higher p_T measurement of the spin alignment will help distinguish these model calculations and thus further improve our understanding of J/ψ production mechanisms.

4. Future upgrades and the related key measurements

STAR and PHENIX recently updated their data acquisition and trigger systems, which will help sample RHIC II luminosity. With several detector upgrades, heavy flavor collectivity and energy loss as well as color screening effects will be studied with better precision at RHIC.

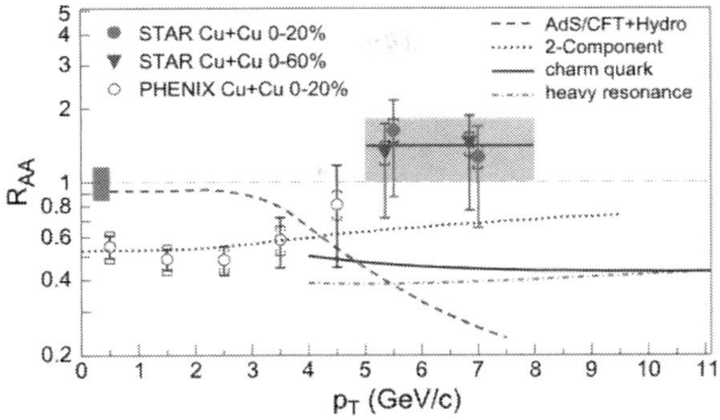

Figure 1. J/ψ R_{AA} versus p_T. The box about unity on the left shows R_{AA} normalization uncertainty, which is the sum in quadrature of p+p normalization and binary collision scaling uncertainties. The solid line and band show the average and uncertainty of the two 0-20% data points. The uncertainty band of 10% for the dotted curve is not shown. Figure is taken from ref. [31].

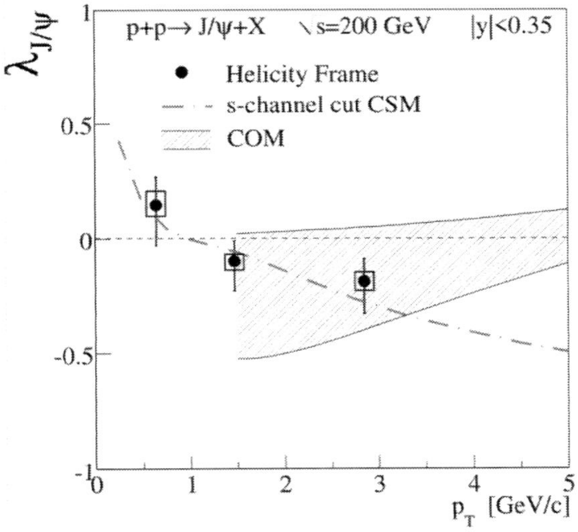

Figure 2. J/ψ spin alignment versus p_T in 200 GeV p+p collisions. Boxes are correlated systematic uncertainties. Figure is taken from ref. [37].

4.1. Heavy flavor collectivity and energy loss

The non-photonic electron analyses suffer from large systematic uncertainties, which are related to photonic background reconstruction and/or subtraction from hadronic decays. In the future, with the Heavy Flavor Tracker upgrade at STAR, the direct topological reconstruction of heavy flavor hadron decays will be feasible and direct charmed hadron measurements will be obtained with good precision [40]. With the Silicon Vertex Detector upgrades, PHENIX will be able to measure non-photonic electrons from charm and bottom decay separately [41]. These measurements are crucial to understand heavy flavor energy loss thus further constrain the details of jet quenching. The collectivity measurements from heavy flavor will be important to understand the thermalization for light flavor.

4.2. Quarkonia production mechanisms, color screening, collectivity and energy loss

To further understand the production mechanisms of quarkonia, color screening effects and medium properties, the precise measurements of the following are needed: nuclear modification factors of J/ψ from low to high p_T in Au+Au and d+Au collisions, J/ψ v_2, forward and backward J/ψ production to address intrinsic charm contributions at large x_F [42], $J/\psi-h$ correlations to access the feeddown contribution, the spin alignment of J/ψ from low to high p_T, higher charmonia states and different Υ states [36, 43]. The Time of Flight system [44], fully installed in the summer of 2009, will enhance the J/ψ capability at low p_T significantly at STAR. The Υ states are also ideal tools to study the effect of color screening in hot and dense QCD matter since its ground state and excited states melt at different temperatures and all of them decay to dileptons [23]. Furthermore, since the $b\bar{b}$ cross section at RHIC energy is expected to be much smaller compared to $c\bar{c}$ cross section from FONLL calculations [45], the recombination contribution from QGP phase might be negligible to bottomonia production. This makes the Υ even a better probe for studying the color screening effect in QGP if sufficient statistics can be achieved experimentally. RHIC II luminosity enables the Υ R_{AA} measurements with good precision. With the possible Muon Telescope Detector upgrade, STAR can cleanly separate the ground state from the excited states even with the additional material from the upgraded inner tracker since the muons in $\Upsilon \rightarrow \mu^+\mu^-$ do not suffer from Bremsstrahlung radiation [46]. With the Silicon Vertex Detector upgrade, PHENIX can measure different upsilon states as well through $\Upsilon \rightarrow e^+e^-$ since the silicon vertex detector brings better mass resolution to quarkonia measurement. The different state Υ measurements will shed more light on the study of the temperature of the QGP created at RHIC.

5. Summary

The recent results on heavy flavor at RHIC are reviewed. The results on charm cross section, heavy flavor collectivity and energy loss, color screening effect and quarkonia production mechanisms are highlighted. Precise measurements with future detector upgrades are discussed.

6. Acknowledgements

The author would like to thank the organizers for the invitation. Thank Prof. M. Calderon for presenting the talk for me at the conference. This work was supported in part by the U. S. Department of Energy under Contract No. DE-AC02-98CH10886.

References

[1] J. Adams *et al.*, Nucl. Phys. A **757**, 102 (2005).
[2] I. Arsene *et al.*, Nucl. Phys. A **757**, 1 (2005).
[3] K. Adcox *et al.*, Nucl. Phys. A **757**, 184 (2005).
[4] B.B. Back *et al.*, Nucl. Phys. A **757**, 28 (2005).
[5] J.C. Collins, D.E. Soper, Annu. Rev. Nucl. Part. Sci. 37 (1987) 383; J.C. Collins, D.E. soper, G. Sterman, Adv. Ser. Direct. High Energy Phys. 5 (1988) 1.
[6] A. D. Frawley, T. Ullrich and R. Vogt, Phys. Rept. **462**, 125 (2008).
[7] Electromagnetic Probes at RHIC II (Working Group Report), G. David, R. Rapp and Z. Xu, Phys. Rept. **462**, 176 (2008).
[8] J. Adams *et al.*, Phys. Rev. Lett. **94**, 62301 (2005); B.I. Abelev *et al.*, Phys. Rev. Lett. **98**, 192301 (2007).
[9] B.I. Abelev *et al.*, arXiv:0805.0364.
[10] H. Liu *et al.*, Phys. Lett. B **639**, 441 (2006).
[11] S.S. Adler *et al.*, Phys. Rev. Lett. **96**, 032301 (2006); A. Adare *et al.*, Phys. Rev. Lett. **98**, 172301 (2007).
[12] A. Adare *et al.*, Phys. Lett. B **670**, 313 (2009).
[13] W. Xie for the STAR Collaboration, contribution to DIS 2010 conference proceedings.
[14] S. Wicks *et al.*, Nucl. Phys. A **784**, 426 (2007).
[15] N. Armesto et al., Phys. Lett. B **637**, 362 (2006).
[16] S.S. Adler *et al.*, Phys. Rev. Lett. **103**, 082002 (2009); M. M. Aggarwal *et al.*, arXiv:1007.1200.
[17] A. Adil and I. Vitev, Phys. Lett. B **649**, 139 (2007).

[18] H. v. Hess, V. Greco and R. Rapp, Phys. Rev. C **73**, 034913 (2006).
[19] W. Liu and C. M. Ko, nucl-th/0603004.
[20] T. Matsui and H. Satz, Phys. Lett. B **178**, 416 (1986).
[21] A. Adare *et al.*, Phys. Rev. Lett. **98**, 232301 (2007).
[22] M. C. Abreu *et al.*, Phys. Lett. B **499**, 85 (2001).
[23] H. Satz, J. Phys. G **32**, R25 (2006); F. Karsch, D. Kharzeev and H. Satz, Phys. Lett. B **637**, 75 (2006).
[24] P. Braun-Munzinger and J. Stachel, Phys. Lett. B **490**, 196 (2000).
[25] L. Grandchamp and R. Rapp, Phys. Lett. B **523**, 60 (2001).
[26] M. I. Gorenstein *et al.*, Phys. Lett. B **524**, 265 (2002).
[27] R. L. Thews, M. Schroedter and J. Rafelski, Phys. Rev. C **63**, 054905 (2001).
[28] M. Asakawa *et al.*, Prog. Part. Nucl. Phys. 46, 459 (2001); M. Asakawa and T. Hatsuda, Phys. Rev. Lett. 92, 012001 (2004); S. Datta *et al.*, Phys. Rev. D 69, 094507 (2004).
[29] H. Liu, K. Rajagopal and U.A. Wiedemann, Phys. Rev. Lett. **98**, 182301 (2007).
[30] T. Gunji *et al.*, J. Phys. G **35**, 104137 (2008).
[31] B.I. Abelev *et al.*, Phys. Rev. C **80**, 041902 (2009), arXiv:0904.0439; Z. Tang, Ph.D. thesis, University of Science and Technology of China, 2009.
[32] A. Adare *et al.*, Phys. Rev. Lett. **101**, 122301 (2008).
[33] W.A. Horowitz private communication; I. Vitev private communication.
[34] P. Artoisenet et al., Phys. Rev. Lett. 101, 152001 (2008), and J.P. Lansberg private communication.
[35] X. Zhao and R. Rapp, Phys. Lett. B **664**, 253 (2008); Y. Liu *et al.*, Phys. Lett. B **678**, 72 (2009).
[36] M. Leitch *et al.*, Nucl. Phys. A **830**, 27c-34c (2009).
[37] A. Adare *et al.*, Phys. Rev. D **82**, 012001 (2010).
[38] H. Haberzettl and J. P. Lansberg, Phys. Rev. Lett. **100**, 032006 (2008).
[39] H. S. Chung, S. Kim, J. Lee, and C. Yu, arXiv:0911.2113.
[40] STAR Heavy Flavor Tracker proposal, http://rnc.lbl.gov/hft/docs/hft_final_submission_version.pdf; S. Kleinfelder *et al.*, Nucl. Instr. Meth. A **565**, 132 (2006).
[41] PHENIX Silicon Vertex Detector proposal.
[42] C. Perkins *et al.*, Nucl. Phys. A **830**, 231C (2009).
[43] H. Liu *et al.*, Nucl. Phys. A **830**, 235c-238c (2009); B.I. Abelev *et al.*, Phys. Rev. D **82**, 012004 (2010).
[44] B. Bonner *et al.*, Nucl. Instr. Meth. A **508**, 181 (2003); M. Shao *et al.*, Nucl. Instr. Meth. A **492**, 344 (2002); STAR Time-of-Flight Proposal: http://www.star.bnl.gov/STAR/tof/publications/TOF_20040524.pdf.
[45] R. Vogt, M. Cacciari and P. Nason, Nucl. Phys. A 774, 661 (2006).
[46] Z. Xu, BNL LDRD project 07-007; L. Ruan *et al.*, J. Phys. G **36**, 095001 (2009); L. Ruan *et al.*, WWND2008 proceedings, nucl-ex/0805.4638; STAR Muon Telescope Detector Proposal: http://drupal.star.bnl.gov/STAR/system/files/MTD_proposal_v14.pdf.

Journal of Physics: Conference Series **270** (2011) 012026 doi:10.1088/1742-6596/270/1/012026

Υ production in p+p, d+Au, Au+Au collisions at $\sqrt{s_{NN}}$ =200 GeV in STAR

Rosi Reed (For the STAR Collaboration)

University of California, Davis

rjreed@ucdavis.edu

Abstract. Quarkonia production can be used to investigate the properties of the dense matter produced in heavy-ion collisions at RHIC because the production is theorized to be suppressed due to the Debye color screening of the potential between the heavy quarks. Lattice calculations indicate that a sequential suppression of quarkonia states in heavy-ion collisions will give us a model dependent measurement of the temperature of the Quark Gluon Plasma (QGP). Suppression is determined by calculating R_{AA}, which is the ratio of the production in p+p scaled by the number of binary collisions to the production in Au+Au. The Υ states are of particular interest because at 200 GeV the effects of recombination and co-mover absoprtion are smaller than for J/ψ, which lowers the systematic uncertainty of the interpretation of the R_{AA} calculation. It is also important to consider cold nuclear effects, which can be determined from d+Au collisions. We will present our results for mid-rapidity Υ production in p+p, as well as our preliminary results in d+Au and Au+Au at $\sqrt{s_{NN}}$ = 200 GeV. These results will then be compared with theoretical QCD calculations.

1. Introduction

Suppression of quarkonia due to the Debye screening of the potential between the two heavy quarks was thought to be a distinct signature of QGP formation [1]. J/ψ suppression was observed in heavy-ion collisions at both SPS and RHIC [2]. This indicates that the strongly interacting matter formed in heavy-ion collisions (sQGP) does modify the production of quarkonia. The magnitude of J/ψ suppression in both systems was similar despite the different energy densities, indicating that other physics processes besides Debye screening needed to be accounted for. These processes include statistical recombination of $Q\overline{Q}$ pairs, co-mover absorption and feed-down from higher states. All of these processes have a smaller effect on the Υ family than for the ψ family at 200 GeV. However, measuring the Υ family is difficult because it has a very small cross-section which requires a large amount of integrated luminosity. An important measurement that can be made by studying the Υ 1S, 2S, and 3S states is a model dependent determination of the sQGP temperature. This is determined by examining the ratio of the 2S and 3S states to the 1S state in heavy-ion collisions compared to the ratio in p+p collisions. At $\sqrt{s_{NN}}$ =200 GeV, lattice calculations indicate that the 3S should be completely dissociated, while the 2S state may dissociate and the 1S state should survive [3,4]. In order to measure the medium modification of the Υ(1S, 2S, 3S) states at STAR, a baseline p+p measurement was required. To understand whether medium modification in Au+Au collisions is due to the hot nuclear matter effects caused by the sQGP, cold nuclear matter effects need to be quantified. This is done by measuring the ϒ(1S, 2S, 3S) yield in d+Au collisions.

2. STAR detectors and trigger

Upsilons are reconstructed at STAR through the decay channel $\Upsilon \rightarrow e^+e^-$. The Barrel Electromagnetic-Calorimeter (BEMC) [5] and the Time Projection Chamber (TPC) [6] are the two main detectors used in the STAR ϒ analyses. The BEMC has a coverage of $|\eta| < 1$ and $0 < \phi < 2\pi$, and is used to measure the energy of the ϒ daughters and for the trigger. The TPC has the same coverage as the BEMC and is used to determine the momentum and the ionization energy loss for particle identification. The two are used in conjunction to calculate *E/p*, which is also used for particle identification. In order to utilize the full luminosity seen at STAR an efficient two stage trigger was developed. The first stage is a hardware trigger (L0) which fires if a single tower in the BEMC is above the necessary threshold. The threshold for the p+p (d+Au, Au+Au) data presented in this paper is ~3.5 (4.3) GeV. This trigger can fire on every bunch crossing. The second stage is a topological software trigger which requires two three-tower BEMC clusters to be above two different thresholds. These clusters must also satisfy an invariant-mass and an opening-angle requirement [7,8,9].

3. Results

The baseline p+p analysis presented in this paper is from $\int \mathcal{L}\, dt = 8.8 \pm 0.8$ (syst.) pb^{-1} taken during run 6 (2006). The electrons were selected using the ionization energy loss and the E/p ratio as explained in section 2. The signal is a combination of the ϒ(1S, 2S, 3S), Drell-Yan and $b\bar{b}$ yield. This is calculated by combining the unlike-charged electron pairs and subtracting the like-sign electron invariant-mass spectrum from it. The ϒ(1S, 2S, 3S) is then determined by subtracting the Drell-Yan and $b\bar{b}$ yield from the signal as shown in figure 1.

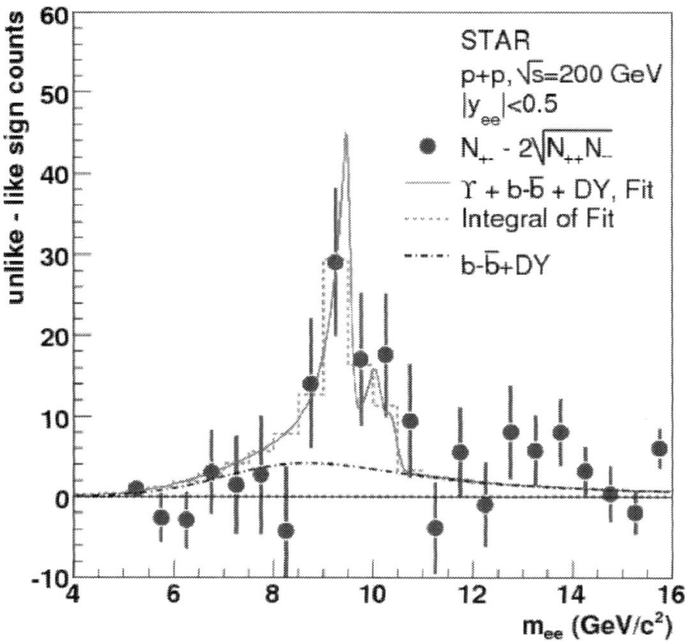

Figure 1. STAR measurement of the like-sign subtracted e+e- invariant mass spectrum from p+p collisions. The solid circles are from data with statistical error bars. The dot-dashed black line is the continuum contribution from Drell-Yan and

uncorrelated $b\bar{b}$ pairs. The solid red curve combines the dot-dashed black curve and the contribution from the three Crystal-Ball functions [11]. The red dotted histogram is the integral of the red curve [10].

The efficiency of the ϒ(1S, 2S, 3S) signal is determined by embedding a GEANT simulated ϒ→e+e- decay into a pythia event which is propagated through the detector geometry and then further embedded into a real zerobias event. A zerobias collision does not require any signals from any of the STAR detectors, only a bunch crossing coincidence. We find the cross section at mid-rapidity in p+p collisions to be: $\sum_{n=1}^{3} B(nS)\times \left.\frac{d\sigma}{dy}\right|_{y=0}(nS) = 114 \pm 38^{+23}_{-24}$ pb [10].

The d+Au analysis presented in this paper is from $\int \mathcal{L}\,dt = 32\ nb^{-1}$ of integrated luminosity from the run 8 (2008) d+Au data. The amount of inner material in the STAR detector was reduced compared to the p+p and Au+Au ϒ analyses. This allows for better resolution. The yield of ϒ(1S, 2S, 3S), Drell-Yan and $b\bar{b}$ was determined by counting the number of like-sign subtracted invariant mass pairs from 7.5 to 11 GeV/c^2. This entire yield was used to calculate R_{dA}, which was found to be $R_{dA} = 0.78\pm0.28\pm0.20$. This is consistent with N_{binary} scaling which means that cold nuclear matter contributions are small [9]. As shown in figure 2, this d+Au result is consistent with a Color Evaporation Model (CEM) calculated which includes anti-shadowing.

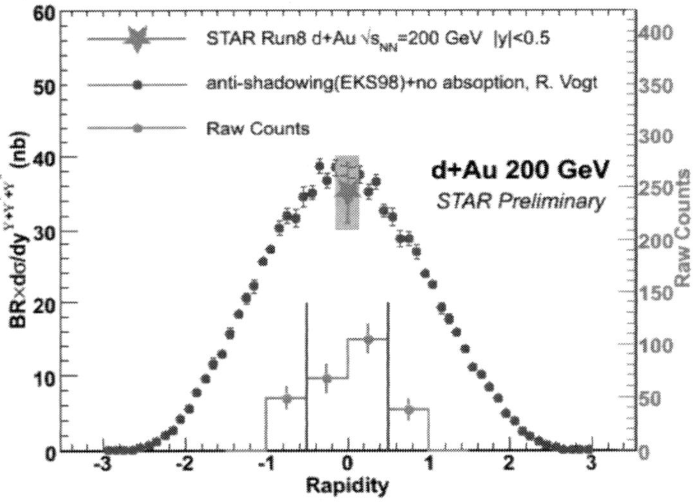

Figure 2. STAR measurement of the ϒ yield in d+Au in the dielectron channel. The left axis determines the scale of the red star and the blue points from theory. The right axis determines the scale for the red histogram. The red star is the calculated d+Au cross section and is agreement with data.

The Au+Au data presented here is from ~500 μb^{-1} of integrated luminosity taken during run 7 (2007). The total yield of ϒ(1S, 2S, 3S), Drell-Yan and $b\bar{b}$ was calculated in the same manner as the p+p and d+Au analysis. In the 0-60% centrality bin, this yield was 95 counts, with a 4.6σ significance. The unlike-sign signal and like-sign background for this calculation are shown in figure 3. In the 0-10% centrality bin, the yield was 47 counts with a significance of 3.5σ. Determination of the corrected ϒ(1S, 2S, 3S) yield from this value is in process.

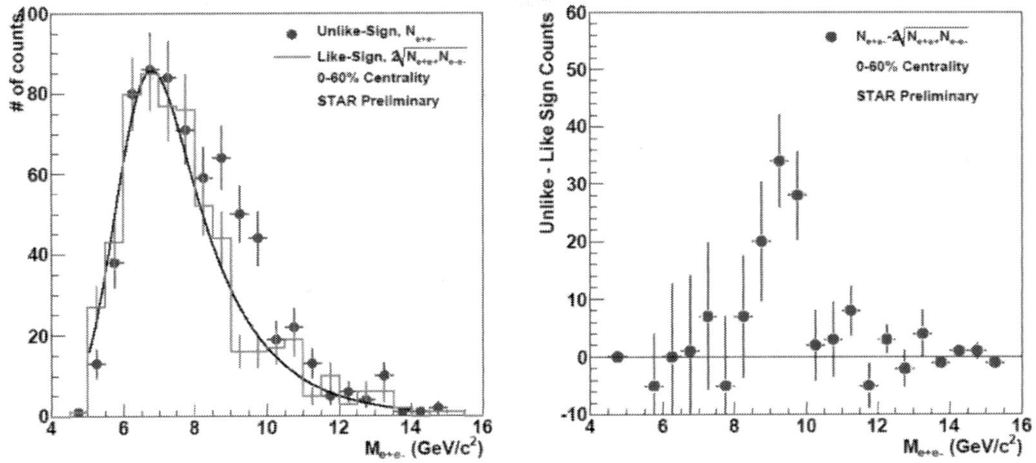

Figure 3. (Left)The STAR measurement of the invariant yield via the dielectron channel for Au+Au collisions in the 0-60% centrality region. The red histogram is the like-sign background yield and the blue circles are the unlike sign yield. The black curve is a power-law multiplied by an erf function used to determine the trigger effect on a continuum process and will be used to model the Drell-Yan and $b\bar{b}$ background yields. (Right)The unlike minus the like sign distributions from the left graph.

4. Conclusions

The STAR experiment has measured the $\Upsilon(1S+2S+3S)\rightarrow e^+e^-$ cross section at midrapidity in p + p to be $\sum_{n=1}^{3} B(nS)\times \left.\frac{d\sigma}{dy}\right|_{y=0} (nS) = 114 \pm 38^{+23}_{-24}$ pb at $\sqrt{S_{NN}} = 200$ GeV. The d+Au cross section for Υ(1S, 2S, 3S) + Drell-Yan + $b\bar{b}$ was measured to be $\sum_{n=1}^{3} B(nS)\times \left.\frac{d\sigma}{dy}\right|_{y=0} (nS) = 35 \pm 4(\text{stat}) \pm 5(\text{syst})$ nb. Combining these gives a R_{dA} = 0.78±0.28±0.2 which is consistent with CEM predictions including anti-shadowing. Preliminary uncorrected yields of Υ(1S, 2S, 3S) + Drell-Yan + $b\bar{b}$ of 95 in the 0-60% centrality bin of Au+Au and 47 in the 0-10% bin was also presented.

References

[1] *T. Matsui and H. Satz*, Phys. Lett. B **178** 416–22 (1986)
[2] *F. Karsch, D. Kharzeev, H. Satz*, Phys. Lett. B **637**, 75 (2006)
[3] *S. Digal, P. Petreczky and H. Satz*, Phys. Lett. B **514** 57–62 (2001)
[4] *C-Y Wong, Phys*. Rev. C **72** 034906 (2005)
[5] *M. Beddo et al.*, Nucl. Instr. Meth. A **499**, 725 (2003)
[6] *M. Anderson et al.*, Nucl. Instr. Meth. A **499**, 659 (2003)
[7] *P. Djawotho et al.*, J. Phys. G: Nucl. Part. Phys. 34 S947–S950 (2007)
[8] *D. Das et al.*, J. Phys. G: Nucl. Part. Phys. 35 104153 (2008)
[9] *H. Liu et al.*, arXiv:0907.4538v2 [nucl-ex] 20 Sep (2009)
[10] *B. Abelev et. al.*, arXiv:1001.2745v2 [nucl-ex] to appear in Phys. Rev. D
[11] *J. Gaiser*, Ph.D. Thesis, SLAC-R-255, Appendix F., p.178.

Charmonium production in pp collisions at LHC with the ALICE Muon Spectrometer

Livio Bianchi for the ALICE Collaboration
Università degli Studi di Torino - INFN di Torino
E-mail: lbianchi@to.infn.it

Abstract. The Forward Muon Spectrometer of the ALICE experiment is designed for the study of quarkonium resonances and open heavy flavour particles coming from the strongly interacting matter produced in nucleus-nucleus collisions at the LHC, where the formation of the Quark Gluon Plasma is expected. Muon pairs are detected at forward rapidity ($-4 < y < -2.5$) and in a wide transverse momentum range (from 0 to tens of GeV/c). The proton-proton physics program of the Muon Spectrometer has the aim to define a proper normalization for nuclear collision studies and to deal with some open issues such as quarkonium production mechanism and polarization. The physics performances of the Muon Spectrometer for charmonium studies are presented and the progress of the data taking with the first LHC high energy run is shown.

1. Quarkonium production in pp collisions

The production of any given quarkonium state [1] is believed to be factorisable into two parts: the first one, where a heavy quark and antiquark pair is produced, is driven by perturbative QCD; the second, concerning the formation and the evolution of the bound state, is governed by non-perturbative QCD.

One convenient way to carry out this separation is through the use of the effective field theory Non-Relativistic QCD (NRQCD), which embeds the non perturbative effects in some universal coefficients. The full expansion of the NRQCD can be truncated in two different ways:

- The Colour Singlet Model (CSM) takes into account only colour singlet elements; it enjoyed some success before CDF measured [2] a J/ψ production rate more than an order of magnitude greater than predicted; moreover the differential p_T distribution provided by the model was wrong. Recently it has received new attention since new calculations at NLO have shown a better description of CDF cross sections and of the observed polarization [3].
- The Colour Octet Mechanism (COM) takes into account the colour octet terms, suggesting that the heavy quark pairs could evolve into quarkonium states through radiation of soft gluons; despite the successes of COM in reproducing the p_T differential cross section, recent measurements at CDF show disagreement with COM predictions on polarization [3].

Another model is the Colour Evaporation Model (CEM), which considers the quarkonium production cross section as a measurable fraction of all $Q\bar{Q}$ pairs below the $H\bar{H}$ threshold (where H is the lowest mass heavy flavour hadron), without any constraint on the colour or spin of the final state. This model, even if purely phenomenological, has the advantage of being relatively easy-to-use once a set of PDFs is chosen for the calculation.

2. The ALICE Forward Muon Spectrometer

ALICE [4] is the dedicated heavy-ion experiment at the LHC. It can be seen as the integration of a central barrel part (which covers mid-rapidity: $|\eta| < 0.9$) with an arm dedicated to the detection of muons, the Forward Muon Spectrometer [5].
The angular acceptance of the Forward Spectrometer covers the range 171°-178°, corresponding to a pseudorapidity range from 2.5 to 4. Since it covers the forward direction, its design is similar to solutions used in fixed target experiments; the important difference is that the forward particle flow cannot be absorbed in a plug because the beam pipe passes through the experiment. Instead, the particles have to be absorbed in a dense beam shield (Pb and W), decreasing the hit rates on the detector planes to acceptable levels.
The front absorber has a total thickness of $10\lambda_I$ and consists of carbon, concrete and steel. The inner and outer surfaces of the absorber are covered by heavy materials (W and Pb), in order to shield the first two tracking stations against background particles. The tracking system consists of five stations with two planes each, made of very thin ($\sim 5\%\ X_0$) cathode pad tracking chambers. To limit the occupation rate to a maximum of 5% the full set of chambers has more than 1 million channels. One tracking station is placed inside a dipole magnet with a total field integral of $B \cdot l = 3$ Tm and a maximum field of 0.7 T. The momentum of the muons is determined from a combination of deflection angle and sagitta measurements. The muon filter, a $7\lambda_I$ Fe absorber wall, suppresses the hadron rate on the 4 planes of single-gap RPCs of the trigger system, which has a high time resolution and provides a level-0 muon and dimuon trigger to the ALICE central trigger system.

3. Performances for Quarkonia studies

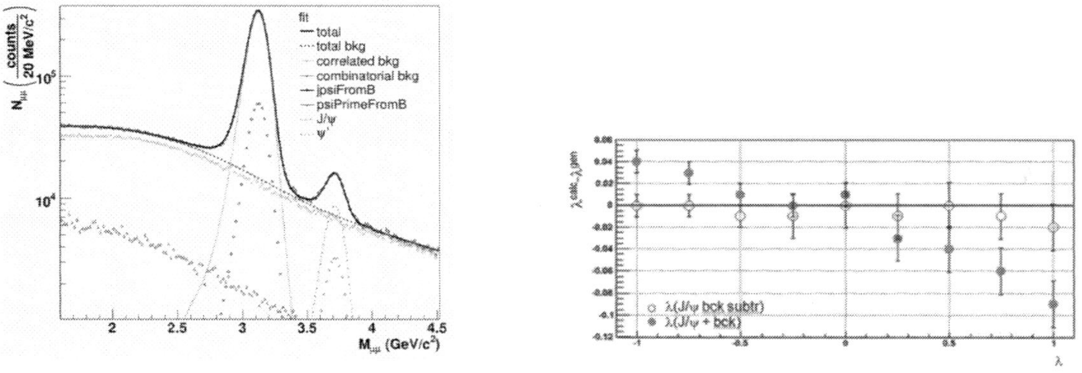

Figure 1. Physics performances of the ALICE Muon Spectrometer in one year of data taking in pp at nominal LHC running conditions: muon pairs invariant mass spectrum in the region 1.5-4.5 GeV/c^2 (left) and polarization parameter extraction capability (right) with and without background subtraction (open and solid points respectively).

The physics performances of the spectrometer [6] have been studied through simulations performed using the CEM predictions at $\sqrt{s} = 14$ TeV. The set of PDFs used is the MRST HO [7] and the mass of the charm quark was set to $m_c = 1.2$ GeV/c^2. The resulting J/ψ total cross section is 53.9 μb, including the feed-down from higher mass resonances. The rapidity distribution is a parameterization of CEM predictions, while the p_T distribution was obtained by extrapolating to LHC energies the one measured by the CDF experiment at $\sqrt{s} = 2$ TeV.
If we consider one year of data taking in pp with nominal LHC running conditions (10^7s running

time with $L = 3 \cdot 10^{30}$ cm^{-2}s^{-1}) the integrated luminosity is 30 pb^{-1} and this leads to $\sim 2.8 \cdot 10^6$ J/ψ reconstructed in the muon spectrometer. The background from heavy flavours decay can be correlated (both muons originate from the same heavy quark pair) or uncorrelated (decay muons from different heavy quark pairs). Both have been simulated using PYTHIA, tuned with the CEM cross section predictions. The background under the J/ψ peak is foreseen to be quite low (S/B $\sim$ 12) and is dominated by correlated pairs of muons (see left panel of Figure 1).
With these statistics the J/ψ differential cross section in p_T and y can be extracted (up to $\sim$ 20 GeV/c for p_T and in all the accessible range for y) with negligible statistical errors.
The y differential distribution will be very useful for putting constraints on the gluon PDF for low-x values (at $\sqrt{s} = 14$ TeV a J/ψ with $y > 3$ has $x < 10^{-5}$), since CEM calculations with different PDF sets lead to different shapes in the rapidity spectrum.
Another study that can be performed is the J/ψ polarization, extracted from the analysis of the complete angular distribution of the muons in a given reference frame [8]:

$$\frac{1}{\Gamma}\frac{d\Gamma}{dcos\theta d\phi} = 1 + \lambda cos^2\theta + \mu sin(2\theta)cos\phi + \frac{\nu}{2}sin^2\theta cos(2\phi)$$

where the λ, μ and ν parameters are related to the spin alignment of the mother particle.
In one year of data taking a detailed study of the J/ψ polarization as a function of p_T can be performed, with expected statisical errors of the order of 3%. This study can help to discriminate among the various production models, since they foresee different values of polarization at high p_T. The right panel of Figure 1 shows the λ parameter reconstruction capability for different values of the parameter itself with and without background subtraction.

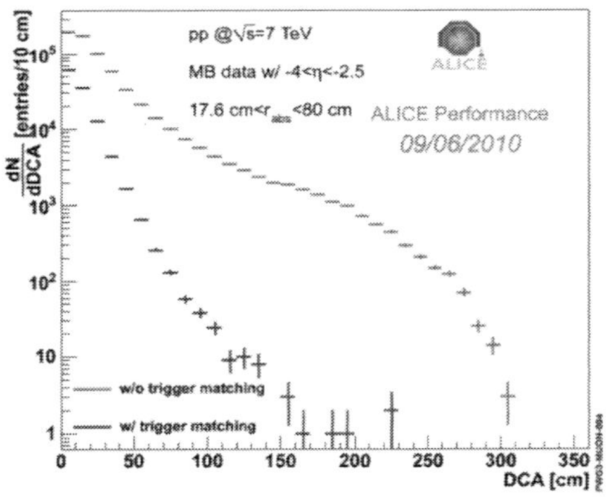

Figure 2. Distance of Closest Approach (DCA) to the interaction vertex in pp collisions at 7 TeV, with (blue points) and without (red points) asking the track to fire the muon trigger.

4. First LHC high energy run: data taking and perspectives

LHC started running in November 2009 at $\sqrt{s} = 900$ GeV, showing excellent performances both of the accelerator and of the experiments and allowing ALICE to publish the first LHC physics article [9]. In December 2009 a few runs were delivered at $\sqrt{s} = 2.36$ TeV, the highest energy value since that time.
In March 2010 the long-term LHC high energy program has started at $\sqrt{s} = 7$ TeV, value that will not change until the end of 2011. The beam intensity has raised up starting from $\sim 10^{10}$ p/bunch and almost reaching the nominal value of 10^{11} p/bunch . The number of bunches per beam is still far from the nominal value (2808), but it is gradually increasing.

The ALICE experiment has shown very good performances collecting more than 10^8 minimum bias events in the first four months of data taking. The Forward Muon Spectrometer was ready to take data since the first run in November 2009 and is currently collecting events both with minimum bias and muon triggers. The effect of the muon trigger is clear when looking at the DCA (Distance of Closest Approach) of single muons in the spectrometer (see Figure 2): the high-DCA wide tail due to residual hadrons in the tracking stations is suppressed when the track is required to fire the muon trigger.

The charmonium analysis started just at the beginning of the 7 TeV period and it is currently

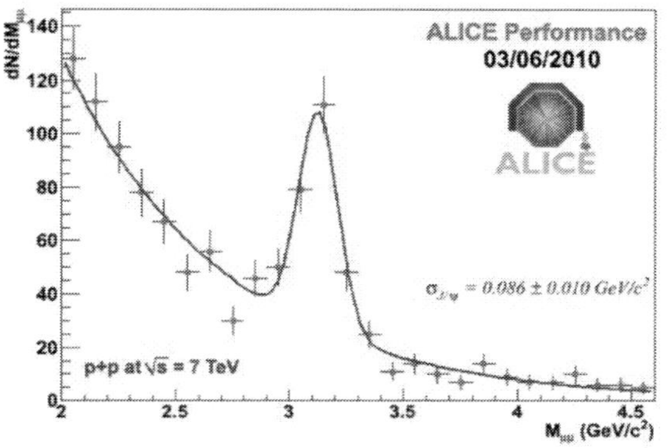

Figure 3. Invariant mass spectrum for muon pairs collected in pp collisions at $\sqrt{s} = 7$ TeV.

proceeding fastly. During the first two months of data taking the spectrometer was not aligned and the J/ψ invariant mass peak appeared to be very wide with respect to the nominal value (~ 200 MeV/c^2 instead ~ 70 MeV/c^2); after a few runs with the dipole magnet off, the offline alignment procedure was performed and the width of the new peak obtained (~ 90 MeV/c^2) is very close to the nominal value, as can be seen in Figure 3.

The acceptances have been obtained through pure signal simulation and reconstruction, whereas the efficiencies are estimated by means of real tracks and are put in the reconstruction procedure in order to properly take into account the performances of the detectors.

The systematical errors are currently under study and they are dominated by the unknown degree of polarization ($> 10\%$).

The amount of statistics for the J/ψ analysis is now quite high and the progresses done in the understanding of all the detector features are promising: the first results, such as the integrated cross section and the differential distributions, are forthcoming.

References

[1] Brambilla N *et al.* 2005 Heavy Quarkonium Physics *Addison-Wesley*
[2] Abe F *et al.* 1997 J/ψ and $\psi(2S)$ Production in $p\bar{p}$ Collisions at $\sqrt{s} = 1.8$ TeV *Phys.Rev.Lett.* **79** 572-7
[3] CDF Collaboration 2007 Polarization of J/ψ and $\psi(2S)$ mesons produced in $p\bar{p}$ collisions at $\sqrt{s} = 1.96$ TeV *Phys.Rev.Lett.* **99** 132001
[4] ALICE Collaboration 2009 The ALICE experiment at the CERN LHC *JINST* 3 S08002
[5] ALICE Collaboration 1999 TDR of the MUON Spectrometer *CERN/LHCC* **99-2**
[6] ALICE Collaboration 2006 Physics performance report - vol. II *Jour. of Phys.* G **32(10)** 1295-2040
[7] Martin A D *et al.* 2009 Parton distributions for the LHC *Eur.Phys.J.* C **63** 189-285
[8] Collins J and Soper D 1977 Angular distribution of dileptons in high-energy hadron collisions *Phys.Rev.Lett.* **D16**
[9] ALICE Collaboration 2010 First proton–proton collisions at the LHC as observed with the ALICE detector: measurement of the charged particle pseudorapidity density at $\sqrt{s} = 900$ GeV *Eur.Phys.J.* C **65** 111-25

Journal of Physics: Conference Series **270** (2011) 012028 doi:10.1088/1742-6596/270/1/012028

Production, elliptic flow and energy loss of heavy quarks in the quark-gluon plasma

Jan Uphoff,[1] Oliver Fochler,[1] Zhe Xu,[2,1] and Carsten Greiner[1]

[1] Institut für Theoretische Physik, Johann Wolfgang Goethe-Universität Frankfurt, Max-von-Laue-Str. 1, D-60438 Frankfurt am Main, Germany

[2] Frankfurt Institute for Advanced Studies, Ruth-Moufang-Str. 1, D-60438 Frankfurt am Main, Germany

E-mail: uphoff@th.physik.uni-frankfurt.de

Abstract. Production, elliptic flow and the nuclear modification factor of charm and bottom quarks are studied in central and non-central heavy-ion collisions at RHIC and LHC using the partonic transport model *Boltzmann Approach of MultiParton Scatterings* (BAMPS). Employing an initial heavy quark yield obtained with PYTHIA the full space-time evolution of charm and bottom quarks in the quark-gluon plasma (QGP) is carried out with BAMPS, taking also secondary production in the QGP into account. Only elastic collisions of heavy quarks with particles from the medium cannot describe the experimentally observed elliptic flow and nuclear modification factor. However, using an improved Debye screening and the running coupling yields a result which is much closer to data.

1. Introduction

Due to their large mass heavy quarks are produced at an early stage of the collision and can cover – depending on their production point – a long distance through the QGP. Interaction along this path and subsequent modifications of the heavy quark can reveal valuable information about the properties of the medium.

The experimentally observed elliptic flow v_2 and nuclear modification factor R_{AA} of heavy quarks [1, 2, 3] hint at an energy loss which is comparable to that of light quarks. This is in contradiction with the expectations which one would draw from the "dead cone effect" [4, 5]. Whether this large energy loss is due to collisional or radiative interactions – or both (or even other effects) – is under investigation (see [3] for a recent overview and comparison with data).

After a short introduction to BAMPS we will briefly address the production of heavy quarks. In Sec. 4 our results on the elliptic flow and nuclear modification factor are discussed and compared to the experimental data.

2. Parton cascade BAMPS

The partonic transport model *Boltzmann Approach of MultiParton Scatterings* (BAMPS) [6, 7] simulates the full space-time evolution of the QGP by solving the Boltzmann equation,

$$\left(\frac{\partial}{\partial t} + \frac{\mathbf{p}_i}{E_i} \frac{\partial}{\partial \mathbf{r}} \right) f_i(\mathbf{r}, \mathbf{p}_i, t) = \mathcal{C}_i^{2\to 2} + \mathcal{C}_i^{2\leftrightarrow 3} + \dots \; , \qquad (1)$$

for on-shell partons and pQCD interactions. So far gluons (g) and heavy quarks (Q) are implemented with the following interactions: $gg \to gg$, $gg \to ggg$, $ggg \to gg$, $gg \to Q\bar{Q}$, $Q\bar{Q} \to gg$, $gQ \to gQ$ and $g\bar{Q} \to g\bar{Q}$. Details of the model and the employed cross sections can be found in [6, 7, 8].

3. Heavy quark production at RHIC and LHC

We study the production of heavy quarks in two stages: initially during hard parton interactions in nucleon-nucleon scatterings and secondary production during the evolution of the QGP. Due to their large mass heavy quark production should be describable in the framework of pQCD. Indeed, next-to-leading order calculations [9] are in good agreement with experimental data from PHENIX [10]. For the present study, however, we use the event generator PYTHIA [11] to determine the initial heavy quark distributions, which agree with the PHENIX data as well. Nevertheless, these distributions have large uncertainties due to their sensitivity on the parton distribution functions in nucleons, the heavy quark masses as well as the renormalization and factorization scales (see [8] for a detailed analysis). For the initial distribution of the gluonic medium we use three different approaches: the mini-jet model, a color glass condensate inspired model and also PYTHIA in combination with the Glauber model.

In the following we give a brief overview of our results on heavy quark production in the QGP. More details concerning this section can be found in [8].

Secondary heavy quark production in the QGP is studied within a full BAMPS simulation of Au+Au collisions at RHIC. According to our calculations the charm quark production in the medium lies between 0.3 and 3.4 charm pairs, depending on the model of the initial gluon distribution, the charm mass and whether a $K = 2$ factor for higher order corrections of the cross section is employed. However, compared to the initial yield these values are of the order of a few percent for the most probable scenarios. Therefore, one can conclude that charm production at RHIC in the QGP is nearly negligible.

At LHC, however, the picture looks a bit different: Here the charm production in the QGP is a sizeable fraction of the initial yield and is even of the same order for some scenarios (with mini-jet initial conditions for gluons with a high energy density). In numbers, between 11 and 55 charm pairs are produced in the QGP.

Bottom production in the QGP, however is very small both at RHIC and LHC and can be safely neglected. As a consequence, all bottom quarks at these colliders are produced in initial hard parton scatterings.

4. Elliptic flow and nuclear modification factor of heavy quarks at RHIC

Commonly used observables for investigating the coupling of heavy quarks with the medium are the elliptic flow

$$v_2 = \left\langle \frac{p_x^2 - p_y^2}{p_T^2} \right\rangle \tag{2}$$

(p_x and p_y are the momenta in x and y direction in respect to the reaction plane) and the nuclear modification factor

$$R_{AA} = \frac{\mathrm{d}^2 N_{AA}/\mathrm{d}p_T \mathrm{d}y}{N_{\mathrm{bin}}\, \mathrm{d}^2 N_{pp}/\mathrm{d}p_T \mathrm{d}y} \tag{3}$$

of heavy quarks at mid-rapidity. A large elliptic flow comparable to that of light partons indicates a strong coupling to the medium. On the other hand a small R_{AA} is a sign for a large energy loss of heavy quarks. Experimental results reveal that both quantities are on the same order as the respective values for light particles [1, 2, 3].

As we have recently shown [12] elastic scatterings of heavy quarks with the gluonic medium using a constant coupling $\alpha_s = 0.3$ and the Debye mass for screening the t channel cannot reproduce the experimentally measured elliptic flow. In order to explain the data one would need a $40-50$ times larger cross section than the leading order one. Of course, this K factor is too large to represent the contribution of higher order corrections. However, as we will show in this section, the discrepancy with the data can be lowered – even on the leading order level – by a factor of 10 by taking the running of the coupling into account and by improving the Debye screening. The remaining factor of 4 difference could then indeed stem from neglecting higher order effects, which, however, must be checked in a future project.

The following calculations are done analogously to [13, 14]. An effective running coupling is obtained from measurements of e^+e^- annihilation and non-strange hadronic decays of τ leptons [15, 13]:

$$\alpha_s(Q^2) = \frac{4\pi}{\beta_0} \begin{cases} L_-^{-1} & Q^2 < 0 \\ \frac{1}{2} - \pi^{-1}\mathrm{atn}(L_+/\pi) & Q^2 > 0 \end{cases} \tag{4}$$

with $\beta_0 = 11 - \frac{2}{3}N_f$, $N_f = 0$ in our case, and $L_\pm = \ln(\pm Q^2/\Lambda^2)$ with $\Lambda = 200\,\mathrm{MeV}$. If $\alpha_s(Q^2)$ is larger than $\alpha_s^{\max} = 1.0$ it is set to $\alpha_s^{\max}$.

Since the t channel of the $gQ \to gQ$ cross section is divergent it is screened with a mass proportional to the Debye mass m_D:

$$\frac{1}{t} \to \frac{1}{t - \kappa m_D^2} \tag{5}$$

The Debye mass is calculated by the common definition $m_D^2 = 4\pi\,(1 + N_f/6)\,\alpha_s(t)\,T^2$ with the running coupling. The prefactor κ in Eq. (5) is mostly set to 1 in the literature without a clear reason. However, one can fix this factor by comparing the $\mathrm{d}E/\mathrm{d}x$ of the born cross section with κ to the energy loss within the hard thermal loop approach to $\kappa \approx 0.2$ [13, 14].

These improvements lead to an enhanced cross section which also increases the elliptic flow. Fig. 1 shows v_2 as a function of the transverse momentum p_T for the leading order cross section without any improvements, with the running coupling, with the corrected Debye screening and with both modifications. The elliptic flow of the latter reproduces the order of magnitude of the data, if the cross section is multiplied with $K = 4$, which is much smaller than the previous employed $K = 40 - 50$ and lies in a region which could account for higher order corrections. However, one has to check if these corrections have indeed a similar effect as a constant K factor of 4. Therefore, the calculation of the next-to-leading order cross section is planned for the near future and will complement $2 \leftrightarrow 3$ interactions for gluons, which are already implemented in BAMPS [6]. The shapes of the theoretical curve and of the data points are, however, slightly different. This is probably an effect of hadronization and decay to electrons, which is neglected in this calculation. That is, in Fig. 1 v_2 and R_{AA} of heavy quarks on the quark level are compared to the heavy flavor electron data. We plan to take these effects into account in future investigations.

On the right hand side in Fig. 1 the R_{AA} of heavy quarks is depicted, which shows for $K = 4$ the same magnitude of suppression as the data. Studies on the R_{AA} of gluons in BAMPS are presented in [16, 17].

5. Conclusions

We have studied the production and space-time evolution of charm and bottom quarks with BAMPS. Charm production in the QGP at RHIC is to a good approximation negligible, but at LHC it can reach values comparable to the initial yield. Bottom production during the evolution of the medium can be neglected at RHIC and LHC.

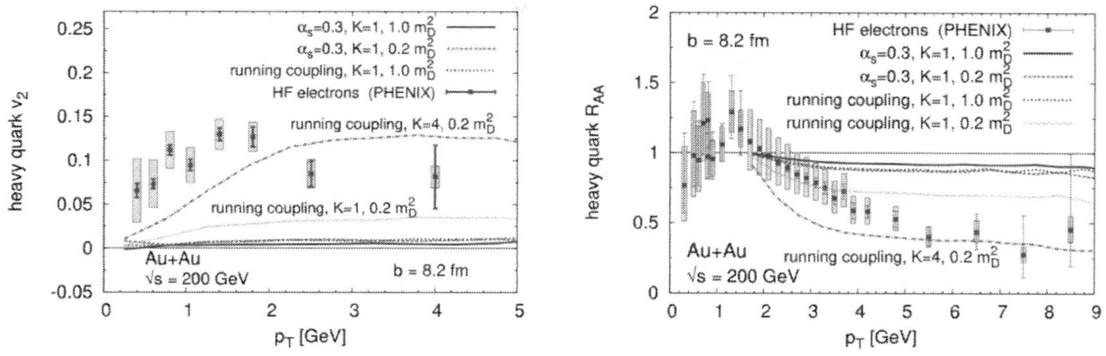

Figure 1. Elliptic flow v_2 (left) and nuclear modification factor R_{AA} (right) of heavy quarks with pseudo-rapidity $|\eta| < 0.35$ at the end of the QGP phase for Au+Au collisions at RHIC with an impact parameter of $b = 8.2$ fm. For one curve the cross section of $gQ \to gQ$ is multiplied with a K factor. For comparison data of heavy flavor electrons [3] is shown (see also text).

The energy loss of heavy quarks in the QGP due to elastic collisions with the leading order cross section is very small. Nevertheless, if one improves the calculation of the cross section by taking the running coupling and a more precise Debye screening into account, the order of magnitude of the experimentally measured elliptic flow and nuclear modification factor can be reproduced, if one multiplies the cross section with a K factor of 4. In order to solve this discrepancy next-to-leading order contributions must be taken into account, which we plan to do in the future.

Acknowledgements

We would like to thank A. Peshier and P.B. Gossiaux for stimulating and helpful discussions.

The BAMPS simulations were performed at the Center for Scientific Computing of the Goethe University Frankfurt. This work was supported by the Helmholtz International Center for FAIR within the framework of the LOEWE program launched by the State of Hesse.

References

[1] Abelev B I *et al.* (STAR) 2007 *Phys. Rev. Lett.* **98** 192301 (*Preprint* nucl-ex/0607012)
[2] Adare A *et al.* (PHENIX) 2007 *Phys. Rev. Lett.* **98** 172301 (*Preprint* nucl-ex/0611018)
[3] Adare A *et al.* (PHENIX) 2010 (*Preprint* 1005.1627)
[4] Dokshitzer Y L and Kharzeev D E 2001 *Phys. Lett.* **B519** 199–206 (*Preprint* hep-ph/0106202)
[5] Zhang B W, Wang E and Wang X N 2004 *Phys. Rev. Lett.* **93** 072301 (*Preprint* nucl-th/0309040)
[6] Xu Z and Greiner C 2005 *Phys. Rev.* **C71** 064901 (*Preprint* hep-ph/0406278)
[7] Xu Z and Greiner C 2007 *Phys. Rev.* **C76** 024911 (*Preprint* hep-ph/0703233)
[8] Uphoff J, Fochler O, Xu Z and Greiner C 2010 (*Preprint* 1003.4200)
[9] Cacciari M, Nason P and Vogt R 2005 *Phys. Rev. Lett.* **95** 122001 (*Preprint* hep-ph/0502203)
[10] Adare A *et al.* (PHENIX) 2006 *Phys. Rev. Lett.* **97** 252002 (*Preprint* hep-ex/0609010)
[11] Sjostrand T, Mrenna S and Skands P 2006 *JHEP* **05** 026 (*Preprint* hep-ph/0603175)
[12] Uphoff J, Fochler O, Xu Z and Greiner C 2010 *J. Phys. Conf. Ser.* **230** 012004 (*Preprint* 1004.4091)
[13] Gossiaux P B and Aichelin J 2008 *Phys. Rev.* **C78** 014904 (*Preprint* 0802.2525)
[14] Peshier A 2008 (*Preprint* 0801.0595)
[15] Dokshitzer Y L, Marchesini G and Webber B R 1996 *Nucl. Phys.* **B469** 93–142 (*Preprint* hep-ph/9512336)
[16] Fochler O, Xu Z and Greiner C 2009 *Phys. Rev. Lett.* **102** 202301 (*Preprint* 0806.1169)
[17] Fochler O, Xu Z and Greiner C 2010 (*Preprint* 1003.4380)

Journal of Physics: Conference Series **270** (2011) 012029 doi:10.1088/1742-6596/270/1/012029

Heavy-quark Langevin dynamics and single-electron spectra in nucleus-nucleus collisions

A Beraudo[1,2,3,4], W M Alberico[3,4], A De Pace[4], A Molinari[3,4], M Monteno[4], M Nardi[4] and F Prino[4]

[1]Centro Studi e Ricerche *Enrico Fermi*, Piazza del Viminale 1, Roma, ITALY
[2]Physics Department, Theory Unit, CERN, CH-1211 Genève 23, Switzerland
[3]Dipartimento di Fisica Teorica dell'Università di Torino, via P.Giuria 1, I-10125 Torino, Italy
[4]Istituto Nazionale di Fisica Nucleare, Sezione di Torino, via P.Giuria 1, I-10125 Torino, Italy

E-mail: `beraudo@to.infn.it`

Abstract. The stochastic dynamics of heavy quarks in the fireball produced in heavy-ion collisions is followed through numerical simulations based on the Langevin equation. The modification of the final p_T spectra (R_{AA}) of c and b quarks, hadrons and single-electrons with respect to pp collisions is studied. The transport coefficients are evaluated treating separately the contribution of soft and hard collisions. The initial heavy-quark spectra are generated according to NLO-pQCD, accounting for nuclear effects through recent nPDFs. The evolution of the medium is obtained from the output of two hydro-codes (ideal and viscous). The heavy-quark fragmentation into hadrons and their final semileptonic decays are implemented according to up-to-date experimental data. A comparison with RHIC data for non-photonic electron spectra is given.

1. Introduction

In a previous paper [1] we evaluated the transport coefficients for heavy quarks in the QGP in the Hard Thermal Loop (HTL) approximation. The above coefficients were then used to solve the relativistic Langevin equation in the case of a static medium, following the approach to thermal equilibrium. Here we improve our previous study in two important aspects. We provide a more accurate microscopic calculation of the transport coefficients and we perform the Langevin simulations for a fully realistic scenario, accounting for the hydrodynamical evolution of the medium. For independent studies based on the Langevin approach see [2, 3, 4, 5].

2. The Langevin equation in a dynamical medium

The relativistic Langevin equation allows to study the stochastic dynamics of heavy quarks in the QGP. The algorithm to follow the evolution of momentum and position of a brownian particle is the following. We focus on a given quark at $(\boldsymbol{x}_n, \boldsymbol{p}_n)$ after n steps of evolution. We move to the local fluid rest-frame and update its position and momentum by the quantities $\Delta\bar{\boldsymbol{x}}_n = (\bar{\boldsymbol{p}}_n/\bar{E}_p)\Delta\bar{t}$ and

$$\Delta\bar{p}^i_n = -\eta_D(\bar{p}_n)\bar{p}^i_n\Delta\bar{t} + \xi^i(\bar{t})\Delta\bar{t} \equiv -\eta_D(\bar{p}_n)\bar{p}^i_n\Delta\bar{t} + g^{ij}(\bar{\boldsymbol{p}}_n)\zeta^j(\bar{t})\sqrt{\Delta\bar{t}}, \qquad (1)$$

where we take – in the fluid rest-frame – $\Delta\bar{t}=0.02$ fm/c. In the above we express the noise term through the tensor (we omit the "bar")

$$g^{ij}(\boldsymbol{p})\equiv\sqrt{\kappa_L(p)}\hat{p}^i\hat{p}^j+\sqrt{\kappa_T(p)}(\delta^{ij}-\hat{p}^i\hat{p}^j), \tag{2}$$

depending on the transverse/longitudinal momentum diffusion coefficients $\kappa_{T/L}(p)$, and the uncorrelated random variables ζ^j, with $\langle\zeta^i(t)\zeta^j(t')\rangle=\delta^{ij}\delta_{tt'}$. Hence, one simply needs to extract three random numbers η^j from a gaussian distribution with $\sigma=1$. Concerning the problem of fixing the friction term $\eta_D(p)$ and further details we refer the reader to Ref. [1]. We then go back to the Lab frame, getting the updated $(\boldsymbol{x}_{n+1},\boldsymbol{p}_{n+1})$. The four-velocity and temperature fields $u^\mu(x)$ and $T(x)$ of the background medium are obtained from the output of two different hydrodynamical codes [6, 7, 8].

3. Evaluation of the transport coefficients

The coefficients $\kappa_{T/L}(p)$ reflect the transverse/longitudinal squared-momentum acquired through the collisions in the medium. Following [9] we introduce an intermediate cutoff $|t|^*\sim m_D^2$ $(t\equiv(P'-P)^2=\omega^2-q^2)$ separating hard and soft scatterings. The contribution of hard collisions $(|t|>|t|^*)$ is evaluated in pQCD, from the diagrams $Qq(\bar{q})\to Qq(\bar{q})$ and $Qg\to Qg$ and reads (employing the notation $\int_k\equiv\int d\vec{k}/(2\pi)^3$ and $P_{\rm tot}\equiv P+K-P'-K'$)

$$\kappa_T^{(g/q){\rm hard}}(p)=\frac{1}{2}\frac{1}{2E}\int_k\frac{n_{B/F}(k)}{2k}\int_{k'}\frac{1\pm n_{B/F}(k')}{2k'}\int_{p'}\frac{1}{2E'}\theta(|t|-|t|^*)(2\pi)^4\delta^{(4)}(P_{\rm tot})\left|\overline{\mathcal{M}}_{g/q}\right|^2q_T^2 \tag{3}$$

and

$$\kappa_L^{(g/q){\rm hard}}(p)=\frac{1}{2E}\int_k\frac{n_{B/F}(k)}{2k}\int_{k'}\frac{1\pm n_{B/F}(k')}{2k'}\int_{p'}\frac{1}{2E'}\theta(|t|-|t|^*)(2\pi)^4\delta^{(4)}(P_{\rm tot})\left|\overline{\mathcal{M}}_{g/q}\right|^2q_L^2, \tag{4}$$

where the squared amplitudes were evaluated in Ref. [10]. On the other hand in soft collisions $(|t|<|t|^*)$ the exchanged gluon has a small virtuality and has thus "time" to feel the presence of the other particles. A resummation of medium effects is required and this is provided by the HTL approximation, leading to the compact formulas $(x\equiv\omega/q)$

$$\kappa_T^{\rm soft}(p)=\frac{C_Fg^2}{8\pi^2v}\int_0^{|t|^*}d|t|\int_0^v dx\frac{|t|^{3/2}}{2(1-x^2)^{5/2}}\overline{\rho}(|t|,x)\left(1-\frac{x^2}{v^2}\right)\coth\left(\frac{x\sqrt{\frac{|t|}{1-x^2}}}{2T}\right) \tag{5}$$

and

$$\kappa_L^{\rm soft}(p)=\frac{C_Fg^2}{4\pi^2v}\int_0^{|t|^*}d|t|\int_0^v dx\frac{|t|^{3/2}}{2(1-x^2)^{5/2}}\overline{\rho}(|t|,x)\frac{x^2}{v^2}\coth\left(\frac{x\sqrt{\frac{|t|}{1-x^2}}}{2T}\right), \tag{6}$$

expressed in terms of the resummed gluon spectral function $\overline{\rho}(|t|,x)\equiv\rho_L(|t|,x)+(v^2-x^2)\rho_T(|t|,x)$ which can be found in [1]. Our results for $\kappa_{T/L}(p)$ are shown in the left panel of Fig. 1 where a very mild dependence on the arbitrary intermediate cutoff is manifest. For large momenta κ_L largely exceeds κ_T.

4. Numerical results

For each explored case we generated an initial sample of $45\cdot10^6$ $c\bar{c}$ and $b\bar{b}$ pairs, using the POWHEG code [11], with CTEQ6M PDFs. In the AA case we introduced nuclear effects in the PDFs according to the EPS09 scheme [12]; the quarks were then distributed in the transverse

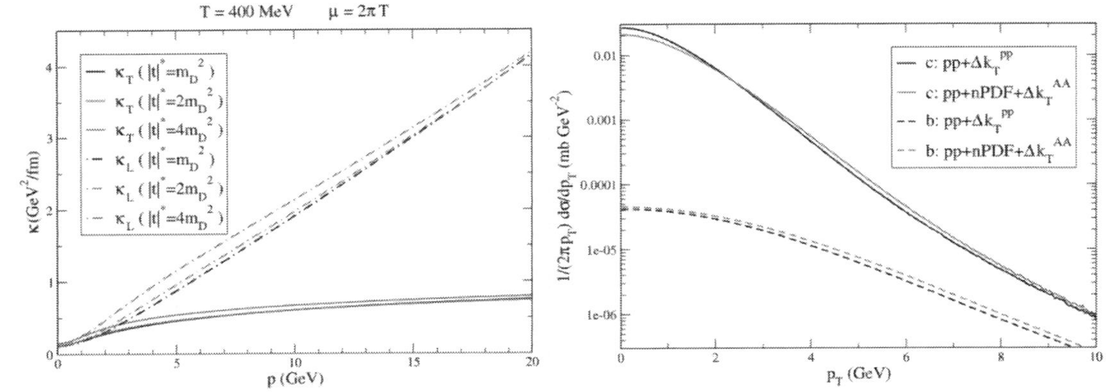

Figure 1. Left panel: the momentum-diffusion coefficients $\kappa_{T/L}(p)$ after summing the soft and hard contributions. The dependence on the intermediate cutoff $|t|^* \sim m_D^2$ is very mild. Right panel: the initial c and b spectra generated by POWHEG for pp and AA (with EPS09 nPDFs and nuclear k_T-broadening) collisions at $\sqrt{s}_{NN}=200$ GeV.

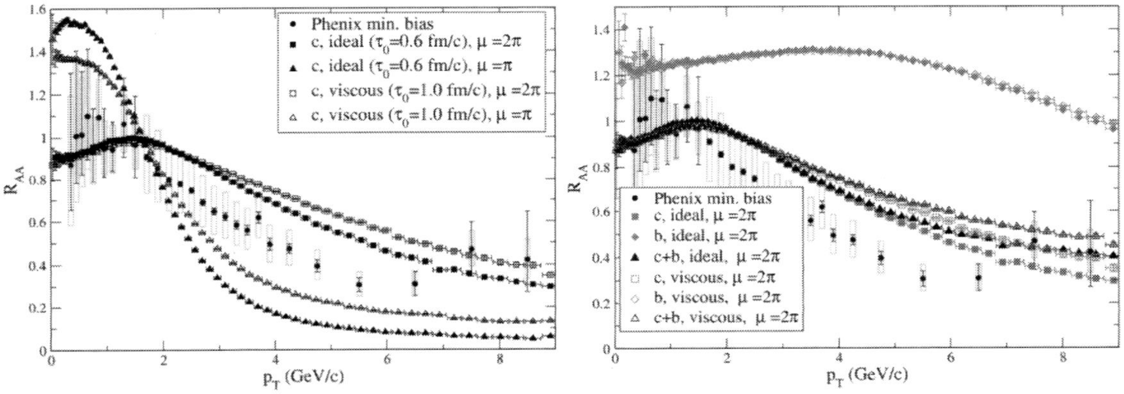

Figure 2. Left panel: $R_{AA}(p_T)$ for the c-quark spectra after the Langevin evolution in the QGP. The sensitivity on the hydrodynamical scenario (ideal vs viscous) appears to be small. On the other hand the dependence on the coupling $g(\mu)$ (at $T=200$ MeV $\alpha_s \approx 0.6/0.34$ for $\mu=\pi T/2\pi T$) is huge. Right panel: $R_{AA}(p_T)$ for c, and b quark spectra, together with their weighted combination.

plane according to the nuclear overlap function $dN/d\boldsymbol{x}_\perp \sim T_{AB}(x,y) \equiv T_A(x+b/2,y)T_B(x-b/2,y)$ and given a further k_T broadening on top of the "intrinsic" one. The initial p_T spectra are displayed in the right panel of Fig. 1. For each quark at the proper-time $\tau \equiv t^2 - z^2 = \tau_0$ we started following its Langevin dynamics until hadronization. The latter was modeled using a Peterson fragmentation function [13], setting $\epsilon = 0.04$ and 0.005 for c and b respectively; the hadron species were then assigned according to the branching fractions taken from [14, 15]. Finally each hadron was forced to decay into electrons with PYTHIA [16], using updated decay tables [17]. The e-spectra from c and b were then combined with a weight given by the respective total production cross-section ($\sigma_{c\bar{c}/b\bar{b}}$).

The results we present refer to minimum-bias Au-Au collisions at $\sqrt{s}_{NN}=200$ GeV. In Fig. 2

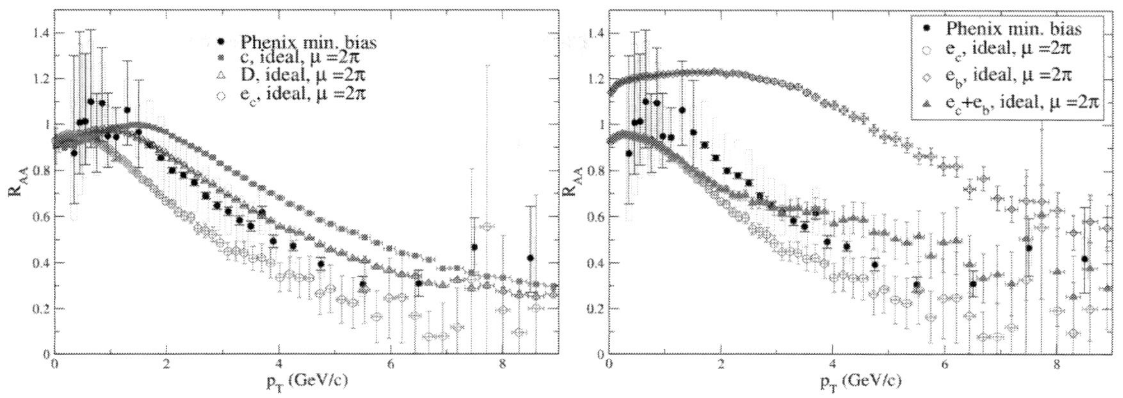

Figure 3. Left panel: the effects of fragmentation and semileptonic decays, leading to a substantial quenching of the electron-$R_{AA}(p_T)$. Right panel: $R_{AA}(p_T)$ for the single-electron spectra from the decays of charm, bottom and their weighted combination. The latter can be compared with the experimental results by Phenix.

we display the $R_{AA}(p_T) \equiv (dN/dp_T)^{AA}/\langle N_{\rm coll}\rangle(dN/dp_T)^{pp}$ for the final quark spectra. The dependence on the hydro scenario appears to be small. On the other hand the results display a huge sensitivity to the value of the coupling (hence, of the transport coefficients): a too strong coupling would overestimate the quenching effect. For the b spectra, the quenching induced by the medium is small: the most relevant effect is the larger production cross section in AA with respect to pp, arising from the anti-shadowing region in the nPDFs. In the left panel of Fig. 3 we consider the effects of fragmentation and decay into electrons: they both give rise to a quenching of the spectrum. Finally in the right panel we display, for a given scenario, the $R_{AA}(p_T)$ of the single-electron spectra and we compare it to the results by Phenix [18]. Our results, obtained in a perturbative setup and referring to a quite moderate value of the coupling, appear in reasonable agreement with the experimental data. This also suggests that it should be worth addressing more carefully the role of collisional energy loss in the quenching of high-p_T spectra, so far mainly attributed to gluon radiation.

References

[1] A. Beraudo, A. De Pace, W.M. Alberico and A. Molinari, Nucl. Phys. A 831, 59 (2009).
[2] Y. Akamatsu, T. Hatsuda, T. Hirano, Phys. Rev. C **79** 054907 (2009).
[3] H. van Hees, V. Greco and R. Rapp, Phys. Rev. C **73**, 034913 (2006).
[4] G.D. Moore, D. Teaney, Phys. Rev. C **71**, 064904 (2005).
[5] P.B. Gossiaux and J. Aichelin, Phys. Rev. C **78**, 014904 (2008).
[6] P.F. Kolb, J. Sollfrank and U. Heinz, Phys. Rev. C **62**, 054909 (2000).
[7] P. Romatschke and U.Romatschke, Phys. Rev. Lett. **99**, 172301 (2007).
[8] M. Luzum and P. Romatschke, Phys. Rev. C **78**, 034915 (2008).
[9] S. Peigné and A. Peshier, Phys. Rev. D **77**, 114017 (2008).
[10] B.L. Combridge, Nucl. Phys. B **151**, 429 (1979).
[11] S. Frixione, P. Nason and G. Ridolfi, JHEP 0709, 126 (2007).
[12] K.J. Eskola, H. Paukkunen and C.A. Salgado, JHEP 0904, 065 (2009).
[13] C. Peterson, D. Schlatter, I. Schmitt and P.M. Zerwas, Phys. Rev. D **27**, 105 (1983).
[14] http://www-zeus.desy.de/zeus_papers/ZEUS_PAPERS/DESY-05-147.ps .
[15] http://www.slac.stanford.edu/xorg/hfag/osc/PDG_2009/#FRAC .
[16] T. Sjostrand, S. Mrenna and P.Z. Skands, JHEP 0605, 026 (2006).
[17] C. Amsler *et al.* (PDG), Phys. Lett. B667, 1 (2008) and 2009 partial update for the 2010 edition.
[18] A. Adare *et al.* (PHENIX Collaboration), Phys. Rev. Lett. **98**, 172301 (2007).

Journal of Physics: Conference Series **270** (2011) 012030 doi:10.1088/1742-6596/270/1/012030

Charm and beauty searches using electron-D^0 azimuthal correlations and microvertexing techniques in STAR experiment at RHIC

Witold Borowski for the STAR collaboration

Laboratoire SUBATECH, 4 rue Alfred Kastler, BP 20722, 44307 Nantes cedex 3, France

E-mail: witold.borowski@subatech.in2p3.fr

Abstract. The energy loss of heavy quarks in the hot and dense matter created at RHIC, can be used to probe the properties of the medium. Both charm and beauty quarks contribute to the non-photonic electrons through their semi-leptonic decays. It is essential to determine experimentally the relative contribution of charm and beauty quarks to understand the suppression of heavy flavors at high p_T in central Au+Au collisions. The azimuthal angular correlations of non-photonic electrons with the reconstructed D^0 allow to disentangle the contribution of charm and beauty and to reduce the background below the D^0 invariant mass as well. We discuss the STAR measurement of non-photonic electron and $D^0 \rightarrow K\pi$ azimuthal correlations in p+p collisions at 200 GeV. Furthermore, we show results from the application of microvertexing techniques for charm and beauty searches in Cu+Cu and Au+Au collisions at 200 GeV using the information of the Silicon tracker of STAR.

1. Introduction

Jet quenching, namely anomalous energy loss of partons in the dense medium formed in early stage of heavy-ion collisions, is one of the most important discoveries at RHIC and allows for an estimate of the initial gluon density. Due to the dead cone effect [1] one expects a hierarchy in the suppression of partons depending on their mass. In particular the suppression is expected to be decreasing for increasing mass, for radiative energy loss models. It is therefore of particular importance to measure the flavor dependence of jet quenching to test theoretical predictions and allow for an estimate of the gluon density, which is in agreement with the jet quenching observed for different flavors.

Charm and beauty at RHIC at high transverse momenta (p_T) are measured through non-photonic electrons (NPE) originating from their semileptonic decays. The suppression of hadrons is quantified through the nuclear modification factor R_{AA}, which is defined as the yield of charm and beauty in heavy-ion collisions, divided by the yield in p+p collisions at same energy scaled by the average number of binary collisions.

Studies on non-photonic electron yields from heavy flavor semi-leptonic decays at RHIC have shown that R_{AA} exhibits suppression for the high values of p_T, which is similar to the one observed for light quark hadrons [2]. These results contradict with the previously mentioned theoretical expectations that predicted heavy flavor spectra to be less suppressed due to the dead cone effect [1] and constitute an important puzzle which needs to be resolved. To understand this phenomenon the contributions of charm and beauty to NPE spectra have to be studied separately. This can be done through the analysis of electron-D^0 and electron-hadron azimuthal angular correlations.

2. STAR Experiment

STAR (Solenoidal Tracker at RHIC) is a versatile detector which consists of many sub-systems dedicated to measurement of different aspects of heavy-ion collisions. Subdetectors used for the purpose of this analysis were: Time Projection Chamber, Electromagnetic Calorimeter with Shower Maximum Detector, Silicon Vertex Tracker and Silicon Strip Detector. The details about the STAR experiment can be found in [3].

2.1. Time Projecting Chamber

TPC is a main tracking and particle identification device of STAR. It is located around the interaction point inside the magnet, which is capable of generating a magnetic field up to 0.5 T.

2.2. Silicon Detectors

In order to improve the pointing resolution, STAR was equipped with two silicon based detectors located at the very center of the detector, between the beam pipe and the inner border of TPC. The inner one, Silicon Vertex Tracker (SVT), is a 3-layered drift detector. Each layer is a cylinder-shaped structure with the radius of 6.85, 10.8 and 14.7 cm. To simplify matching of the tracks reconstructed in TPC with points from SVT, the 4th layer of silicon, Silicon Strip Detector (SSD), with the radius of 23 cm was introduced. Together SSD and SVT can improve the resolution in Au+Au collisions by a factor of 10 for a particle with $p = 1\ GeV/c$ [4], from ~ 0.27 cm for a TPC only track to ~ 0.03 if the track has at least 2 silicon hits (Fig. 3).

2.3. Electromagnetic Calorimeter

The Electromagnetic Calorimeter (EMC), located around TPC barrel, has a form of towers built of lead scintillator with thickness of 21 X_0. This enables a measurement of the energy of electrons within $|\ \eta\ |< 1.0$. The Shower Maximum Detector (SMD) located at 5 X_0 inside the towers is a two-dimensional (ϕ, η) wire proportional detector which enables the analysis of the shower shape with the accuracy of 0.007 in both planes [5].

Figure 1: Schematic view of the fragmentation of a $c\bar{c}$ pair (left) and a $b\bar{b}$ pair (right) [6].

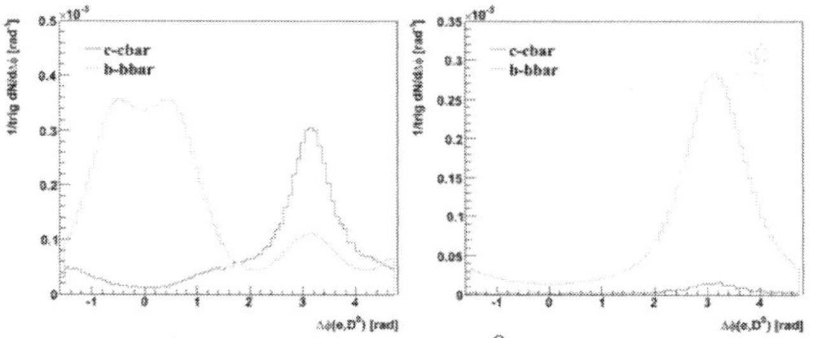

Figure 2: Azimuthal electron-D^0 correlation distribution from PYTHIA for like-sign e-K charge correlation (left plot) and unlike-sign correlation (right plot) [7, 8].

3. Correlation Technique

The D^0 meson can originate either directly from the hadronisation of the $c\bar{c}$ pair or from the B meson decay (as shown on Fig. 1). Thus it can be considered as a good probe for studying charm, but through the B meson, also the beauty. These contributions can be separated by combining the azimuthal correlations $(\Delta\phi)$ between an electron which comes either from the D or B semileptonic decay and the direction of D^0 particle itself with a charge correlation between that electron and kaon which comes from the D^0 hadronic decay. This gives three different possibilities shown on Fig. 2 as the result of a PYTHIA (leading order) simulation. Two of them exist for the e-K like-sign correlation, the beauty only case for $\Delta\phi \sim 0$ and charm dominant case with a small beauty contribution for $\Delta\phi \sim \pi$. In the unlike-sign e-K case only the $\Delta\phi \sim \pi$ correlation is applicable and represents the beauty domination with a small charm contribution. Analogous analysis performed with Next to Leading Order (NLO) simulations gives similar structure [6].

4. Data Analysis and Results

The described technique has been applied to the dataset collected during the 2006 p+p run with the energy of $\sqrt{s} = 200\ GeV$. In the analysis only the events where the primary vertex z coordinate was reconstructed less than 30 cm away from the center of the experiment and for which at least one high energy ($E_t > 5.4\ GeV$) non-photonic electron was registered were taken into consideration.

The electron candidates were selected by taking all particles for which $dE/dx \in (3.5, 5.0)\ keV/cm$ and $p > 1.5\ GeV/c$. Later, in order to exclude hadron contamination, the cut on $p/E \in (0, 2)$ was performed, for which momentum (p) was taken from the track reconstructed in the TPC and the energy (E) was given by the calorimeter. Electrons were expected to give a value close to unity as they are supposed to deposit all their energy inside the calorimeter. Finally a cut on a shower shape was performed with the use of the information gathered by the Shower Maximum Detector. It is known that hadrons give much narrower shower than electrons. Thus only the ones which were covering more than one strip in each plane were considered as electron-originated.

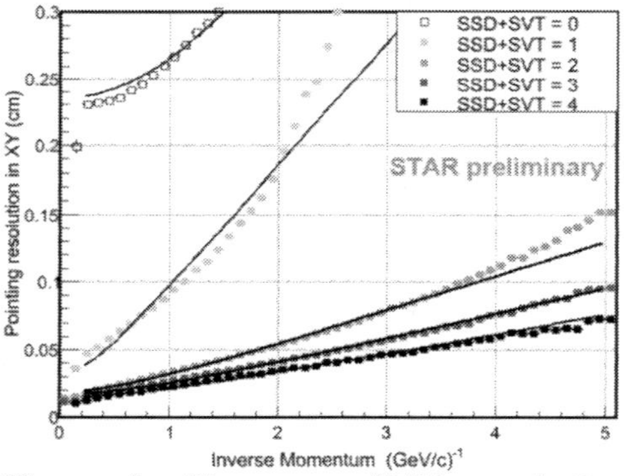

Figure 3: Transverse DCA resolution vs 1/p in Au+Au $\sqrt{s_{NN}} = 200\ GeV$. Tracks with at least two silicon hits have 10 times better pointing resolution than tracks for which only the TPC information is available.

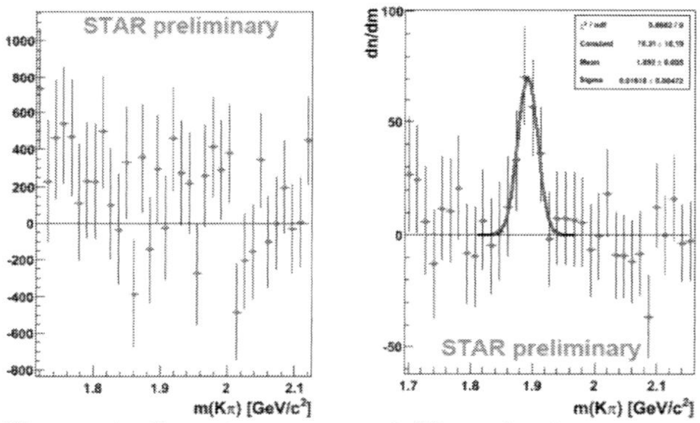

Figure 4: Invariant mass of $K\pi$ pairs in p+p at $\sqrt{s} = 200\ GeV$ for all events (left) and selected by non-photonic electron trigger (right) [9]. Selecting only electron triggered events allows to reduce background by a factor of 100.

The sample of particles selected in the presented way was highly contaminated by photonic electrons which came from neutral meson decays or photon conversions inside the detector. In order to get rid of them for each electron pair the Invariant Mass was calculated and particles for which the obtained value was below 150 MeV/c^2 were rejected. In this way up to 70% of the photonic electrons have been removed.

For events tagged by electrons, selected with the technique described above, the invariant mass method was applied to all $K\pi$ pairs [7, 8]. As a result a D^0 yield was obtained (Fig. 4) and studied in the different azimuthal angle and charge correlation variants. The results shows that charm and beauty contributions are similar as it is presented on Fig. 5 together with PYTHIA and NLO Monte Carlo simulation which have been fitted to the data. The observed effect is also in agreement with the similar analysis done for the electron-hadron correlation and FONLL (Fig. 6).

The studies on electron-D^0 correlations in p+p have provided a baseline for analysis in Au+Au collisions. However, due to much greater multiplicity and particle density of such events another technique of secondary vertex selection (microvertexing) had to be applied in order to reduce the combinatorial background. That method has been tested on minimum bias Au+Au $\sqrt{s_{NN}} = 200\ GeV$ data collected in the run 2007 and provided a preliminary D^0 peak. To improve the pointing precision silicon detectors were used in that run. More details on the D^0 reconstruction in Au+Au collisions with silicon detectors can be found in [10].

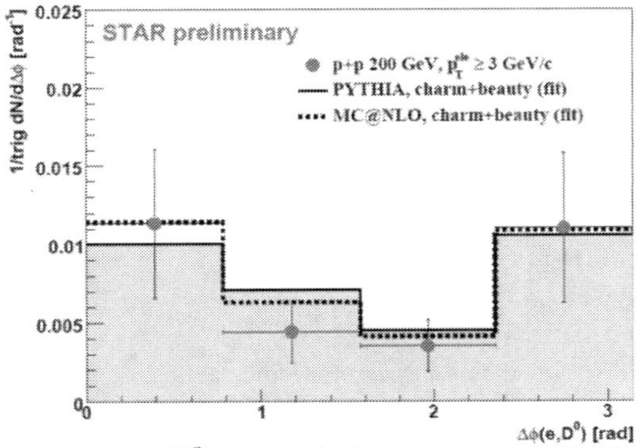

Figure 5: e-D^0 azimuthal angular correlations in p+p collisions at 200 GeV in comparison with PYTHIA and NLO simulations [7, 8].

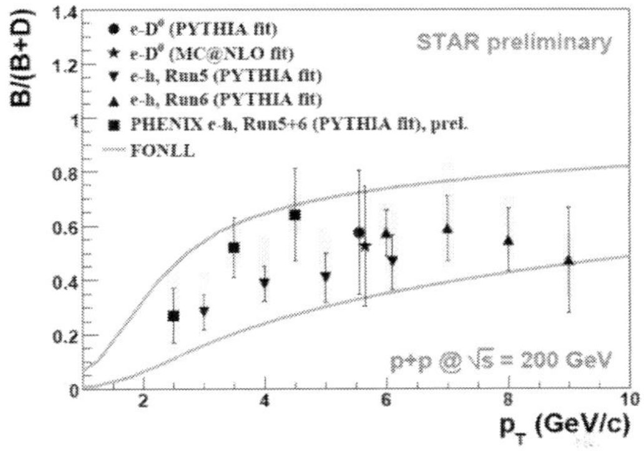

Figure 6: Relative beauty contribution $B/(B+D)$ as a function of trigger electron p_T with a FONLL simulation [7, 8].

Summary

Studies on non-photonic electron yields from heavy flavor semi-leptonic decays at RHIC have shown that R_{AA} exhibits suppression for the high values of p_T, which is similar to the one observed for light quark hadrons. To understand this phenomenon the contribution of charm and beauty to NPE spectra has been studied through the electron-D^0 and electron-hadron azimuthal angular correlations. Due to the use of these methods it has been found that in p+p collisions the beauty contribution increases with p_T and becomes comparable to charm contribution around $p_T = 5.5\ GeV/c$. This effect was found to be compatible with the FONLL simulations within the uncertainties. In order to perform a similar analysis in Au+Au collisions the microvertexing method has been applied to the data collected by STAR with inner silicon subdetectors. The technique has been successfully tested and the analyses on correlations are underway.

References

[1] Dokshitzer Y L and Kharzeev D E 2001 *Phys. Lett.* B **519** 199–206
[2] Adare A *et al.* (PHENIX) 2010 (*Preprint* 1005.1627)
[3] Ackermann K H *et al.* (STAR) 2003 *Nucl. Instrum. Meth.* **A499** 624–632
[4] Fisyak Y V *et al.* 2008 *J. Phys. Conf. Ser.* **119** 032017
[5] Beddo M *et al.* (STAR) 2003 *Nucl. Instrum. Meth.* A **499** 725–739
[6] Mischke A 2009 *Phys. Lett.* B **671** 361–365 (*Preprint* 0807.1309)
[7] Mischke A (STAR) 2008 *J. Phys.* G **35** 044022 (*Preprint* 0710.2599)
[8] Mischke A (STAR) 2008 *J. Phys.* G **35** 104117 (*Preprint* 0804.4601)
[9] Geromitsos A (STAR) 2009 *AIP Conf. Proc.* **1182** 787–790 (*Preprint* 0911.2490)
[10] LaPointe S (STAR) *these proceedings*

Direct Photons from a Hybrid Approach – Exploring the parameter space

B Bäuchle and M Bleicher

Frankfurt Institute for Advanced Studies, Ruth-Moufang-Straße 1, 60438 Frankfurt am Main, Germany

Institut für Theoretische Physik, Max-von-Laue-Straße 1, 60438 Frankfurt am Main, Germany

E-mail: baeuchle@th.physik.uni-frankfurt.de

Abstract.
Direct photon spectra are calculated within a transport+hydrodynamics hybrid approach, in which the high-density part of the transport evolution has been replaced by a 3+1-dimensional hydrodynamic calculation. We study the effects of changing the parameters of the two interfaces between the transport- and hydrodynamic descriptions on the resulting direct photon spectra.

1. Introduction

Heavy-ion physics is the principal tool to investigate the phase diagram of strongly interacting matter. Knowledge about the matter created in high-energy nuclear collisions is inferred by studying the remnants of the violent interaction. Since no more direct observation is possible, the experimental observations need to be complemented by theoretical calculations. Unfortunately, first-principle calculations of the whole collision are not feasible, so effective models need to be employed.

Most remnants from heavy-ion collisions are themselves strongly interacting particles, which therefore are mostly emitted in the late stages of the fireball evolution. While they may carry information about the early stages collectively, such as flow patterns, the individual hadron carries only indirect information about the hot and dense phases. Direct photons are well-suited to probe the early stages of a heavy-ion reaction, since due to their small scattering cross-section they leave the reaction zone essentially unscathed. The experimental task of extracting direct photon spectra is challenging, because the decay of hadrons such as the π^0 create a huge background of decay photons. In this article, we investigate a model for direct photon emission [1] with rates from [2] (emission from hydrodynamic phase) and cross-sections from [3] (emission from transport phase) with respect to the various parameters used therein.

2. The Model

We use the UrQMD v3.3 transport model [4, 5, 6] and its recent extension to substitute the intermediate high-density stage with an ideal 3+1-dimensional hydrodynamic calculation [7]. The early and late stages are still calculated in the transport model, which implements hadron and string degrees of freedom and uses PYTHIA [8] for scatterings at high momentum transfer. In the intermediate stage, hydrodynamic calculations are performed using three different Equations of State (EoS): A hadron gas EoS (HG-EoS [9]) is used, which has no phase transition, but the same degrees of freedom as the transport description. A chiral EoS (χ-EoS [10]) which has a cross-over phase transition to chirally restored and deconfined matter, and a MIT-bag model EoS (BM-EoS [11]) which has a strong first order phase transition with large latent heat to deconfined matter, are also used.

The interfaces between transport and hydrodynamics are the main subject of the present investigations. Transport → hydrodynamics: All particles and their momenta are transformed to baryon number-, energy- and momentum densities. In order to obtain the densities at one specific point, all particle properties are represented by a gaussian distribution centered around the particle's position with the width 1 fm. The sum of the contributions of each particle at the point of interest is the relevant density at that point. Since the gaussian distributions are normalized, this approach inherently conserves baryon number, energy and momentum. In this process, the system is forced into local thermal equilibrium, because in ideal hydrodynamics only perfectly equilibrated matter can be described. The time at which this transition happens t_{start} is, in standard setup, when the initial nuclei have passed through each other. This number depends on the (radii of the) incoming nuclei as well as the incident energy. We investigate the changing of this parameter from $\frac{1}{4}t_0$ to $4t_0$, with t_0 being the standard value. At high energies, this time is too short to allow even for partial thermalization. The minimal time for switching to the hydrodynamic description has been set to $t_{\mathsf{start}} = 0.6$ fm, which is $0.41t_0$ for Pb+Pb-collisions at top SPS-energies ($E_{\mathsf{lab}} = 158$ AGeV) and more than $4t_0$ at $\sqrt{s_{\mathsf{NN}}} = 200$ GeV. Therefore, the latter energy has been omitted from that investigation.

Hydrodynamics → transport: Besides the criterion when the transition from hydrodynamics back to transport happens, also the details of this transition can be changed. Usually, the transition happens when the system has diluted below $\varepsilon = 5\varepsilon_0$, where $\varepsilon_0 = 146$ MeV/fm^3 is the nuclear ground state energy density. This number is varied between $\varepsilon_{\mathsf{crit}} = 2.5\varepsilon_0$ and $\varepsilon_{\mathsf{crit}} = 10$. The scenario for the transition can either be that the criterion must be met by the part of the system that has the same z-coordinate (along the beam axis), whereafter this slice is transferred to the transport calculation. This is the default **gradual** scenario (for more information the reader is referred to [12]). The other scenario, called **isochronous** scenario, requires the criterion to be met by the whole system, after which it is transferred to the transport calculation instantaneously.

We will change the time for the first transition t_{start}, the critical energy density for the second transition $\varepsilon_{\mathsf{crit}}$ and the scenario for the second transition (gradual vs. isochronous). During each of these changes, the other two parameters are kept at their default values. The calculations are done for central Pb+Pb-collisions ($b < 5$ fm) at $E_{\mathsf{lab}} = 8$, 45 and 158 AGeV and $\sqrt{s_{\mathsf{NN}}} = 200$ AGeV.

3. Results

Figure 1 shows the direct photon yield of calculations with varying start time of the hydrodynamic description. At $E_{\mathsf{lab}} = 8$ AGeV and late switching times $t_0 > 2$ fm, the whole system has an energy density below or close to the threshold for switching back to transport calculations $\varepsilon = 5\varepsilon_0$, so that the hydrodynamic calculation only runs for a very short time. The yield of direct photons is maximal in calculations with $t_{\mathsf{start}} \approx 0.75t_0$ for all systems. In calculations with large t_{start} values, most of the high-density evolution of the system takes place in the transport phase. Here, chemical equilibration is reached only after some time, so that especially the $\pi\rho \to \gamma\pi$-channel, which dominates the hadronic sources, is suppressed by the lack of ρ-mesons in the unequilibrated cascade calculation with respect to the equilibrated hydrodynamic calculation.

At the second transition from hydrodynamics to the transport calculation, the system remains in thermal equilibrium. The investigation shown in Figure 2 varies the critical energy density $\varepsilon_{\mathsf{crit}}$ below which the system is described by transport calculations from $\varepsilon_{\mathsf{crit}} = 2\varepsilon_0$, where the

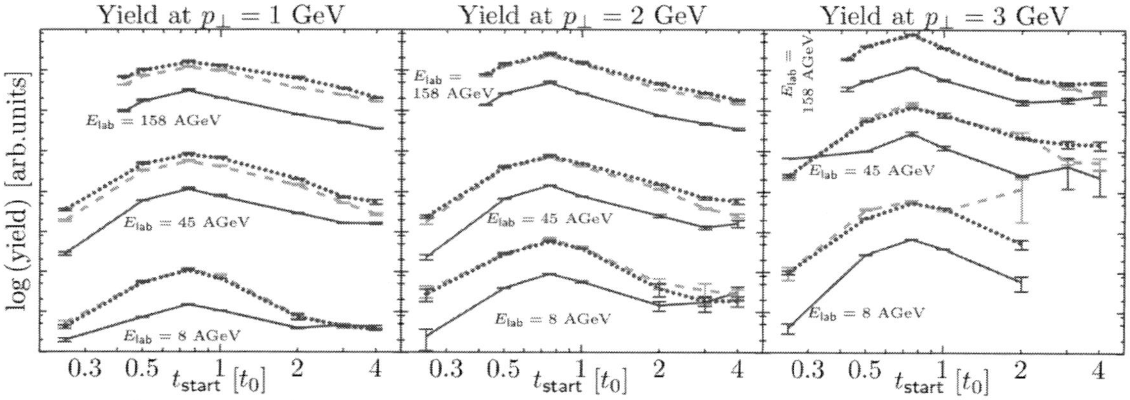

Figure 1. (Color Online) Invariant direct photon yield of calculations with varying start time of the hydrodynamic description. The left panel shows the yield at $0.5 < p_\perp < 1.5$ GeV, the middel panel at $1.5 < p_\perp < 2.5$ GeV and the right panel shows the yield at $2.5 < p_\perp < 3.5$ GeV. Each panel shows the yields for $E_{lab} = 158$ AGeV, scaled by 10 (upper curves), $E_{lab} = 45$ AGeV (middle curves) and $E_{lab} = 8$ AGeV, scaled by 1/10 (lower curves) for HG-EoS calculations (blue solid lines), χ-EoS calculations (orange dashed lines) and BM-EoS calculations (purple dotted lines). The scale is consistent within each panel.

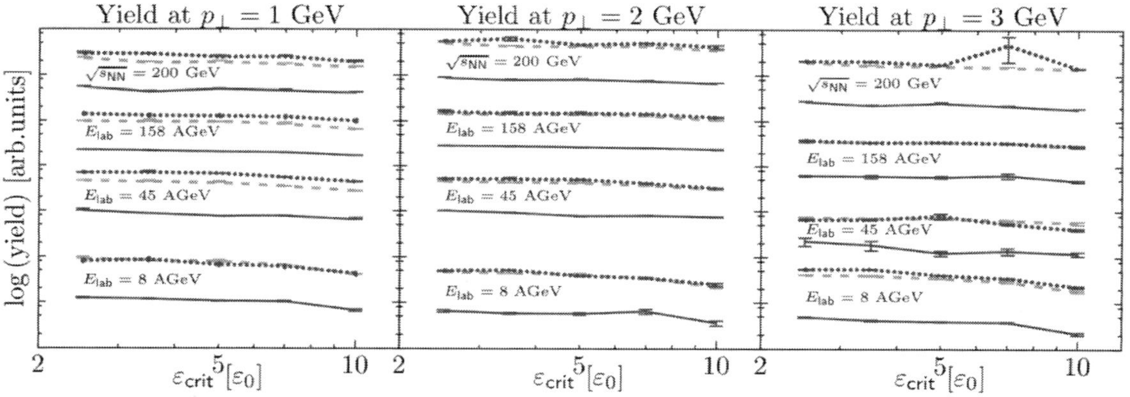

Figure 2. (Color Online) Invariant direct photon yield of calculations with varying critical energy density for the transition from hydrodynamic description to transport description. For explanation of the panels and lines, see the description of Figure 1.

applicability of hydrodynamics seems very questionable to $\varepsilon_{crit} = 10\varepsilon_0$, where the use of a transport calculation might be questioned. We find a slight decrease of the photon spectra as the threshold is raised, consistent with the findings above.

The influence of changing the transition scenario is investigated in Figure 3. The yield in the gradual transition scenario is very similar to the yield in the isochronous transition scenario, albeit a small decrease is visible in most of the calculations. Since in the isochronous transition scenario the fraction of the system calculated with hydrodynamics is larger than in the gradual transition scenario, this is also consistent with the investigations on ε_{crit} and t_{start}.

4. Summary

The parameter space of the UrQMD-hybrid model has been explored with respect to direct photon emission. The two major model parameters – the transition time from the initial transport calculation to the hydrodynamic calculation t_{start} and the critical energy density for

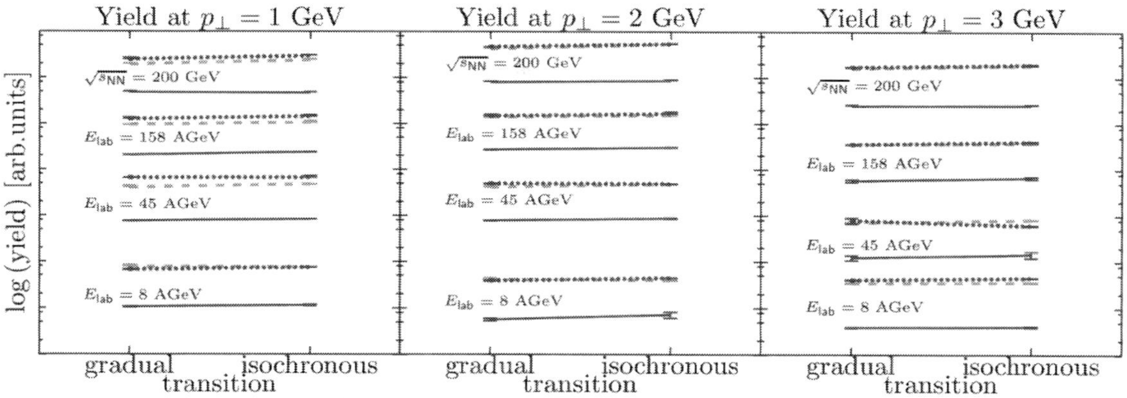

Figure 3. (Color Online) Invariant direct photon yield of calculations with varying the transitin scenario for the transition from hydrodynamic description to transport description. For explanation of the panels and lines, see the description of Figure 1.

the transition time from hydrodynamics to the transport calculation ε_{crit} – have been varied over a range that includes all reasonable values. In all cases, we find that if a larger part of the evolution is calculated in hydrodynamics, the yield of direct photons increases. Only when the initial transition is done too early (small t_{start}), the direct photon yield is reduced.

The increase of the photon spectra with longer hydrodynamic phase is expected in calculations with a Quark-Gluon-Plasma, which is only present in the hydrodynamic phase and emits more photons than the hadronic phase. In the hadronic phase, the increase is attributed to the higher number of ρ-mesons in the chemically equilibrated hydrodynamic calculation.

Acknowledgments

This work has been supported by the Frankfurt Center for Scientific Computing (CSC), the GSI and the BMBF. B. Bäuchle gratefully acknowledges support from the Deutsche Telekom Stiftung, the Helmholtz Research School on Quark Matter Studies and the Helmholtz Graduate School for Hadron and Ion Research. This work was supported by the Hessian LOEWE initiative through the Helmholtz International Center for FAIR. The authors thank Elvira Santini and Rene Bellwied for valuable discussions.

References

[1] B. Bäuchle and M. Bleicher, Phys. Rev. C **81** (2010) 044904 [arXiv:0905.4678 [hep-ph]].
[2] P. B. Arnold *et al.*, JHEP **0112** (2001) 009, S. Turbide *et al.*, Phys. Rev. C **69** (2004) 014903.
[3] J. I. Kapusta, P. Lichard and D. Seibert, Phys. Rev. D **44** (1991) 2774 [Erratum-ibid. D **47** (1993) 4171] [Phys. Rev. D **47** (1993) 4171].
[4] S. A. Bass *et al.*, Prog. Part. Nucl. Phys. **41** (1998) 255 [Prog. Part. Nucl. Phys. **41** (1998) 225] [arXiv:nucl-th/9803035].
[5] M. Bleicher *et al.*, J. Phys. G **25** (1999) 1859 [arXiv:hep-ph/9909407].
[6] H. Petersen, M. Bleicher, S. A. Bass and H. Stocker, arXiv:0805.0567 [hep-ph].
[7] H. Petersen, J. Steinheimer, G. Burau, M. Bleicher and H. Stöcker, Phys. Rev. C **78** (2008) 044901 [arXiv:0806.1695 [nucl-th]].
[8] T. Sjöstrand, S. Mrenna and P. Z. Skands, JHEP **0605** (2006) 026 [arXiv:hep-ph/0603175].
[9] D. Zschiesche, S. Schramm, J. Schaffner-Bielich, H. Stöcker and W. Greiner, Phys. Lett. B **547** (2002) 7 [arXiv:nucl-th/0209022].
[10] J. Steinheimer, S. Schramm and H. Stöcker, arXiv:0909.4421 [hep-ph].
[11] D. H. Rischke, Y. Pursun and J. A. Maruhn, Nucl. Phys. A **595** (1995) 383 [Erratum-ibid. A **596** (1996) 717] [arXiv:nucl-th/9504021].
[12] J. Steinheimer, V. Dexheimer, H. Petersen, M. Bleicher, S. Schramm and H. Stöcker, Phys. Rev. C **81** (2010) 044913 [arXiv:0905.3099 [hep-ph]].

Two particle correlations: a probe of the LHC QCD medium

Yaxian Mao[1,2], for the ALICE Collaboration

[1] Laboratoire de Physique Subatomique et de Cosmologie, Grenoble 38026, France
[2] Key Laboratory of Quark & Lepton Physics (Huazhong Normal University), Ministry of Education, Wuhan 430079, China

E-mail: maoyx@iopp.ccnu.edu.cn

Abstract. The properties of γ–jet pairs emitted in heavy-ion collisions provide an accurate mean to perform a tomographic measurement of the medium created in the collision through the study of the medium modified jet properties. The idea is to measure the distribution of hadrons emitted on the opposite side of the direct photon. The feasibility of such measurements is studied by applying the approach on the simulation data, we have demonstrated that this method allows us to measure, with a good approximation, both the jet fragmentation and the back-to-back azimuthal alignment of the direct photon and the jet. Comparing these two observables measured in pp collisions with the ones measured in AA collisions reveals the modifications induced by the medium on the jet structure and consequently allows us to infer the medium properties. In this contribution, we discuss a first attempt of such measurements applied to real proton-proton data from the ALICE experiment.

1. Introduction

Quantum ChromoDynamics (QCD) [1] is a theory of the strong interaction, the fundamental force describing the interactions of quarks and gluons making up hadrons. The QCD calculations performed on a lattice, indicate that a phase transition from normal hadronic matter to partonic matter, the Quark-Gluon Plasma (QGP), will occur beyond a critical temperature of $T_c \sim$ 170 MeV [2]. By colliding heavy ions at ultra relativistic energies, this new state of matter can be created and its properties, such as the equation of state, the degrees of freedom and the transport properties can be measured.

The phase diagram has been explored in various regions with heavy-ion collisions at continuously increasing kinetic energies. Experiments at CERN's Super Proton Synchrotron (SPS) [3] concluded on the indirect evidence of a "new state of matter". Current experiments at Brookhaven National Laboratory's Relativistic Heavy Ion Collider (RHIC) [4] have found that matter does not behave as an ideal gas of free quarks and gluons predicted by theory, but, rather, as an almost perfect fluid. The new experiment ALICE [5] at CERN's Large Hadron Collider (LHC), will push further the study of the QCD medium. Thanks to the huge step in collision energy ($\sqrt{s_{NN}} = 5.5\ TeV$ in Pb-Pb collisions), LHC will open new avenues for the exploration of matter under extreme conditions of temperature and density. Since the hot QCD medium will be formed at higher temperatures than at RHIC, the deconfined phase will last longer and more readily modify our experimental probes, allowing for a more accurate study of this new state matter.

Hard scattered partons produced in initial stage of the collisions, have been identified as a valuable probe of the medium. Indeed, medium properties can be inferred from the modifications experienced by the partonic shower inside the medium. Partons are only observed indirectly, as a collimated jet of hadrons coming from the fragmentation of the partonic shower [6]. Comparing the measurements of the jet fragmentation in proton-proton and heavy-ion collisions will reveal the modifications produced by the medium on the hard scattered partons. Ideally, one needs to know the 4-momentum of the parton when it has been produced in the hard scattering and after it has been modified by the medium. This can be achieved by selecting particular hard processes in which there is a photon in the final state. Since the photon does not interact with the medium, its 4-momentum is not modified and thus provides a measure of the hard scattered parton emitted back-to-back with the photon. Measuring the hadrons opposite to the photon is thus a promising way to measure the jet fragmentation and misalignment between photon and hadrons to quantify the modifications due to the medium.

2. Approach Validation with Monte-Carlo Data

The experimental technique consists in tagging events with a well identified high energy direct photon and measuring the distribution of hadrons emitted oppositely to the photon as a function of the parameter $x_E = -\vec{p}_T^{\,h} \cdot \vec{p}_T^{\,\gamma} / \mid p_T^{\gamma} \mid^2$. Such a measurement requires an excellent direct photon identification and the measurement of charged and neutral hadrons with good p_T resolution. In ALICE, the electromagnetic calorimeters, PHOS ($|\Delta\eta| < 0.12$ and $\Delta\phi = 100^o$) and EMCal ($|\Delta\eta| < 0.7$ and $\Delta\phi = 100^o$) [7, 8], are capable to measure photons with high efficiency and resolution [9]. The central tracking system (ITS and TPC), covers the pseudorapidity $-0.9 \leq \eta \leq +0.9$ and the full azimuth, is helpful for direct photon extraction with the isolation technique.

We have first established the feasibility of γ–hadrons correlation measurement with ALICE detectors using Monte-Carlo data. As a first result [11] of this study, PYTHIA [10] generator is used to simulate pp collisions at $\sqrt{s} = 14$ TeV containing a 2→2 process with a direct photon inside PHOS acceptance. we have demonstrated that this measurement allows us to determine, both the jet fragmentation distribution and the back-to-back azimuthal alignment of the direct photon and the jet. However because of the limited acceptance covered by the calorimeters, the measurement is restricted by statistics to photon with energies below 50 GeV. This kinematic region is particularly interesting because jets of such low energy loose a large fraction of their energy while traversing the medium, rendering the medium modification most visible. In addition, because jets with energy below 50 GeV can hardly be reconstructed in the heavy-ion environment, the photon tagging technique provides a sensitive measurement of jets in this kinematic range. Systematic errors due to the improper identification of direct photons remain, within this kinematic range, lower than statistical errors from our study [11].

To quantify the medium modification, the photon–hadrons correlation distribution has been studied with events generated in pp collisions at $\sqrt{s} = 5.5$ TeV containing a 2→2 process with a direct photon inside EMCal acceptance. They have been generated by PYTHIA [10] generator and qPYTHIA, which includes a parton energy loss model [12] with the medium transport parameter $\hat{q} = 50$ GeV^2/fm for photon energies between 5 and 200 GeV. At this stage of the study, the heavy-ion collision background has not yet been taken into account. Direct photons are identified with the isolation technique requiring no hadronic activity around the direct photon candidate inside a given cone size [13]. Hadrons detected in the azimuthal range $\pi/2 < \Delta\phi < 3\pi/2$ relative to the photon were used to construct the correlation function. The contribution of hadrons from the underlying event was calculated from the hadrons emitted in the same azimuthal hemisphere as the photon.

The relative azimuthal angle, $\Delta\phi = \phi_\gamma - \phi_h$, between the direct photon and charged hadrons is strongly peaked at π as expected for the 2→2 process (Fig. 1). When medium effects are simulated (qPYTHIA), the $\Delta\phi$ distribution becomes broader. The broadening can be related to the medium transport parameter $\hat{q}$. However, the effect is quite small which will make the measurement in the heavy-ion environment quite challenging. A stronger signal is expected to be observed in the photon–hadrons distribution from heavy-ion collisions when compared to the distribution from pp collisions. The resulting photon-triggered hadrons distributions, after subtraction of underlying events, are shown in Fig. 2, normalized to the number of trigger particles found in corresponding generation. The statistical errors are estimated from the annual yield of photon events with p_T larger than 30 GeV we anticipate to collect during one PbPb run at nominal luminosity [5]. The distribution exhibits the expected suppression at high x_E, due to the enegy loss of the hard scatered parton and the enhancement at low x_E due to the fragmentation of soft gluons radiated in the medium.

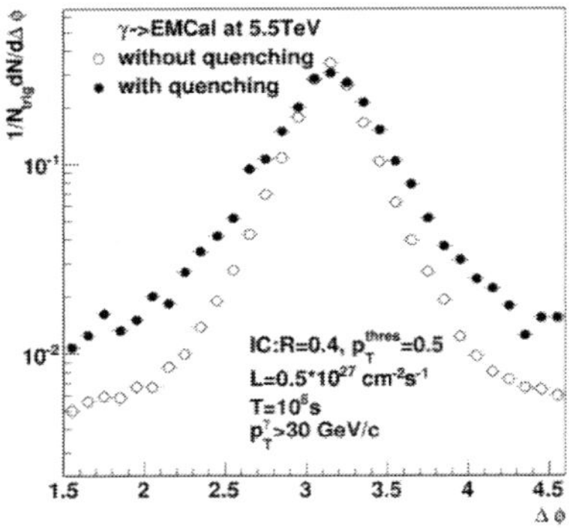

Figure 1. Relative azimuthal angle distribution $\Delta\phi = \phi_\gamma - \phi_{hadron}$ for γ-jet events in pp collisions at $\sqrt{s} = 5.5$ TeV.

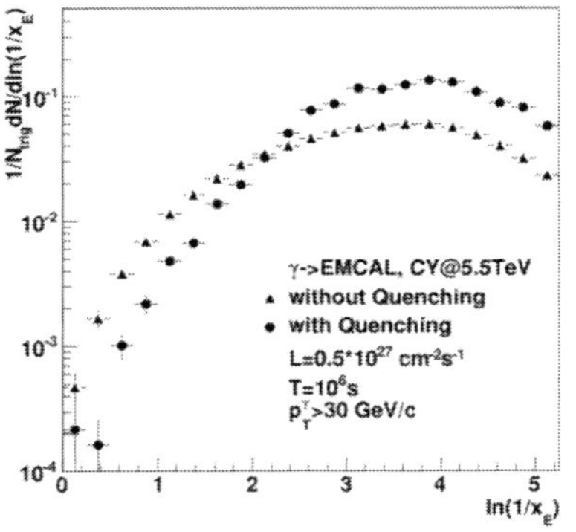

Figure 2. γ-hadron correlation distributions in quenched and unquenched PYTHIA events as a function of $ln(1/x_E)$.

3. Two Particle Correlations in pp@7TeV

Minimum bias data have been collected in pp collisions at center of mass $\sqrt{s} = 7$ TeV. We have analyzed about 35 million events in a first attempt to measure γ-hadrons correlations. The trigger particle is selected as the one with the highest transverse momentum measured either in the central tracking system or in the electromagnetic calorimeters. In the calorimeters, electromagnetic particles are detected as clusters of hit calorimeter cells. Roughly we have identified π^0 candidate as a pair of clusters which invariant mass matches the π^0 mass range, 135 ± 15 MeV, and single clusters (which do no pair with another cluster) as direct photon candidates. No particle identification has been applied yet so that the single cluster sample contains a sizable fraction of charged particles which develop a shower in the calorimeters.

The azimuthal correlation between the trigger particle (charged particle, π^0 candidate, single cluster) and the charged hadrons are shown in Fig. 3. The near side ($\Delta\phi = 0$) and away side ($\Delta\phi = \pi$) peak are clearly observed. Note, however, that these distributions have not been corrected for efficiency. It is interesting to remark that at this very preliminary stage of the analysis we find that the underlying event background level, outside the peaks region, is independent of the type of trigger particles, giving some confidence in the measurement. By applying an isolation selection on the trigger candidate, where hadron activity carries less than 30 % transverse momentum of the trigger candidate inside a cone with size $R = 0.4$ required, the probability of direct photon or single particle jets in the sample enhances. Comparing the azimuthal correlation with and without isolation selection, obviously, a suppression of the near side peak is observed, but the away side peak is almost unaffected, as expected (Fig. 4). However this preliminary analysis does not allow to draw any conclusion other that these results indicate the expected behaviour. The isolation parameters are not well adjusted and especially in our case only charged tracks are considered in our isolation cone due to the limited calorimeter acceptance (40 % EMCAL and 60 % PHOS have been installed so far).

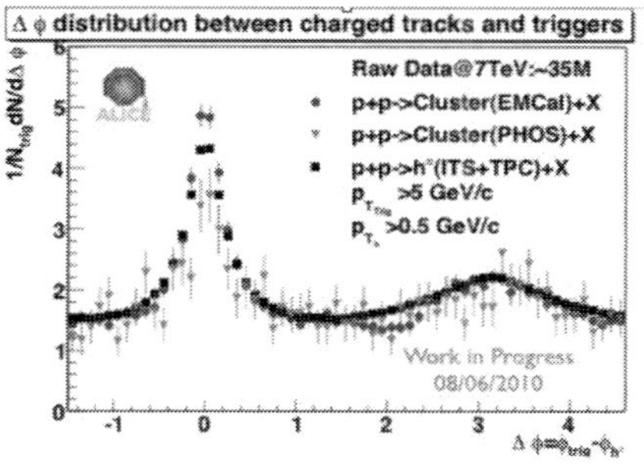

Figure 3. Relative azimuthal angle distribution $\Delta\phi = \phi_{trigger} - \phi_{hadron}$ in pp collisions at $\sqrt{s} = 7$ TeV.

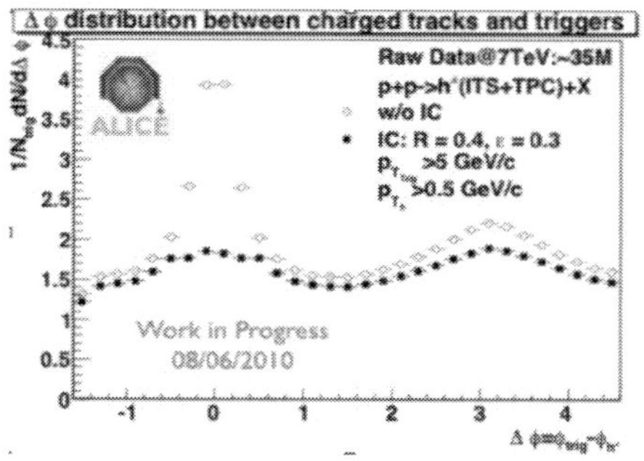

Figure 4. Azimuthal correlation distributions before and after isolation cut (IC) on the trigger particles with $p_T > 5GeV/c$

4. Summary and outlook

The feasibility to measure γ–hadrons correlation in pp collisions and medium modification effect in PbPb collisions with ALICE has been evaluated. Such a measurement provides an exclusive observable sensitive to the properties of the medium formed in heavy-ion collisions. So far only a preliminary analysis has been performed on a small fraction of the data collected by ALICE. Exciting physics will certainly come with the final analysis of large statistics with well calibrated detectors.

Acknowledgement

The work is partially supported by the NSFC (10875051, 10635020 and 10975061), the Key Project of Chinese Ministry of Education (306022), the Program of Introducing Talents of Discipline to Universities of China (QLPL200909, B08033 and CCNU09C01002).

5. References

[1] D. J. Gross and F. Wilczec 1973, *Phys. Rev. D* **8** 3633.
[2] E. V. Shuryak 1980, *Phys. Rep.* **61** 71.
[3] U. W. Heinz and M. Jacob 2000, arXiv:nucl-th/0002042.
[4] RHIC White paper 2005, *Nucl. Phys. A* **757** 1.
[5] K. Aamodt et al 2008, *JINST* **3** S08002.
[6] A. Morsch 2007,*Nucl. Phys. A* **783** 427.
[7] ALICE Collaboration 1999, http://alice.web.cern.ch/Alice/TDR/.
[8] T. M. Cormier 2004, *Eur. Phys. J C* **34** s333.
[9] Y. X. Mao et al., 2008, *Chinese Physics C* **32(07)**.
[10] Y. X. Mao et al. 2008, *Eur. Phys. J. C* **57** 613.
[11] T. Sjostrand et al. 2001, *Comput. Phys. Commun.* **135** 238.
[12] N. Armesto et al. 2009, hep-ph/0906.0754; *Eur. Phys. J C* **63** 679.
[13] G. Conesa et al. 2007 *Nucl. Instr. and Meth. Nucl. Res.* A **580** 1446.

Journal of Physics: Conference Series **270** (2011) 012033 doi:10.1088/1742-6596/270/1/012033

Studies of isolated photon production in simulated proton-proton collisions with ALICE-EMCal

Raphaëlle Ichou for the ALICE Collaboration
Subatech, 4 rue Alfred Kastler, BP 20722 44307 NANTES cedex 3, France
E-mail: raphaelle.ichou@cern.ch

Abstract. The production of prompt photons at high transverse momentum in proton-proton collisions (p-p) is a useful tool to study perturbative Quantum-Chromo-Dynamics (pQCD). In particular, they yield valuable information about parton distribution functions in the proton. The experimental measurement of prompt γ is a difficult task due to the large background of decay photons from neutral mesons, mainly π^0. We present a full simulation and reconstruction study of prompt γ identification in p-p at $\sqrt{s}$ = 14 TeV in the ALICE electromagnetic calorimeter EMCal, giving details on the methods developed to separate them from decay photons with the help of shower-shape and isolation cuts. We present Monte Carlo predictions for signal and background. The method used to extract the final isolated γ corrected cross-section is presented and the calculation of various experimental corrections is outlined.

1. Physics motivations

The study of photon production at large transverse momenta ($p_T \gg \Lambda_{QCD} = 0.2$ GeV) in hadronic interactions is a valuable testing ground of the perturbative regime of Quantum Chromodynamics (pQCD) [1]. At the LHC, photons will allow one to confront the data with pQCD predictions at energies never reached before. Since these photons come directly from parton-parton hard scatterings, they allow one to constrain the gluon distribution function in the proton at small parton momentum fraction $x = p_{parton}/p_{proton}$, down to 10^{-3} at mid-rapidity. Also, photons produced in p-p collisions provide a vacuum baseline reference for the study of their production rates in nucleus-nucleus collisions.
At lowest order, three partonic mechanisms produce prompt γ in hadronic collisions: (i) quark-gluon Compton scattering $qg \rightarrow \gamma q$, (ii) quark-antiquark annihilation $q\bar{q} \rightarrow \gamma g$, and (iii) the collinear fragmentation of a final-state parton into a photon, e.g. $qq \rightarrow qq \rightarrow \gamma X$. The photons produced in the two first point-like processes are called *direct*, the latter *fragmentation* photons. The Compton channel is particularly interesting as it provides information on the proton gluon distribution [2], which is otherwise only indirectly constrained via the "scaling violations" of the proton structure function in deep-inelastic scattering (DIS) e-p collisions [3].

The measurement of high-p_T photon production is complicated by a large γ background from hadrons, specially from π^0 mesons, which decay into two photons. At high p_T there are between 10 and 100 (at 100 GeV/c and 10 GeV/c respectively) times more π^0 than prompt γ. Furthermore, above 30 GeV/c, the two π^0 decay photons merge into a single cluster in EMCal and thus cannot be identified as a π^0 via γ-γ invariant mass analysis. In order to measure the prompt γ signal out of the overwhelming background of π^0, which are produced in the fragmentation of jets, one requires the photon candidate to be isolated from any hadronic activity within a given distance around its direction. The corresponding measurements are then dubbed *isolated* photons. A standard isolation requirement is that within a cone around the γ direction defined in pseudo-rapidity η and azimuthal angle ϕ by $R = \sqrt{(\eta - \eta_\gamma)^2 + (\phi - \phi_\gamma)^2}$, the accompanying hadronic transverse energy is less than a fixed fraction ε (e.g. often 10%) of the photon's p_T. R is usually taken between 0.4 and 0.7. Isolation enables one to reject a large fraction of the π^0 decay photons, without supressing too many photons of interest. The bigger R and the smaller ε, the more selective is the isolation. But the choice of optimal isolation criteria also needs to consider detector acceptance and Underlying Event (UE) limitations. The values choosen are $R = 0.2$ (for the first year of running at $\sqrt{s} = 7$ TeV, with only 4 Super-Modules of EMCal installed), $R = 0.4$ and $\varepsilon = 0.1$ (which e.g. removes UE-hadrons with p_T <2 GeV/c in events with a 20-GeV γ).

2. Monte Carlo predictions

The `JETPHOX` [4] program (for JET-PHOton/hadron X-sections) allows one to calculate perturbatively the γ cross-sections at Next to Leading Order accuracy. It includes both the fragmentation and the direct components and isolation cuts can be applied at the parton level. Figure 1 shows the expected isolated photons (R =0.2, ϵ=0.1) cross-sections at 7 TeV

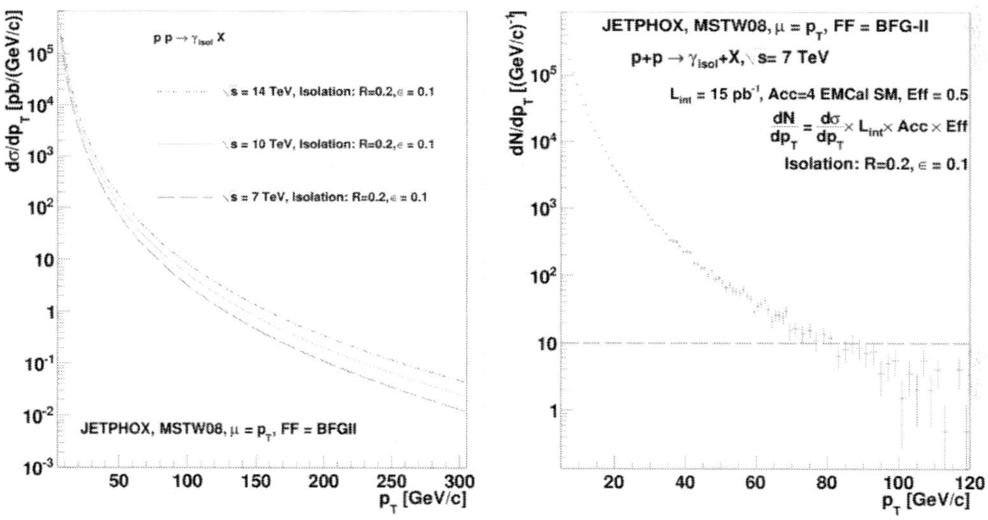

Figure 1. Differential cross-section as a function of p_T for isolated photons (R =0.2, ϵ=0.1) in p-p collisions at $\sqrt{s} = 7$, 10 and 14 TeV (left) in one unit of rapidity at mid-rapidity and expected isolated yields and p_T reach at $\sqrt{s} = 7$ TeV, for an integrated luminosity of L = 15 pb^{-1} (right).

and, for comparison, at 10 and 14 TeV (left) and the corresponding yields in p-p at 7 TeV (right). The expected p_T reach, which will be accessible in the ALICE EMCAL acceptance ($\Delta\eta$ = 1.4, $\Delta\phi$=1.1), during the first year of LHC running at $\sqrt{s} = 7$ TeV, considering an integrated luminosity[1] of L = 15 pb^{-1}, is found to be around 80 GeV/c. `PYTHIA 6.420` [5], with the Perugia hard and soft tunes [6], has been used to study the signal and the background, for p-p collisions at $\sqrt{s} = 14$ TeV. We have generated the different photon subprocesses: Compton

[1] We assume fully-efficient high-p_T EMCal cluster triggers with no scale-down reduction.

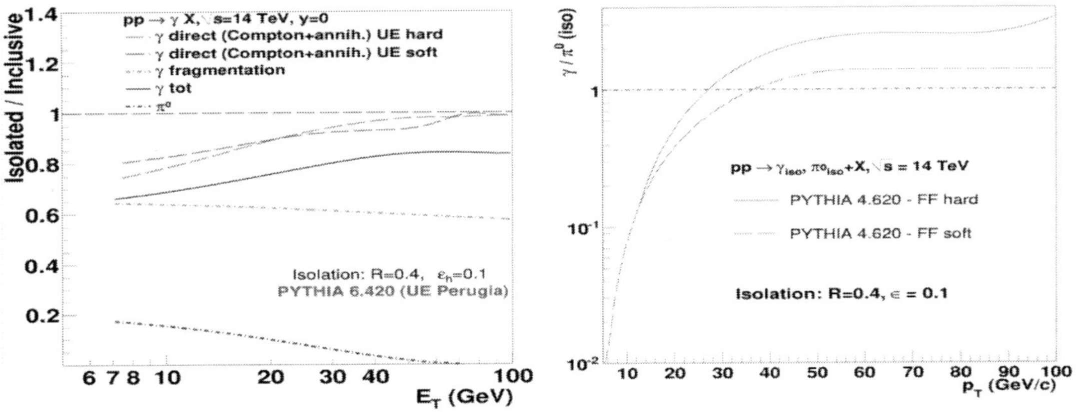

Figure 2. Fraction of isolated ($R = 0.4$, $\varepsilon = 0.1$) over inclusive photons and π^0 for p-p collisions simulated by PYTHIA at $\sqrt{s}$ =14 TeV (left), and signal (γ) over background (π^0) after isolation cuts for both the FF hard and soft PYTHIA tunes (right).

(MSUB = 29), annihilation (MSUB = 14) and fragmentation (MSEL = 14), as well as π^0 (MSEL = 1) events, over one unit of rapidity at $y = 0$, for different p_T bins: [5-20], [20-50], [50-100], [100-250], [250-500] and [500-1000] GeV/c, with 10 millions events in each. Isolation cuts ($R = 0.4$, $\varepsilon = 0.1$) have been applied on them. Figure 2 (left) shows the fraction of isolated over inclusive particles, for the different photon subprocesses and for π^0, at the MC level. For the direct γ, which are theoretically all isolated, we have studied the effect of the UE on the isolation by using two different PYTHIA tunes, hard and soft, which differ by their proportions of initial and final-state radiation, multiple interactions, beam remnants and parton-to-hadron fragmentation functions (FF). The fraction of isolated direct photons ($R = 0.4$, $\varepsilon = 0.1$) goes from 80% to 100% from 10 to 100 GeV. Thus, the UE leads to a maximum of 20% (±5%) of isolated direct photons loss at 10 GeV. The fraction of isolated fragmentation photons with PYTHIA is around 60%. This leads to a total amount of isolated over inclusive γ of about 70-80% over the full E_T range. The isolated π^0 fraction is represented by high-z π^0 which carry a large fraction of the jet energy ($z = p_{hadron}/p_{parton}$) and are thus isolated from accompanying hadronic activity. With PYTHIA we find a fraction of isolated π^0 going from 20% to 3% from 10 GeV to 60 GeV. High-z isolated π^0 are thus considered as the main background in our prompt isolated photon analysis.

Figure 2 (right) shows the MC-level predictions for the signal over background (S/B) with isolation cuts, for both PYTHIA tunes, in order to assess the systematic uncertainty linked to the choice of the FF. Isolation allows one to strongly enhance the signal over background ratio. Yet, with isolation cuts, the S/B remains smaller than 1 up to ~ 30 GeV/c. In order to suppress the remaining background due to isolated π^0, one thus needs to employ shower-shape cuts which improve the photon over π^0 ratio, and subtract statistically any final remaining π^0 background.

3. Analysis

The goal of this analysis is to obtain a fully corrected isolated prompt photon spectrum at 14 TeV. The isolated photon cross-section can be written as :

$$E\frac{d^3\sigma}{dp^3} = \frac{1}{2\pi p_T}\frac{1}{p(\gamma|\gamma)}\left(\frac{\text{N(clusters id. }\gamma\text{ and iso.)}}{A \times L \times \epsilon \times \Delta p_T \times \Delta\eta} - p(\gamma|\pi^0)\frac{d\sigma^{\pi^0_{iso}}}{dp_T}\right) \tag{1}$$

with N(clusters id. γ and iso.) the number of clusters identified as photons and isolated, which represents the Signal+Background, $p(\gamma|\gamma)$ the gamma identification efficiency [2] and $p(\gamma|\pi^0)$ the identification contamination due to π^0 identified as photon, L the integrated luminosity, A the acceptance term, and ε all the other corrections (reconstruction and UE isolation, shown in Fig. 2 left). For the acceptance term, we consider a fiducial cut corresponding to an isolation radius of $R = 0.4$, which represents 1/4 of EMCal acceptance. The acceptance with respect to full coverage for one unit of rapidity at mid-rapidity is thus found to be around 10%. Full simulation and reconstruction events for signal (γ-jet) and background (di-jets) at 14 TeV have been used to obtain all correction factors in Eq. (1). The clusters are identified by the shower shape

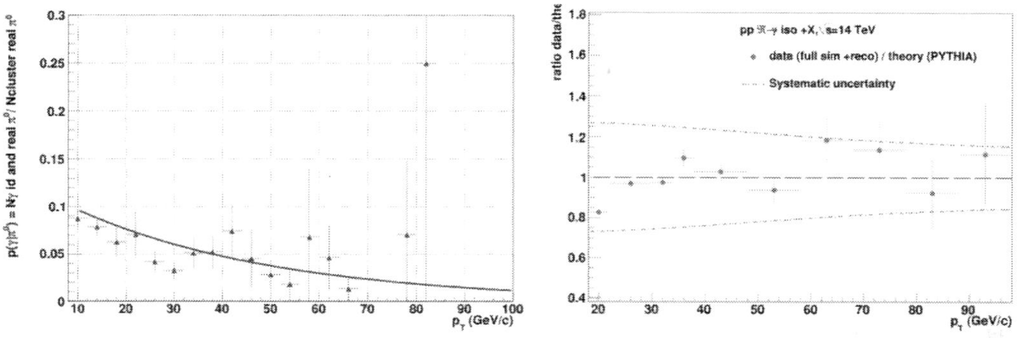

Figure 3. PID contamination of the π^0 identified as γ, $p(\gamma|\pi^0)$ (left), and ratio data/theory (full sim reco vs PYTHIA) for isolated γ ($R = 0.4$, $\varepsilon = 0.1$) in p-p at $\sqrt{s}$ =14 TeV (right).

Bayesian method [7]. A track-matching cluster–charged particle cut, using information from the central tracker, is also applied on the photon candidates. The photon identification efficiency, $p(\gamma|\gamma)$, is found to be constant at ~80% up to 45 GeV/c. The background contamination due to π^0's identified as γ, $p(\gamma|\pi^0)$, is found to be less than 10% above 10 GeV/c and decreasing with p_T (Fig. 3 left). Our shower-shape cuts remove, thus, a significant fraction of neutral pions. The reconstruction efficiency is ~90% up to ~80 GeV/c. The signal/background after photon reconstruction, identification and isolation is found to be greater than one above $p_T \approx 15$ GeV/c.

Finally, we use the PYTHIA spectrum of isolated π^0 to statistically remove any remaining contamination. This π^0 cross-section, weighted by the term $p(\gamma|\pi^0)$, is subtracted from the fully-corrected left term of Eq. (1) to obtain the final corrected isolated γ spectrum at 14 TeV. The systematic errors propagated on the final yields are dominated by the EMCal energy scale (about 15%) and the (theoretical) uncertainty in the subtracted isolated-π^0's (~20%). Fig. 3 (right) shows the final data/theory ratio, i.e. the ratio of the full sim+reco spectrum over the original PYTHIA distribution. The fully corrected spectrum is in good agreement with the input MC LO spectrum within the expected statistical (for 24 pb^{-1}) and systematic uncertainties.

References

[1] P. Aurenche, M. Fontannaz, J. P. Guillet, E. Pilon and M. Werlen, Phys. Rev. D **73** (2006) 094007
[2] R. Ichou and D. d'Enterria, Phys. Rev. D **82** (2010) 014015
[3] M. Klein and R. Yoshida, Prog. Part. Nucl. Phys. **61** (2008) 343
[4] JetPHOX, P. Aurenche et al., (http://lappweb.in2p3.fr/lapth/PHOX_FAMILY/jetphox.html)
[5] T. Sjostrand, S. Mrenna and P. Z. Skands, JHEP **0605**, 026 (2006)
[6] P. Z. Skands, arXiv:1005.3457 [hep-ph].
[7] Bellwied, R. et al., arXiv:1008.0413.

[2] Defined as a conditionnal probability of identifying a reconstructed cluster as a γ knowing that it is indeed a γ.

Hot Quarks 2010 IOP Publishing
Journal of Physics: Conference Series **270** (2011) 012034 doi:10.1088/1742-6596/270/1/012034

Medium-induced multi-photon radiation

Hao Ma, Carlos A. Salgado
Departamento de Física de Partículas, Universidade de Santiago de Compostela, E-15782 Santiago de Compostela, Spain

Konrad Tywoniuk
Lund University

Abstract. We study the spectrum of multi-photon radiation off a fast quark in medium in the BDMPS/ASW approach. We reproduce the medium-induced one-photon radiation spectrum in dipole approximation, and go on to calculate the two-photon radiation in the Molière limit. We find that in this limit the LPM effect holds for medium-induced two-photon ladder emission.

1. Introduction

To date, the medium-modification of the jet fragmentation function has only been calculated for the single-particle emission spectrum. In QED, the one-photon radiation off an energetic quark/electron propagating a dense, finite-sized medium was calculated as a multiple scattering process in [1]. In the context of QCD, the corresponding gluon radiation process, including multiple gluon interactions, was calculated in [2, 3, 4, 5]. In general, since it does not interact, the photon spectrum is hardly suppressed by formation time effects, in contrast to the gluon case. Thus, while the suppression of high-$p_\perp$ hadrons serves as a probe [6] of the hot and dense medium created in the wake of a heavy-ion collision [7, 8, 9, 10], photon production is used as a calibration due to its lack of interaction with the surrounding QCD matter [11].

On the other hand, medium-modifications of more exclusive observables, e.g., two-particle emission, are hitherto unknown. Particularly in vacuum, the latter process comprise key features such as the energy and angular ordering of QCD radiation which arise due to interference between emitters. In QED, the lack of ordering of the radiation can be traced back to the fact that the photon does not carry charge. On the contrary, intensive medium-interactions of the quark could induce unprecedented interferences between subsequent emissions which, in principle, could be measured by experiment. Furthermore, the two-photon spectrum is expected to be less involved than the corresponding QCD process due to the sterile nature of the photons, and thus serves also as a first attempt at computing multi-particle radiation in medium.

Following the BDMPS/ASW approach [1, 2, 3, 4, 5], we focus on a highly energetic quark traversing the medium with energy E_q, which radiates a soft photon with energy ω and transverse momentum $k_\perp$ ($E_q \gg \omega \gg k_\perp$). The medium is modeled as a collection of static scattering centers, given by the gauge field $A^a(x)t^a$ described by a Yukawa potential with the Debye mass serving as the inverse screening length, where t^a is the color matrix in the fundamental representation. Due to the interaction with the medium, the hard quark acquires an eikonal phase which is described by the Wilson line. Additionally, due to the Bloch-Nordsieck structure of the QED radiation vertex, we need to take into account corrections from the Brownian motion of the quark in the transverse plane.

2. Medium-induced one-photon radiation

Here we are interested in the leading contribution to the spectrum in the high-energy approximation. In QED, one has to keep both order $\mathcal{O}((1/p_+)^0)$ and order $\mathcal{O}(1/p_+)$ terms in the radiation vertex, where $p_+ \simeq \sqrt{2}E_q$ is the quark light cone momentum. Accordingly, we derive the spin non-flip medium-induced one-photon radiation spectrum for a finite medium of length L, as in [1].

For the purpose of illustration, let us consider spectrum for a fixed number of medium scatterings in the Bethe-Heitler and factorization limits. In the first case, the N scattering centers are well-separated, thus the radiation off these scattering centers is the sum of N Bethe-Heitler spectra, given by

$$\frac{\mathrm{d}^5 I^{\mathrm{BH}}}{\mathrm{d}^2\mathbf{p}_{f\perp}\mathrm{d}^2\mathbf{k}_\perp \mathrm{d}(\ln x)} \propto \frac{\alpha_{em}}{\pi^3}\sum_i \frac{x^2\,\mathbf{q}_{i\perp}^2}{\mathbf{k}_\perp^2(\mathbf{k}_\perp - x\,\mathbf{q}_{i\perp})^2}, \tag{1}$$

where $\mathbf{p}_{f\perp} = \sum_i \mathbf{q}_{i\perp} - \mathbf{k}_\perp$ and $\mathbf{k}_\perp$ are the transverse momentum of the outgoing quark and emitted photon, respectively, while x is given by $\omega = xE_q$, $\mathbf{q}_{i\perp}$ is the momentum transfer from the ith scattering center, and finally α_{em} is the electromagnetic coupling. In the factorization limit, on the other hand, the scattering centers are not resolvable, thus the resulting spectrum is given by

$$\frac{\mathrm{d}^5 I^{\mathrm{fact}}}{\mathrm{d}^2\mathbf{p}_{f\perp}\mathrm{d}^2\mathbf{k}_\perp \mathrm{d}(\ln x)} \propto \frac{\alpha_{em}}{\pi^3}\frac{x^2\,(\sum_i \mathbf{q}_{i\perp})^2}{\mathbf{k}_\perp^2(\mathbf{k}_\perp - x\sum_i \mathbf{q}_{i\perp})^2}. \tag{2}$$

When $i = 1$, we get the expression of the opacity expansion at first order from both of these two limits. The full medium-induced one-photon radiation spectrum interpolates between the above two limits. This is how one incorporates the Landau-Pomeranchuk-Migdal (LPM) effect [12, 13], which accounts for the fact that the scattering centers in medium interfere destructively during the formation time of an on-shell photon.

The coherent nature of the spectrum becomes most pronounced in the so-called Molière limit, i.e. i) for soft and small angle photon emission keeping the transverse momentum of the photon much larger than the total transverse momentum transfer from the medium, ii) in case of large length and iii) small density of the medium at hand. In this approximation, we can perform the sum over all multiple soft scatterings and the spectrum then reads

$$\frac{\mathrm{d}^5 I^{\mathrm{Mol}}}{\mathrm{d}^2\mathbf{p}_{f\perp}\mathrm{d}^2\mathbf{k}_\perp \mathrm{d}(\ln x)} \approx \frac{\alpha_{em}}{\pi^3}\frac{x^2}{\mathbf{k}_\perp^4}\exp\left\{-\frac{(\sum_i \mathbf{q}_{i\perp})^2}{2\,n_0\,C\,L_+}\right\} = \frac{\alpha_{em}}{\pi^3}\frac{x^2}{\mathbf{k}_\perp^4}\exp\left\{-\frac{(\mathbf{p}_{f\perp}+\mathbf{k}_\perp)^2}{2\,n_0\,C\,L_+}\right\} \tag{3}$$

where n_0 is the density of the medium, C is the probability of the hard quark scattering and $L_+ = \sqrt{2}L$ is the size of the medium. In eq. (3), we still find the characteristic x^2 rapidity dependence, accompanied by the $1/\mathbf{k}_\perp^4$ drop, as for the Bethe-Heitler and factorization spectra in the limit $k_\perp \gg q_\perp$. Yet, where as the two latter vanish in the limit $q_\perp \to 0$, the Molière spectrum peaks at zero momentum transfer. This illustrates the fact that the radiation spectrum for multiple soft scattering retains features which cannot be obtained from a calculation at fixed order in opacity.

3. Medium-induced multi-photon radiation

We go on to calculate the radiation of two photons inside the medium in both the amplitude and the conjugate amplitude for the situation of identical sequence of radiations in the amplitude and conjugate amplitude, respectively. We denote this particular contribution as a *ladder* emission.

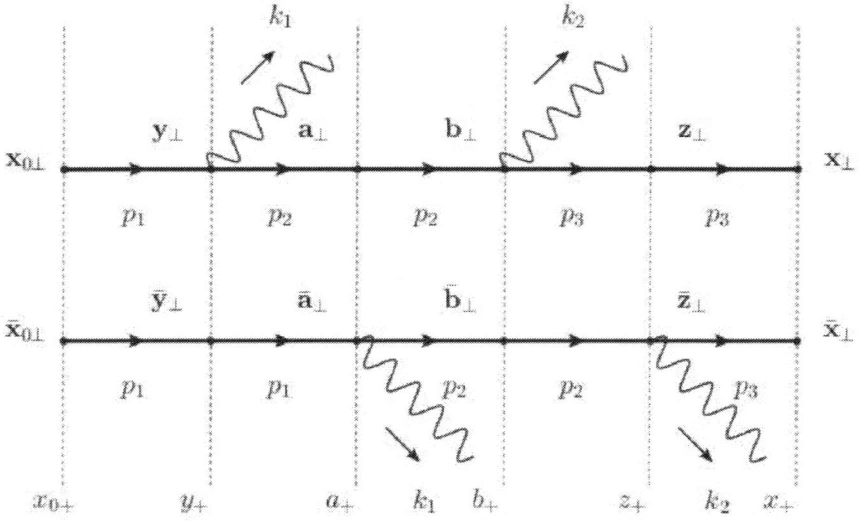

Figure 1: The radiations are inside the medium in both amplitude and conjugate amplitude. Here $p_1 = p_+$, $p_2 = (1-x)\,p_+$ and $p_3 = (1-y)\,(1-x)\,p_+$.

The corresponding diagram is depicted in Figure 1. In particular, first a photon is radiated with transverse momentum $\mathbf{k}_{1\perp}$ and carrying a momentum fraction x of the parent quark, and subsequently another photon with transverse momentum $\mathbf{k}_{2\perp}$ and carrying a momentum fraction y of the quark. In this case, the expression reads

$$\begin{aligned}
\frac{\mathrm{d}^8 I}{\mathrm{d}^2\mathbf{p}_{f\perp}\mathrm{d}^2\mathbf{k}_{1\perp}\mathrm{d}^2\mathbf{k}_{2\perp}\mathrm{d}(\ln x)\mathrm{d}(\ln y)} &= \frac{(\alpha_{em})^2}{(2\,\pi)^6}\,\mathrm{Re}\,\frac{2}{x\,y\,(1-x)^2\,(p_+)^4} \\
&\times \int_{x_{0+}}^{b_+}\mathrm{d}y_+\int_{y_+}^{b_+}\mathrm{d}a_+\int_{y_+}^{x_+}\mathrm{d}b_+\int_{b_+}^{x_+}\mathrm{d}z_+\int \mathrm{d}^2\rho_{1\perp}\mathrm{d}^2\rho_{2\perp}\mathrm{d}^2\rho_{4\perp} \\
&\times \exp\left\{-\int_{x_{0+}}^{y_+}\mathrm{d}\xi_1\,\Sigma(\xi_1,x\,\rho_{1\perp}) - \int_{a_+}^{b_+}\mathrm{d}\xi_3\,\Sigma(\xi_3,x\,\rho_{2\perp}) - \int_{z_+}^{x_+}\mathrm{d}\xi_5\,\Sigma(\xi_5,y\,\rho_{4\perp})\right\} \\
&\times \exp\left\{-i\,y\left(\mathbf{p}_{f\perp} - \frac{1-x}{x}\,\mathbf{k}_{1\perp} - \frac{1-y}{y}\,\mathbf{k}_{2\perp}\right)\cdot\rho_{4\perp} - i\,x\left(\frac{\mathbf{k}_{1\perp}}{x} + \frac{\mathbf{k}_{2\perp}}{y}\right)\cdot\rho_{1\perp}\right\} \qquad (4)\\
&\times \exp\left\{i\left(\bar{q}_1 - \frac{\mathbf{k}_{1\perp}\cdot\mathbf{k}_{2\perp}}{y\,(1-x)\,p_+}\right)(y_+ - a_+)\right\}\exp\left\{i\left(\bar{q}_2 - \frac{\mathbf{k}_{1\perp}\cdot\mathbf{k}_{2\perp}}{x\,p_+}\right)(b_+ - z_+)\right\} \\
&\times \left(\frac{\partial}{\partial\rho_{1\perp}} - i\,\frac{x}{y}\,\mathbf{k}_{2\perp}\right)\cdot\left(\frac{\partial}{\partial\rho_{2\perp}} + i\,\frac{x}{y}\,\mathbf{k}_{2\perp}\right)\mathcal{K}(\rho_{1\perp},y_+;\rho_{2\perp},a_+|\mu_1) \\
&\times \left(\frac{\partial}{\partial\rho_{2\perp}} - i\,(1-x)\,\mathbf{k}_{1\perp}\right)\cdot\left(\frac{\partial}{\partial\rho_{4\perp}} + i\,\frac{(1-x)\,y}{x}\,\mathbf{k}_{1\perp}\right)\mathcal{K}\left(\frac{x}{y}\,\rho_{2\perp},b_+;\rho_{4\perp},z_+|\mu_3\right),
\end{aligned}$$

where $\bar{q}_1 = \frac{x\,m_q^2}{2\,(1-x)\,p_+}$ and $\bar{q}_2 = \frac{y\,m_q^2}{2\,(1-y)\,(1-x)\,p_+}$ are the reciprocals of the photon formation lengths, m_q is the mass of the quark and $\Sigma(\xi_1, x\,\rho_{1\perp}) = n(\xi_1)\sigma(x\,\rho_{1\perp})/2$ is the product of the elastic

Mott cross section and the density of scattering centers in the medium. Finally, $\mu_1 = (1-x)x\,p_+$ and $\mu_3 = (1-y)y(1-x)p_+$. The path integrals are given by

$$\mathcal{K}(\rho_{1\perp}, y_+; \rho_{2\perp}, x_+|\mu_1) = \int \mathcal{D}\mathbf{r}_\perp \exp\left\{\frac{i\,\mu_1}{2}\int_{y_+}^{x_+} \mathrm{d}\xi_2\left[\dot{\mathbf{r}}_\perp^2 - \Sigma(\xi_2, x\,\mathbf{r}_\perp)\right]\right\}, \quad (5)$$

where $\mathbf{r}_\perp(y_+) = \rho_{1\perp}$ and $\mathbf{r}_\perp(x_+) = \rho_{2\perp}$ are the boundary conditions. In the dipole approximation, where $\sigma(x\,\rho_{1\perp}) \approx Cx^2\rho_{1\perp}^2$, the path integral is given as a solution to the harmonic oscillator with imaginary frequency $\Omega_1 = (1-i)\sqrt{n_0\,C\,x^2/(2\,\mu_1)}$ [1], namely

$$\begin{aligned}\mathcal{K}_{osc}(\rho_{1\perp}, y_+; \rho_{2\perp}, a_+|\mu_1) &= \frac{\mu_1\,\Omega_1}{2\,\pi\,i\,\sin(\Omega_1\,(a_+ - y_+))} \\ &\times \exp\left\{\frac{i\,\mu_1\,\Omega_1\left[(\rho_{1\perp}^2 + \rho_{2\perp}^2)\cos(\Omega_1\,(a_+ - y_+)) - 2\,\rho_{1\perp}\cdot\rho_{2\perp}\right]}{2\,\sin(\Omega_1\,(a_+ - y_+))}\right\}. \end{aligned} \quad (6)$$

In this case, and furthermore in the Molière limit, one can perform all integrations in eq. (4) analytically. Thus, the spectrum reads

$$\frac{\mathrm{d}^8 I^{\mathrm{Mol}}}{\mathrm{d}^2\mathbf{p}_{f\perp}\mathrm{d}^2\mathbf{k}_{1\perp}\mathrm{d}^2\mathbf{k}_{2\perp}\mathrm{d}(\ln x)\mathrm{d}(\ln y)} \propto \exp\left\{-\frac{(\mathbf{k}_{1\perp} + \mathbf{k}_{2\perp} + \mathbf{p}_{f\perp})^2}{2\,n_0\,C\,L_+}\right\}, \quad (7)$$

where $|\mathbf{k}_{1\perp} + \mathbf{k}_{2\perp} + \mathbf{p}_{f\perp}| = \sum_i \mathbf{q}_{i\perp}$ is the total momentum transfer from the medium. One can draw several important conclusions from this, seemingly simple, final result. The appearance of the Molière factor in eq. (7) signals that the LPM effect still holds for the medium induced two-photon ladder emission. In particular, eq. (7) is not simply a superposition of two one-photon spectra, as one naively would have expected from independent radiation, e.g., in vacuum. Thus, the medium induces some coherence on the multi-emission spectrum. The details of these coherence effects are still under investigation.

4. Summary and outlook

We present the study of the spectrum of one- and two-photon radiation off a quark in medium. In the Molière limit, we find similar features for both, namely the appearance of Gaussian distribution of the total momentum transfer from the medium. In the latter case, this is quite surprising as the corresponding vacuum spectrum is a superposition of two independent emissions.

The above conclusions hold, strictly speaking, only for the ladder emission diagram, depicted in Figure 1. For the two-photon case, there appears Feynman diagrams with much more complicated emission sequences, in addition to the diagrams including interference with vacuum emissions. We are still working on a complete, analytical solution of all diagrams.

References

[1] U. A. Wiedemann and M. Gyulassy, Nucl. Phys. B **560** (1999) 345-382.
[2] R. Baier, Yu. L. Dokshitzer, S. Peigné and D. Schiff, Phys. Lett. **345B** (1995) 277.
[3] U. A. Wiedemann, Nucl. Phys. B **582**, 409 (2000).
[4] C. A. Salgado and U. A. Wiedemann, Phys. Rev. D **68**, 014008 (2003).
[5] N. Armesto, C. A. Salgado and U. A. Wiedemann, Phys. Rev. D **72**, 064910 (2005).
[6] C. A. Salgado, Prog. Theor. Phys. Suppl. **168**: 355-363, 2007.
[7] K. Adcox *et al.* [PHENIX Collaboration], Nucl. Phys. A **757**, 184 (2005).
[8] B. B. Back *et al.* [PHOBOS Collaboration], Nucl. Phys. A **757**, 28 (2005).
[9] I. Arsene *et al.* [BRAHMS Collaboration], Nucl. Phys. A **757**, 1 (2005).
[10] J. Adams *et al.* [STAR Collaboration], Nucl. Phys. A **757**, 102 (2005).
[11] S. S. Adler *et al.* [PHENIX Collaboration], Phys. Rev. Lett. **94**, 232301 (2005).
[12] L. D. Landau and I. Ya. Pomeranchuk, Dokl. Akad. Nauk. SSSR **92** (1953) 535, 735.
[13] A. B. Migdal, Phys. Rev. **103** (1956) 1811.

Journal of Physics: Conference Series **270** (2011) 012035 doi:10.1088/1742-6596/270/1/012035

π^0 and η reconstruction from photon conversions in ALICE for first p-p collisions at the LHC.

Kenneth Aamodt, for the ALICE collaboration
Department of Physics, University of Oslo, Sem Sælandsvei 24, N-0316 Oslo
E-mail: Kenneth.Aamodt@cern.ch

Abstract. Measurements of π^0 and η inclusive spectra provide reference data for upcoming heavy ion runs, as well as a check on the applicability of perturbative QCD calculations at LHC energies. The high-resolution central tracking system of ALICE can be used to reconstruct π^0 and η through photon conversions, as an alternative to direct measurement in the ALICE Calorimeters. Knowledge of the ALICE material budget is crucial for the extraction of the absolute yield from the conversion technique. The statistics availible from this technique are comparable to that of the photon spectrometer (PHOS), and the reconstruction method can also be applied for Pb-Pb collisions. Moreover, the implementation of a photon conversion trigger in the High Level Trigger (HLT) framework gives the possibility of identifying π^0 and η candidates online, and increasing the statistics at higher momentum. The status of the π^0 meson reconstruction from photon conversions from p-p collisions at $\sqrt{s}$ = 900 GeV and $\sqrt{s}$ = 7 TeV is presented in this article.

1. Introduction

Measurements of π^0 and η inclusive p_t spectra in proton-proton collisions are important since they provide the essential reference for jet quenching measurements in heavy ion collisions via inclusive hadron suppression. In addition, they provide a test of the applicability of perturbative QCD (pQCD) calculations at the highest energies ever recorded. Precise knowledge of these spectra is also essential to extract direct photon yields from the large background of decay photons from mainly π^0 and η decays. In heavy-ion collisions direct photons carry thermodynamical information on the quark-gluon plasma (QGP). Results from RHIC [1] have shown that π^0s and ηs are supressed by a factor 5 in heavy-ion collisions compared to the binary scaled proton-proton yield in the nuclear modification factor R_{AA}, while direct photons show no suppression. The suppression is an indication that a dense medium is created in heavy ion collisions, the QGP, in which colored partons (jets) interact while the electromagnetically interacting direct photons escape without modification. The measurements of the π^0 and η inclusive p_t spectra in proton-proton collisions will provide a baseline for heavy-ion collisions, where their suppression can be studied through the nuclear modification factor.

ALICE [2] has two calorimeters, PHOton Spectrometer (PHOS) and ElectroMagnetic CALorimeter (EMCAL), designed to measure photons, neutral mesons, electrons and jets. ALICE also has a high resolution central barrel tracking system capable of reconstructing photons that convert in the detector material. The tracking system consists of Inner Tracking System (ITS) [3], Time Projection Chamber (TPC) [4], Transition Radiation Detector (TRD) and Time Of Flight (TOF). The ITS and TPC are used in the work presented here. Figure 1 depicts the full ALICE detector.

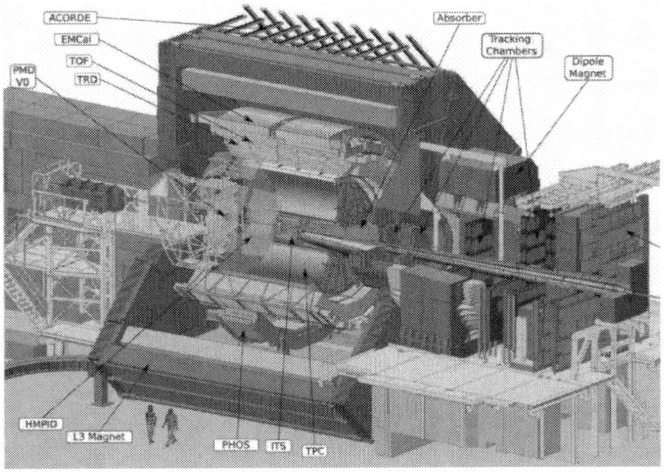

Figure 1. The ALICE detector, the only LHC experiment designed for heavy-ion collisions. The central barrel detectors used in this work are situated around the interaction point and consist of ITS and TPC.

2. Reconstruction

This work presents results from the two channels $\pi^0 \rightarrow \gamma\gamma \rightarrow e^+e^-e^+e^-$ and $\eta \rightarrow \gamma\gamma \rightarrow e^+e^-e^+e^-$. Both π^0 and η have the two photon decay channel as their most abundant one, with branching ratios of 0.99 and 0.39, respectively. Both decay photons are required to convert to e^+ e^- pairs in the detector material, support structures or gas. The acceptance of ALICE for reconstructing conversions is $|\eta| < 0.9$ and radius $r < 180$ cm, corresponding to full tracking in η for conversion daugthers. For $r > 180$ cm the e^+ and e^- tracks are too short to reconstruct the γ with high presicion.

Reconstruction of the photons is done with a general $V0$ reconstruction package which reconstructs Λ, $\overline{\Lambda}$, and K^0_S, as well as conversion photons. Several cuts need to be applied to select the photons from the total $V0$ sample. A series of quality cuts are applied to the tracks and $V0$ properties in order to obtain a clean sample of conversion photons. The most efficient cuts are listed in Table 1.

Table 1. Most significant cuts used to extract photons from $V0$ sample. The numbers in square brackets corresponds to the cut value used for the 900 GeV and 7 TeV data, respectively.

$\chi^2 < [30, 50]$ χ^2 of $V0$ reconstruction package fit function
$dE/dx(e) - [3, 10]\sigma_{(dE/dx)} < dE/dx < dE/dx(e) + [5, 10]\sigma_{(dE/dx)}$ dE/dx of tracks used in the $V0$ combination is required to be higher than $[3, 10]\sigma$ below the dE/dx electron line, and lower than $[5, 10]\sigma$ above the dE/dx electron line
$dE/dx > dE/dx(\pi)$, $p > 1\text{GeV}/c$ dE/dx of tracks must be higher than the dE/dx pion line for $p > 1$ GeV/c

3. Results

Using the reconstructed conversion point locations, it is possible to make a map of where the material of ALICE is located. Comparison between real data and simulated data shows very good agreement on the material budget. Studies using simulated data have been performed to determine the conversion probability, which is 8% within the acceptance. The simulated data also allow for checking the reconstruction efficiency. It is defined as the number of reconstructed photons divided by number of converted photons. For the cuts used in this analysis the photon reconstruction efficiency is found to be 69% for p_t larger than 1.1 GeV/c. This gives a maximum π^0 and η reconstruction efficiency of 0.0030.

In Figure 2 the two-photon invariant mass distribution is shown for 900 GeV minimum bias p-p collisions, using a sample of 3.5×10^6 events. The black points are the data, and a clear π^0 peak can be seen. In blue the combinatorial background, determined by a mixed event technique in which photons from different events are combined in the same z-vertex and charged track multiplicity classes, is scaled to the total distribution in the region [0.6 GeV/c^2, 0.8 GeV/c^2]. The background describes the data very well. Figure 3 shows the same distribution for 7 TeV minimum bias p-p collisions, where both a clear π^0 and η peak can be seen.

Less strict cuts were used to extend the measurement to higher p_t, also resulting in more background. Investigation is ongoing to understand why the background is not describing the data between the π^0 and η peaks, currently believed to be an effect of the less strict cuts.

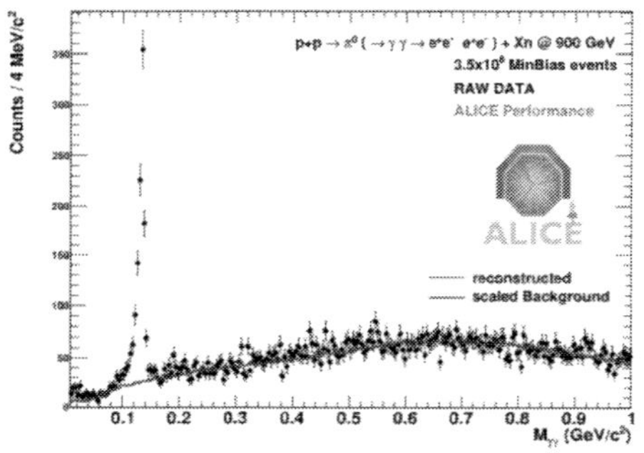

Figure 2. Invariant mass distribution for 3.5×10^6 minimum bias p-p events at 900 GeV.

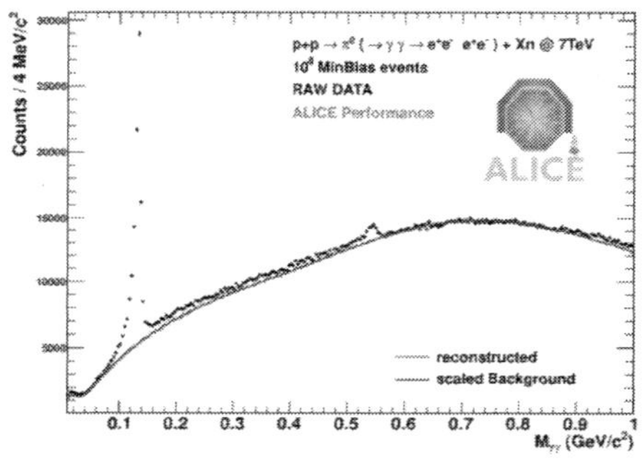

Figure 3. Invariant mass distribution for 10^8 minimum bias p-p events at 7 TeV.

The invariant mass distributions have also been divided into several p_t bins to extract the raw yield of the π^0. A clear π^0 peak was seen in p_t bins ranging from p_t of 0.4 GeV/c to 3.4 GeV/c for the 900 GeV data, and 0.4 GeV/c to 10 GeV/c for the 7 TeV data. For the 900 GeV/c data the signal-to-background (S/B) ranges between 2 and 6 for the four p_t bins, with significance between 35 and 60. The 7 TeV data were divided into 13 p_t bins where the S/B ranged between 1.5 for the lowest and highest p_t bins up to 4.5 for the midrange bins. The significance was between 10 and 350, where for most of the p_t bins it was above 100.

To fit the peaks in the different p_t bins Equation 1 is used. It is a Gaussian function modulated with an exponential to account for the bremsstrahlung tail.

$$y = G + \exp\left(\frac{M_{\gamma\gamma} - m_{\pi^0}}{\lambda}\right)(1-G)\theta(m_{\pi^0} - M_{\gamma\gamma}), \quad G = \exp\left(-\frac{1}{2}\left(\frac{M_{\gamma\gamma} - m_{\pi^0}}{\sigma}\right)^2\right) \quad (1)$$

The peak position is systematically around 0.5 MeV/c^2 below the PDG value of 135 MeV/c^2 due to calibration. The resolution of the peaks is very good, ranging from 3-5 MeV.

In Figure 4 the raw yield of π^0 is shown for both the 900 GeV data in blue points, and for the 7 TeV data in red points. The error bars are statistical. The p_t range achieved for the 900 GeV data is from 0.4 GeV/c to 3.4 GeV/c. For the 7 TeV data it ranges from 0.4 GeV/c to 10 GeV/c.

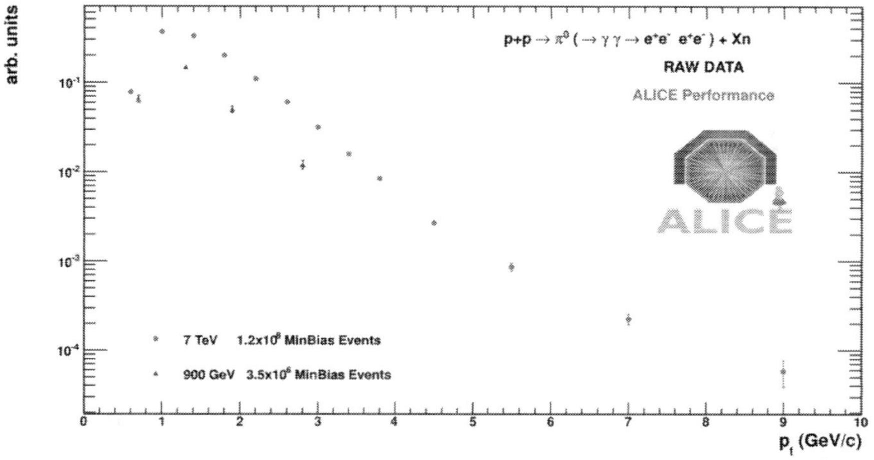

Figure 4. Raw yield of π^0 in arbitrary units versus p_t for 900 GeV data in blue, and 7 TeV data in red. Different sets of cuts have been used for the 900 GeV and the 7 TeV data.

4. Summary and outlook

It has been shown that measuring photons using reconstructed e^+e^- pairs from conversions can be carried out with high precision. Reconstructed conversion points provide information on the material budget of ALICE. Very good knowledge of the material budget is achieved so far, and further discussions with individual subdetector groups are ongoing to improve this agreement further. π^0 and η mesons can be reconstructed via converions with a mass resolution of 3-5 MeV, and very high significance. These measurements extend the measurement to low p_t, 0.4 GeV/c, below the reach of the calorimeter approach. A raw yield for the π^0 meson has been shown as a function of p_t for 900 GeV and 7 TeV data.

In the near future statistics at high p_t will be enriched using a conversion trigger that has been developed and is undergoing testing in the High Level Trigger (HLT), the online reconstruction framework of ALICE. In addition, the implementation of a conversion trigger in the TRD at hardware level is planned. PID can be improved further by using information from TOF and TRD, and using cuts in q_t from Armenteros-Podolanski plots. Fully corrected π^0 and η inclusive spectra will soon be available. Reconstruction using photon conversion from Dalitz decay is ongoing, and a study in which calorimeter and photon conversion measurements are combined has been started.

References

[1] RHIC white papers: Nuclear Physics A Volume 757, Issues 1-2 , 8 August 2005, Pages 1-27 (BRAHMS), Pages 28-101 (PHOBOS), Pages 102-183 (STAR) and Pages 184-283 (PHENIX).
[2] ALICE TDR, URL http://www.cern.ch/Alice/TDR
[3] Alignment of the ALICE Inner Tracking System with cosmic-ray tracks(March 6, 2010) ALICE collaboration 2010. Jinst 5 P03003
[4] The ALICE TPC, a large 3-dimensional tracking device with fast readout for ultra-high multiplicity events. J. Alme et al. Jan 2010. arXiv:1001.1950

Journal of Physics: Conference Series **270** (2011) 012036 doi:10.1088/1742-6596/270/1/012036

Quarkonia propagation in QGP: study of elastic and inelastic scattering processes

H. Berrehrah*, **J. Aichelin**, **PB. Gossiaux****
SUBATECH, Université de Nantes, EMN, IN2P3/CNRS, 4 rue Alfred Kastler, 44307 Nantes cedex 3, France.
E-mail: *berrehra@subatech.in2p3.fr, **gossiaux@subatech.in2p3.fr

Abstract. We propose to study the quarkonia (ϕ) propagation in the QGP. We are especially interested in the elastic and inelastic scattering process of these quarkonia in the medium. We developed the Bethe-Salpeter formalism to calculate the elastic cross section (σ_{elas}) for $\phi - gluon/hadron$. Results obtained in this work show that $\sigma_{elas}(\phi - gluon/hadron)$ might have non negligible effects in the study of $Q\bar{Q}$ propagation.

1. Introduction

Among the possible probes of the quark gluon plasma (QGP) properties, quarkonia (ϕ) is an unavoidable observable. Indeed, since 20 years, the suppression of quarkonia created in the first moments of the nucleus-nucleus collision has been studied, either by screening mechanism "Debye mechanism" (Matsui and Satz [1]) or via the gluo-dissociation of ϕ states that remain strongly bound (Kharzeev [2]). Nevertheless, the comparison between experimental data and scenarios including suppression mechanisms is not very conclusive.

Rather than focusing only on these "suppressed" quarkonia, an alternative track is to focus also on the physical properties of quarkonia survivors in QGP. In fact, recent results of NA50 and NA60 experiments of nucleus-nucleus collisions at the SPS showed that these quarkonia indeed carry of a non-zero (positive) elliptic flow. This could lead us to reconsider the propagation of these quarkonia in the QGP, especially the elastic scattering process of ϕ with the QGP partons.

In the present study, we investigate the elastic and inelastic scattering of quarkonia in the QGP and in hadronic phase. First, we review briefly the historical models for calculating $\phi - gluon/hadron$ elastic and inelastic scattering cross sections. Then, the calculation of $\phi + g/q \rightarrow \phi' + g'/q'$ elastic cross section using the formalism equivalent to 1st order of pQCD (Bethe-Salpeter formalism) is described. Finally, we discuss the results.

2. σ_{elas}, σ_{inel} calculation formalisms

Several theoretical models were developed to calculate $\sigma_{elas}(\phi - g/h)$ and $\sigma_{inel}(\phi - g/h)$. Bhanot and Peskin [3] developed a formalism to evaluate quarkonia-partons interactions, based on the application of pQCD. The interactions are taken as a field-dipole point of view. Hadron dependent models were also used to calculate $\sigma_{inel}(\phi - hadron)$ (meson exchange model [4], QCD sum rules [5] ...). For $\sigma_{elas}(\phi - g/h)$ calculation, one can cite for instance, Hüfner et al [6], [7] who treat the elastic scattering of quarkonia in the medium as an interaction between the $Q\bar{Q}$ and a background of gluonic field, i.e interactions with successive color field.

We show in this paper that Bhanot and Peskin formula can be rederived using Bethe-Salpeter (BS) formalism [8]. This method is equivalent to the 1st order in pQCD and is based on the factorization theorem combined with the knowledge of the BS amplitudes of the bound states. Oh et al [9] have used this BS formalism to calculate $\phi - gluon/hadron$ inelastic cross section. The main motivation of our calculation is that the predictions for the value of $\sigma_{elas}(\phi - g/h)$ differ significantly depending on the models, without being systematically related to its counterpart $\sigma_{inel}(\phi - g/h)$, and that a common framework and a more precise knowledge of both of those cross section is primordial to study the quarkonia propagation and $Q\bar{Q}$ evolution in the environment.

3. Bethe-Salpeter Formalism for $\sigma_{elas}(\phi - g/h)$ and $\sigma_{inel}(\phi - g/h)$ calculation

The BS formalism allows us to determine the BS amplitudes for the bound states vertices, which are incorporated in the Feynman diagrams of $\phi - g/h$ elastic and inelastic scattering process.

The BS formalism is guided by the BS equation. This equation describes the diffusion processes as an infinite sum of Feynman diagrams contributing to the reaction. The diffusion amplitude $\mathcal{M}$ of these Feynman diagrams is given as a product of a kernel $\mathcal{V}$ and a propagator $\mathcal{G}$ (1). If $\mathcal{V}$ is small, $\mathcal{M}$ can be written as a convergent geometric series of Born, where each term in the series corresponds to a Feynman diagram.

A bound state produces a pole in the matrix diffusion $\mathcal{M}$. The condition for the existence of a pole in $\mathcal{M}$ is that the matrix $\mathcal{VG}$ has an eigenvalue equal to 1, the corresponding eigenvector "Γ" satisfies $\Gamma = \mathcal{VG}\ \Gamma$. From this condition, we deduce the BS equation for the bound state ϕ (2) (see also Fig.1), where Γ is the BS vertex function, $C_{color} = (N_c^2 - 1)/2N_c$. $p_1(-p_2)$ is the 4-momentum of q ($\bar{q}$), $q = p_1 + p_2 (p = (p_1 - p_2)/2)$ is the CM (relative) 4-momentum of ϕ, with $q_0 = M_\phi = 2m + \epsilon_0$ ($\epsilon_0 < 0$: binding energy), and finally $\Delta(p) = 1/(\not{p} - m + i\varepsilon)$, $V(k) = -1/(k^2 + i\varepsilon)$.

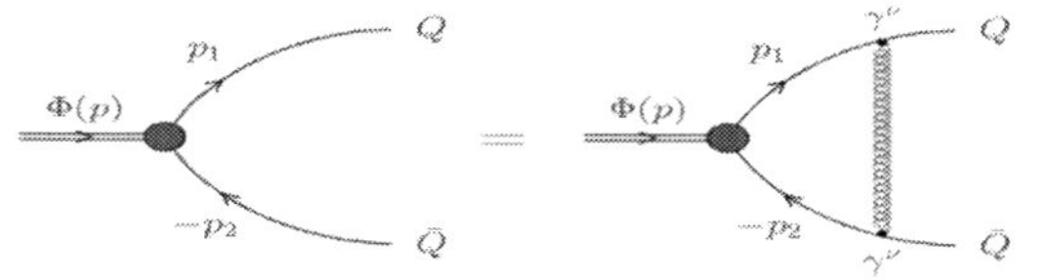

Figure 1. The diagrammatic representation of the bound state BS equation.

$$\mathcal{M} = \mathcal{V} + \int \mathcal{VGV} + \int\int \mathcal{VGVGV} + \cdots + \left(\int \mathcal{VG}\right)^n \mathcal{V} + \cdots = \frac{\mathcal{V}}{1 - \int \mathcal{VG}} \tag{1}$$

$$\Gamma_\mu(\vec{q} = \vec{0}, p) = iC_{Color} \int \frac{d^4k}{(2\pi)^4} g^2 V(k)\gamma^\nu \Delta(p_1 + k)\ \Gamma_\mu(\vec{q} = \vec{0}, p + k)\Delta(-p_2 + k)\gamma_\nu \tag{2}$$

In equation (2), the physical content is hidden in the term $\gamma^\nu\ V(k)\ \gamma_\nu$. It gives the complex structure of the bound state of $Q\bar{Q}$. To treat this vertex, one can decompose it into two contributions: a dominant contribution related to the instantaneous interactions, and a sub-dominant part, which contains retardation interaction and hyperfine effects. To the order of accuracy we want, as a first study, we consider only the instantaneous interactions.

Using the consistent BS equation satisfied by BS amplitudes, the basic idea is to compute the elastic scattering of quarkonia of different velocities but undergoing diffusions at most semi-hard; for this, we rely on the Feynman diagrams displayed at Fig. 4. For ϕ in the rest frame, the BS vertex for a vector particle is given by (3), in function of the wave function $\psi(\vec{p}\,)$. By introducing the BS amplitude in the BS equation, it becomes similar to a Schrödinger equation for a Coulombic bound state written in the momentum space, allowing for the evaluation of the normalized wave function $\psi(\vec{p})$ of ϕ. By applying Lorentz boost on the BS vertex in the rest frame, we can next obtain the vertex of ϕ at finite velocity.

$$\Gamma^\mu(p, P) = \left(\frac{\vec{p}^{\,2}}{m} - \epsilon\right)\sqrt{\frac{M_\phi}{N_c}}\ \psi(\vec{p})\frac{1 + \gamma^0 + \not{\phi}}{2}\gamma^\mu \frac{1 - \gamma^0 + \not{\phi}}{2} \tag{3}$$

Once the BS vertices have been mastered, we can evaluate $\sigma_{elas}(\phi - g/h)$ and $\sigma_{inel}(\phi - g/h)$.

Journal of Physics: Conference Series **270** (2011) 012036 doi:10.1088/1742-6596/270/1/012036

4. $\sigma_{inel}(\phi - g/h)$: calculation and principal results

The cross section for the gluon dissociation process of the bound state ϕ by the gluons $\sigma_{inel}(\phi - g/h)$ in the case of J/ψ ($c\bar{c}$) is given in function of gluon energy k by (4) [9].

$$\sigma_{J/\psi g} = \frac{128g^2}{3N_c}\, a_0^2\, \frac{(k/\epsilon - 1)^{(3/2)}}{(k/\epsilon_0)^5} \tag{4}$$

To calculate the cross section of the hadron dissociation process $\sigma_{inel}(\phi - h)$, one can use the factorization theorem (5), where $\sigma_{\phi h}(\nu)$ ($\sigma_{\phi g}(\nu)$) is the cross section of $\phi - hadron$ ($\phi - gluon$) interaction, $g(x)$ is the gluons distribution function in the hadron, it's parametrization by BP [9] is given in (5).

$$\sigma_{\phi h}(\nu) = \int_0^1 dx\ \sigma_{\phi g}(x\nu)\ g(x),\ with:\ g(x) = 0.5(\eta + 1)\frac{(1-x)^\eta}{x},\ \eta = 5(BP). \tag{5}$$

The plot of $\sigma_{inel}(\phi - g)$ versus the energy k of the absorbed gluon is given by Fig.2 for different values of the coupling constant α. The plot Fig.3 gives the variation of $\sigma_{inel}(\phi - h)$ versus $\sqrt{s}$. From these plots, one sees that these cross sections have a threshold. We note also that these inelastic cross sections depend on the binding energy ϵ_0 and on the c-quark mass m_c.

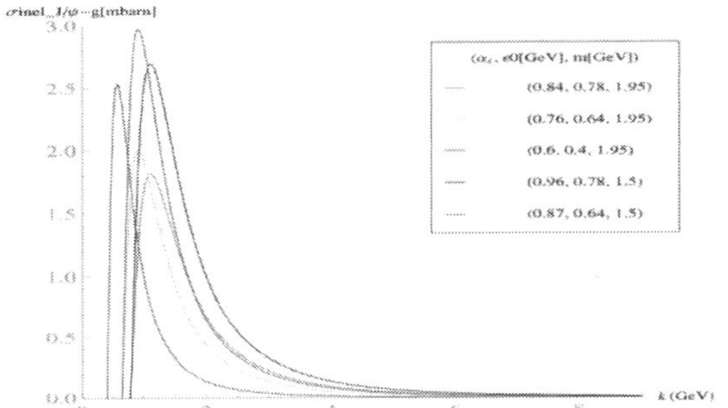

Figure 2. $\sigma_{inel}(\phi - g)$ versus k; the various couples (α, m_c) correspond to several choices compatible with the experimental values of the 1S and 2S charmonia states [10].

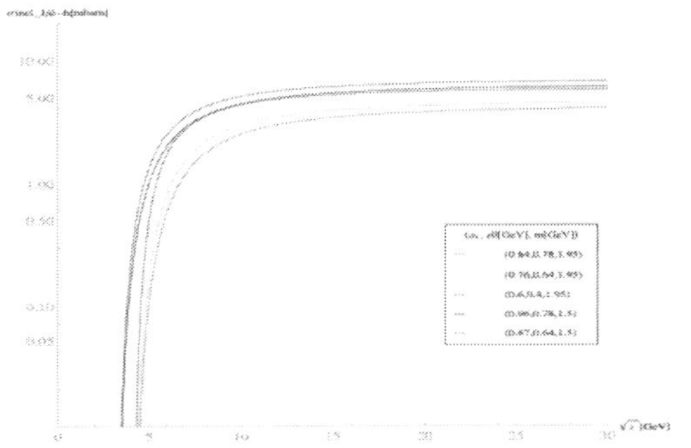

Figure 3. $\sigma_{inel}(\phi - h)$ versus $\sqrt{s}$; same parameters as Fig. 2.

5. $\sigma_{elas}(\phi - g)$: calculation and principal results

This section summarizes the calculation of the Compton diffusion of ϕ. At the Leading Order in coupling constant α, with absorbed and emitted gluons attached to the heavy quark lines, there are 6 diagrams corresponding to the topology of Fig.4-(a). If we look at the NLO and to NNLO, with the external gluons attached to the exchanged gluons between Q and $\bar{Q}$, we have 4 diagrams for the topology Fig.4-(b), 7 diagrams for Fig.4-(c), and finally 1 diagram with a 4 gluons vertex Fig.4-(d). In this contribution, we restricted to the diagrams of the class 4-(a).

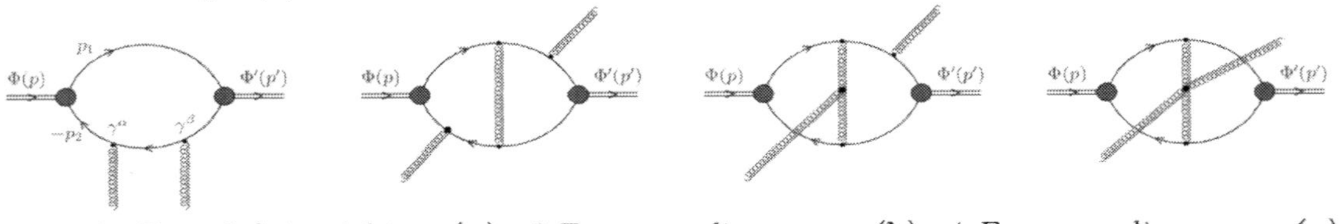

Figure 4. *From left to right :* ***(a)****: 6 Feynman diagrams,* ***(b)****: 4 Feynman diagrams,* ***(c)****: 7 Feynman diagrams,* ***(d)****: 1 Feynman diagram.*

For the case of J/ψ, we calculate the transition amplitude $\mathcal{M}$ for different energies of gluons. First, with soft gluons ($k \sim \epsilon_0 \approx mg^4$, g : coupling cst), the plot of $\mathcal{M}$ (Fig.5) shows that for gluons with $k > \epsilon_0$, $\mathcal{M}$ acquires a non-vanishing imaginary part which corresponds to the opening of the inelastic channel. For semi-hard gluons ($k \approx mg^2$) and for a Coulombic wave function, $\mathcal{M}$ is proportional to a form factor, $\sigma_{elas}(\phi - g)$ is then given by (6).

$$\sigma_{elas}(\phi - g) \approx \frac{g^2}{\pi m^2 N_c} \times \frac{1 + \frac{k^2}{8m|\epsilon_0|} + \frac{1}{2}\left(\frac{k^2}{8m|\epsilon_0|}\right)^2}{\left(1 + \frac{k^2}{8m|\epsilon_0|}\right)^3} \tag{6}$$

6. $\sigma_{elas}(\phi - g)$, $\sigma_{inel}(\phi - g)$: comparison and brief discussion

To illustrate the interest of elastic processes in the study of the quarkonia propagation in the QGP, we compare $\sigma_{elas}(\phi - g)$ and $\sigma_{inel}(\phi - g)$. For this purpose, we plot in the same graph Fig.6, $J/\psi - g$ gluon dissociation and $J/\psi - g$ Compton diffusion and the parton distribution density in the QGP (n_{mb}), where we take the Maxwell-Boltzmann distribution for a temperature $T = 300$ MeV (RHIC) and $T = 500$ MeV (LHC). By examining these curves Fig.6, one concludes that $\sigma_{inel}(\phi - g)$ is certainly larger than the $\sigma_{elas}(\phi - g)$, in agreement with the general understanding found in the literature. However, this $\sigma_{inel}(\phi - g)$ has a threshold, and it does not involve the majority of the partons in the plasma. On the other hand, the $\sigma_{elas}(\phi - g)$ has no threshold, so any parton can diffuse elastically in the plasma. Moreover, the calculation of reaction rates and various transport coefficients imply a folding of the cross sections with the distributions n_{mb}. Thus, we see that the overlap between $\sigma_{elas}(\phi - g)$ and n_{mb} is much higher than between $\sigma_{inel}(\phi - g)$ and n_{mb}.

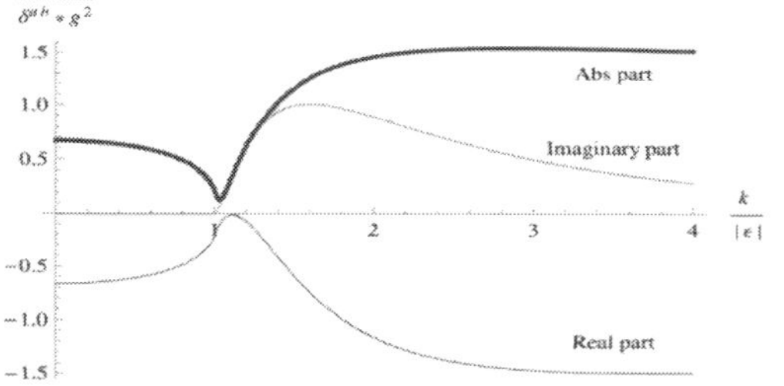

Figure 5. $\mathcal{M}$ for soft gluons ($k \sim \epsilon_0 \approx mg^4$).

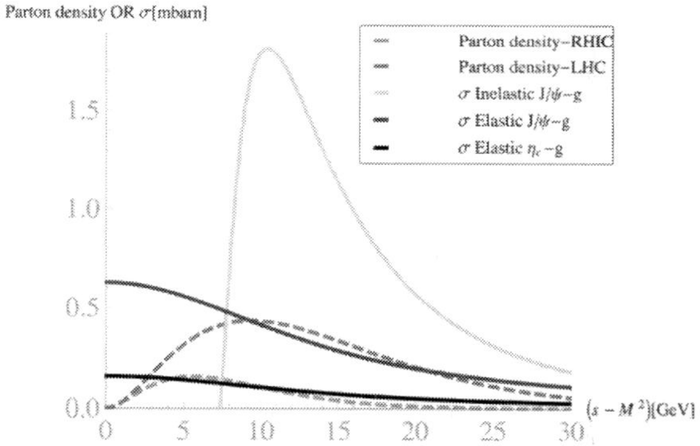

Figure 6. $\sigma_{elas}(\phi - g)$, $\sigma_{inel}(\phi - g)$ comparaison.

7. Conclusion

The present study focus on quarkonia propagation and scattering in the QGP and in a hadronic phase. Our preliminary results reveal that the elastic process in the plasma cannot be neglected. On the contrary we have pointed out the importance of such process in the medium. In this context, we developed a formalism to calculate $\phi - g/h$ elastic cross sections. These results provide some aspects of the $Q\bar{Q}$ pairs propagation in the QGP. The following step will be the development of the propagation model of quarkonia, based on our cross sections calculations.

References

[1] T. Matsui and H. Satz. *Phys. Lett B.*, **178**:4, 1986.
[2] D. Kharzeev and H. Satz. *arXiv:hep-ph/9505345v1*, 1995.
[3] G. Bhanot and M.E.Peskin. *Nucl. Phys. B.*, **156**:391, 1979.
[4] S.G Matinyan and B. Mller. *Phys. Rev. C*, **58**:2994, 1998.
[5] F.S Navarra et al. *Phys. Lett. B.*, **529**:87, 2002.
[6] S. Gardner J. Hufner, B. Povh. *Phys. Lett. B.*, **238**:103, 1990.
[7] M. C. Nemes C.H. Lewenkope, J. Hufner. *Nucl. Phys. A.*, **518**:297, 1990.
[8] E.E Salpeter and H.A Bethe. *Phys. Rev.*, **84**:1232, 1951.
[9] S.H.Lee Y.Oh, S.Kim. *Phys. Rev. C*, **65**:067901, 2002.
[10] F. Arleo et al. *Phys. Rev. D*, **65**:014005, 2001.

Di-muon measurements in Pb+Pb and p+p collisions with CMS

Catherine Silvestre, on the behalf of the CMS Collaboration
Los Alamos National Laboratory, P.O. Box 1663, Los Alamos 87545, U.S.A.
E-mail: catherine.silvestre@cern.ch

Abstract.
Di-muons are especially relevant to study the properties of the strongly interacting QCD matter created in Pb+Pb collisions at the LHC, since they are produced at early times and propagate through the medium, mapping its evolution. Simulations of CMS di-muon measurements in such an environment are presented in this paper. In particular, we show that CMS has very good detection conditions for the studies of J/ψ and Υ production, with an excellent di-muon mass resolution and a rather good acceptance. CMS will also be able to measure Z^0 production in heavy ion collisions for the first time. Early corresponding p+p measurements are reviewed as they will serve as the baseline for the heavy ion measurements.

1. Introduction

The CMS experiment [1] has been recording events for only a few months and has already produced high precision measurements suggesting that incredible physics will be accomplished in the coming years. The paragraphs below detail CMS' capability of measuring di-muons, emphasizing first quarkonia and weak boson measurements in Pb+Pb and p+p collisions at the LHC.

2. Muon reconstruction

The relatively compact CMS experiment [1] is designed to precisely measure muons by utilizing a 3.8 tesla superconducting solenoid containing, from the inside out, a pixel detector, a silicon tracker, an ECAL and a HCAL. Muon chambers are embedded in the magnetic iron yoke and the muon system consists of three independent subsystems: Drift Tubes at mid-rapidity, Cathode Strip Chambers at forward rapidity and Resistive Plate Chambers covering $|\eta| < 2.1$. The muon system is crucial for the CMS physics program because muons provide clean signatures for many physics processes.

CMS is able to measure muons with very good momentum resolution, $\sigma_p^\mu \sim 1(10)\%$ for $p_T = 100(1)$ GeV(TeV), due to the precise silicon tracking to which tracks in the muon chambers are matched. In addition the large rapidity coverage of the chambers makes it ideal to study di-muons. CMS is able to measure di-muons with a mass resolution of $\sigma_{J/\psi} \approx 20(40)$ MeV in $|\eta| \approx 0$ ($|\eta| < 2.2$). This resolution varies as a function of η due to increasing material and to the different lever arm a track has above $\eta \sim 1.6$.

Reconstructing muons in heavy ion collisions requires extra care in order to handle the dense environment. Pb+Pb collisions are simulated using HYDJET [2], an event generator that simulates jet production, jet quenching and flow effects in ultra relativistic heavy ion collisions. The reconstruction algorithm used in p+p collisions is modified for heavy ion collisions because it was too cpu- and time-intensive for high occupancy events: tracks are formed inside-out using three pixel seeds instead of pairs, and segments for the muons stations are matched to tracks made in the pixel layers with a regional algorithm.

The heavy ion run at the LHC is expected to start in November 2010. Pb+Pb collisions at $\sqrt{s_{NN}} = 2.8$ TeV will be recorded at 150 Hz with a minimum bias trigger. This allows the recording of an integrated luminosity of 1 to 10 μb^{-1} which represents 7 to 70 million minimum bias events for 7 b cross-section.

3. Quarkonia

Past experiments at the SPS and RHIC have been measuring quarkonia after it was suggested 20 years ago [3] to be a prime signal for studying the quark gluon plasma (QGP) : quarkonium production would be suppressed by Debye screening of the heavy quark binding potential.

3.1. J/ψ measurement

J/ψ measurements at lower energies demonstrated that this meson is not such a clean probe of the QGP. Indeed, at RHIC energy, PHENIX measured that J/ψs are more suppressed at forward rapidity than mid-rapidity as a function of the number of participants in the collision [4], a result which defies models based on energy density. Explanations for this behavior have since then been proposed. In particular, there may be hints that cold nuclear matter effects may play a more important role than expected in this behavior [5]. On the other hand, the profusion of c and $\bar{c}$ quarks in the medium, more abundant at mid-rapidity than forward rapidity, may constitute a bigger source from recombination at mid-rapidity. Measuring J/ψ production at higher energy is therefore very interesting in order to try to disentangle models since the amount of charm quarks produced is expected to be ten times higher at the LHC energies.

The first heavy ion run at the LHC should provide 250 to 2500 J/ψs with an expected signal over background ratio of 1.12 (and up to 5 in the barrel region) and a signal significance of 10 (to 30), without considering any shadowing or recombination effect, for a multiplicity of $dN_{ch}/dy = 2500$ and a luminosity of 1 (to 10) μb^{-1}. The p_T reach of the first year measurement is expected to be 20 GeV/c and will go up to 40 GeV/c for a nominal year of 0.5 nb^{-1} when 80K J/ψ should be measured. Fig. 1 shows the statistics expected for the J/ψ invariant mass distribution that CMS will measure with the first run in 2010.

The first p+p measurements at 7 TeV will be used as a baseline for Pb+Pb measurements as well as to study J/ψ production mechanisms, which consist of three contributions: direct J/ψ production, prompt decays of excited states and non-prompt J/ψs from B hadron decays. CMS will be able to disentangle the prompt and non-prompt contributions to the J/ψ production as well as measure directly the feed-down contributions from higher states. The dominant systematics of the differential production will come from uncertainties in the luminosity, the polarization, and the fit of the signal. For a nominal year at $\sqrt{s} = 14$ TeV, CMS will be competitive with the Tevatron after only 3 pb^{-1} and will probe cross sections beyond 20 GeV/c after 3 pb^{-1} for the first time.

Fig. 2 shows the invariant mass spectra of the inclusive J/$\psi \rightarrow \mu\mu$ production in p+p collisions at 7 TeV for an integrated luminosity of 15 nb^{-1}. Details of the analysis can be found in [6]. An extended maximum-likelihood fit has been used with an exponential for the background and a Crystal Ball function for the signal, which is a gaussian plus power-law on the low mass side to account for the radiative tail. The shape parameters of the exponential are fixed from a fit to the side bands. All parameters in the Crystal Ball function are free.

3.2. Υ measurement

Measuring Υs at the LHC will be very exciting since they will be produced for the first time in abundance in heavy ion collisions. Υs will give us insight into how their production compares

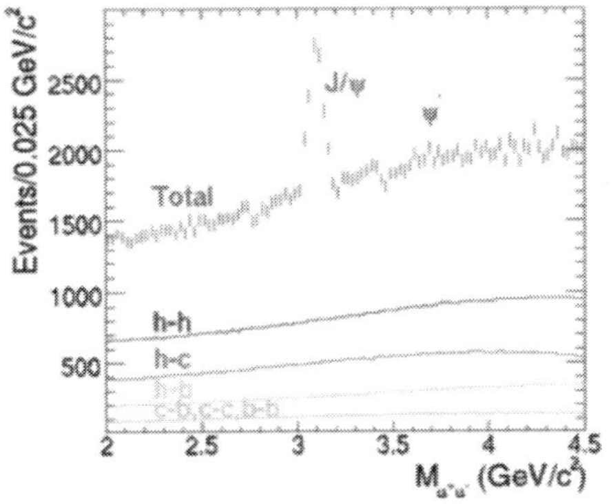

Figure 1. Invariant mass distribution from simulation of Pb+Pb collisions for $dN_{ch}/dy = 5000$ from [7] rescaled to 1 μb^{-1} and 2.8 TeV, reconstructed with CMS in $|\eta| < 2.4$, and background contributions.

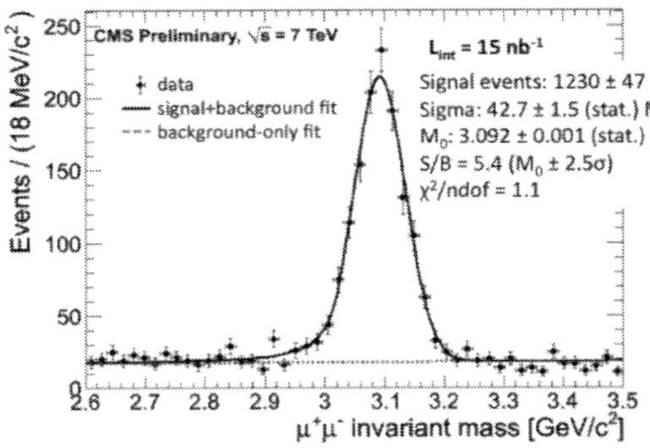

Figure 2. Invariant mass distribution measured by CMS in p+p collisions at 7 TeV for 15 nb^{-1}, fitted with a Crystal Ball function and and exponential for the background.

to the lower mass vector meson production and thus help understanding the overall picture of quarkonia as probes the QGP. In addition, CMS' high resolution, $\sigma_\Upsilon = 54$ MeV/c^2 at $|\eta| < 0.8$ and $\sigma_\Upsilon = 90$ MeV/c^2 at $|\eta| < 2.4$, will allow separation of the Υ states. Υ could follow a first order-like transition giving indication of the various melting temperatures of the QGP. For the first heavy ion run at the LHC, CMS expects to measure between 30 and 300 Υs (if no shadowing or recombination effects are taken into account) with a signal over background ratio of ~ 0.12 and a signal significance of 1 to 4. We expect to measure Υs up to 20 GeV/c the first year, and 30 GeV/c when LHC reaches nominal luminosity.

In p+p, CMS has shown its ability to separate the 1S state from the higher states (2S+3S) [8] already with only a few weeks of data taken at 7 TeV.

4. Z^0 measurement

At the LHC energies, measuring the Z^0 production in heavy ion collisions will be feasible for the first time. Since Z^0 are weakly interacting bosons, they should not be sensitive to the final state QGP. Therefore by measuring Z^0 production, initial state effects like shadowing and multi parton scattering will be probed, which will help understanding the interplay of such effects in quarkonia production.

In addition, when associated with an opposite energetic jet, measuring Z^0 production can give insight into PDF modifications and energy loss in the medium. Jets tagged with Z^0 are a unique possibility to calibrate jet energy loss and reconstruct jet fragmentation functions. Similar to γ+jet measurements, the initial transverse momentum of the hard jet can be determined in Z^0 +jet events since at leading order the jet momentum is balanced by the Z^0 momentum, thus the energy lost by the parton in the QCD medium can be directly estimated [9]. Even though the expected yield of Z^0 +jet is lower than for γ+jet, it is a complementary measurement since it can access low p_T regions where distinguishing γ from pions is hard. New NLO development of such measurement may however indicate that it could be harder than initially expected [10].

Measuring Z^0 with CMS during the heavy ion run in 2010 will be done with a fast track analysis which will allow quick results, especially since the signal over background ratio at such high mass is expected to be high (~ 20). Between 10 to 100 Z^0 are expected with a signal significance of 3 to 10. The background is composed of heavy quarks and $D\bar{D}/B\bar{B}$ decays and dominated by heavy quark di-muons, which can be rejected by a secondary vertex cut of 50 mm in radius. Fig. 3 shows the expected high mass di-muons measured by CMS for a nominal year.

Z^0s have already been measured by CMS in p+p collisions at 7 TeV. Fig. 4 shows the triggered di-muon high invariant mass spectrum, for muons above 20 GeV/c, one muon being detected in $|\eta| < 2.1$, for 16 nb^{-1} integrated luminosity as decribed in [11]. Monte Carlo contributions to the spectra have been simulated with normalized MCFM NLO predictions [12] using MSTW08 NLO PDFs [13]. The EWK backgrounds include $W \to \mu\nu$ and $t\bar{t}$ and the QCD background is PYTHIA at LO with a k factor of 1.

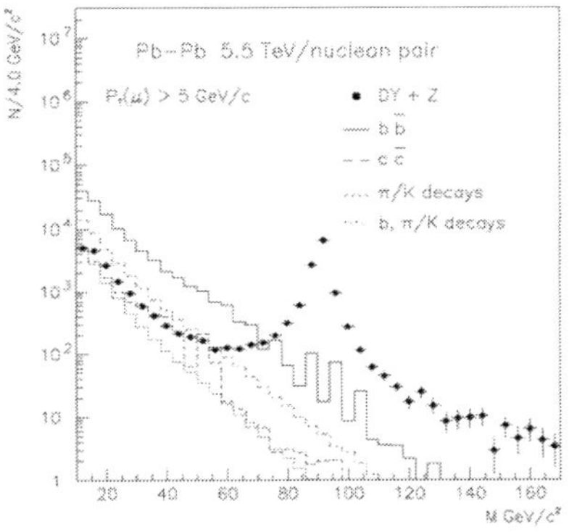

Figure 3. Invariant mass distribution for Pb+Pb collisions and 0.5 nb^{-1} with background contributions. Cuts: $p^{\mu} >$ 5 GeV/c and $\eta^{\mu} < 2.4$

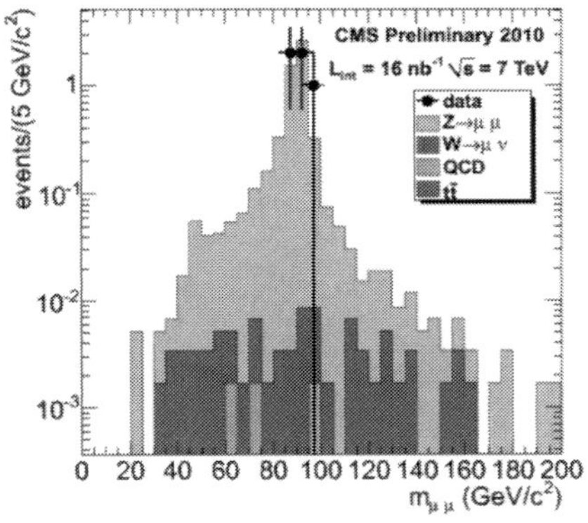

Figure 4. Measured invariant mass distribution in p+p collisions at 7 TeV by CMS with 16 nb^{-1} and different background contributions from simulations.

5. Conclusion

CMS is ready to measure di-muons in Pb+Pb collisions. For the first heavy ion run at the LHC in 2010 at $\sqrt{s_{NN}} = 2.8$ TeV, Z^0 production will be studied quickly thanks to a good signal over background ratio at such high mass. It will give insight into initial state effects. Measuring J/ψ at the LHC energies is very much awaited to test recombination model predictions. Studying the Υ suppression as a signal of QGP formation will be done for the first time with significant statistics and should give indication of the temperature of the QGP.

Acknowledgments

I would like to thanks CMS and the organizers of HQ2010 for giving me the opportunity to give this talk. This work was in part supported by LANL LDRD grant ER20090303.

References

[1] CMS Collaboration *JINST* **3** (2008) S08004. doi:10.1088/1748-0221/3/08/S08004.
[2] I. P. Lokhtin and A. M. Snigirev *Eur. Phys. J.* **C45** (2006) 211–217, arXiv:hep-ph/0506189.
[3] T. Matsui and H. Satz *Phys. Lett.* **B178** (1986) 416. doi:10.1016/0370-2693(86)91404-8.
[4] PHENIX Collaboration *Phys. Rev. Lett.* **98** (2007) 232301. doi:10.1103/PhysRevLett.98.232301.
[5] PHENIX Collaboration *Phys. Rev. C* **79** (May, 2009) 059901. doi:10.1103/PhysRevC.79.059901.
[6] CMS Collaboration *CMS PAS* **BPH-10-002** (2010).
[7] CMS Collaboration *J. Phys.* **G34** (2007) 2307–2455. doi:10.1088/0954-3899/34/11/008.
[8] CMS Collaboration *CMS PAS* **BPH-10-003** (2010).
[9] G. J. Kunde et al. *Eur. Phys. J.* **C61** (2009) 785–788. doi:10.1140/epjc/s10052-009-0955-9.
[10] R. B. Neufeld et al. arXiv:1006.2389.
[11] CMS Collaboration *CMS PAS* **EWK-10-002** (2010).
[12] J. M. Campbell and R. K. Ellis *Phys. Rev.* **D65** (2002) 113007. doi:10.1103/PhysRevD.65.113007.
[13] A. D. Martin et al. *Eur. Phys. J.* **C63** (2009) 189–285. doi:10.1140/epjc/s10052-009-1072-5.

Measurement of the η and ω Dalitz decays transition form factors in p-A collisions at 400 GeV/c with the NA60 apparatus

Antonio Uras for the NA60 Collaboration
University of Cagliari and INFN, Cagliari. Complesso Universitario di Monserrato - S.P. per Sestu Km 0.700 09042 Monserrato (Cagliari) - Italy

E-mail: antonio.uras@ca.infn.it

Abstract. The NA60 experiment has studied low-mass muon pairs production in proton-nucleus collisions at 400 GeV/c at the CERN SPS. The mass spectrum is well described by the superposition of the two-body and Dalitz decays of the light neutral mesons η, ρ, ω, η' and ϕ. The large collected statistics allows to isolate the contributions due to the Dalitz decays of the η and ω mesons, from which the electromagnetic transition form factors of the two mesons have been extracted. The found values agree with the previous available measurements, improving their uncertainty thanks to the higher statistics. The results thus confirm the discrepancy with the prediction of the Vector Meson Dominance model in the case of the electromagnetic form factor of the ω meson.

Proton-nucleus studies have been part of the heavy-ion experimental program at the CERN SPS, of which the NA60 experiment represents the most recent and one of the most fruitful efforts. Beyond serving as a robust reference baseline for the heavy-ion data, proton-nucleus data also stimulate a genuine physics interest, allowing measurements in an environment of cold nuclear matter. This is in fact the case of the electromagnetic transition form factors of the η and ω mesons, which have been studied here through their Dalitz decays $\eta \to \mu^+\mu^-\gamma$ and $\omega \to \mu^+\mu^-\pi^0$. In these $A \to Bl^+l^-$ processes, the mesons decay electromagnetically into a virtual photon with mass M – in turn converting into a lepton pair – and a third body. Assuming point-like particles, the mass distribution of the emitted lepton pair (that is, the decay rate of the process as a function of M) can be exactly described by QED as firstly shown by N.M. Kroll and W. Wada [1] in the 1950s.

However, the observed mass distribution for the lepton pairs emitted in an $A \to Bl^+l^-$ decay differs from the point-like one, because of the internal electromagnetic structure of the decaying meson, which reveals at the vertex of the transition $A \to B$. This difference can be formally described by a function of M, usually expressed as $|F_{AB}(M)|^2$, known as *transition form factor*. Experimentally, $|F_{AB}(M)|^2$ is directly accessible by comparing the measured invariant mass spectrum of the lepton pairs from the Dalitz decays, with the point-like QED prediction. The transition form factors are usually given in terms of the pole parametrization:

$$|F|^2 = (1 - M^2/\Lambda^2)^{-2} \ . \qquad (1)$$

The Vector Meson Dominance (VMD) model gives theoretical predictions both for Λ_η^{-2} and Λ_ω^{-2}. These predictions have been tested by the Lepton-G experiment [3] using pion beams, and

Journal of Physics: Conference Series **270** (2011) 012038 doi:10.1088/1742-6596/270/1/012038

recently by the NA60 experiment in In-In peripheral collisions [4]. While the experimental values agree among each other both for the η and ω mesons, only the VMD model prediction for the η meson is confirmed by the data, while the one for the ω meson is not, strongly underestimating the observed form factor. The NA60 p-A data can now give a further significant contribution to the measurement, thanks to the high quality and the large statistics available.

A description of the NA60 apparatus can be found for example in [5]. Here, only few relevant details are given. The produced dimuons are identified and measured by the muon spectrometer, composed of a set of tracking stations, trigger scintillator hodoscopes, a toroidal magnet and a hadron absorber. The stopping of hadrons (provided by more than 20 interactions lengths) allows to select the highly rare dimuon events. On the other hand, the material which stops the hadrons also induces multiple scattering and energy loss on the muons, degrading the mass resolution of the measurement made in the spectrometer. To overcome this problem, NA60 already measures the muons before the absorber with a tracking telescope, made of pixel silicon detectors [6]. The muon tracks are found among the many other charged particle tracks by a correct matching, both in angular and momentum coordinates, with the reconstructed tracks in the muon spectrometer.

The preliminary results presented here are obtained from the analysis of the data collected during three days within the 2004 proton run, with a system of nine sub-targets of different nuclear species, simultaneously exposed to an incident 400 GeV proton beam. The first step of the analysis is a fit of the reconstructed mass spectrum (see Figure 1). The fit is performed comparing the data with the superposition of the expected sources (estimated via the MC). These sources are the Dalitz decays of the η, ω and η' mesons, the 2-body decays of the η, ρ, ω and ϕ and the contribution of the open charm process. Before the comparison between data and MC is performed, the small component of combinatorial background (originating from π and K decays) is subtracted from the real data: its shape is estimated with an event mixing technique, while its normalization is established fixing the LS (like-sign) component coming out from the mixing, to the LS component of the data (containing no signal from correlated pairs at the SPS energies). The background accounts for less than 10% of the integrated mass spectrum below 1.4 GeV/c^2. The comparison between mixed and real sample gives an average uncertainty of 10%, for both the $(++)$ and the $(--)$ components: given the rather low level of the combinatorial background contribution, this uncertainty hardly affects the evaluation of the shape for the Dalitz decays under study.

The processes which are not of interest for the measurement of the form factors of the η and ω are then subtracted from the real data, using the normalizations fixed by the fit. Only the Dalitz decays of η and ω are retained, together with the two-body decay of the ρ: this latter is left in, making it much easier to control the systematics due to the small contribution coming from the low-mass tail of the ρ in the mass region of interest here. The mass spectrum resulting after the subtraction is then corrected for the effects of geometrical acceptance and reconstruction efficiency. In order to do this, a correction profile as a function of the mass is built, weighting the profiles obtained from the MC for the three processes separately. After the correction for the acceptance × efficiency effects, the resulting mass spectrum – now free from any acceptance distortion – is compared with the superposition of the expected line shapes of the three processes $\eta \to \mu^+\mu^-\gamma$, $\omega \to \mu^+\mu^-\pi^0$ and $\rho \to \mu^+\mu^-$.

Figure 2 shows the fit on the acceptance-corrected mass spectrum (black triangles), having isolated the contribution of the three processes $\eta \to \mu^+\mu^-\gamma$, $\omega \to \mu^+\mu^-\pi^0$ and $\rho \to \mu^+\mu^-$. The fit function is a superposition of the expected line shapes for the three processes, described by the following functions:

$$\left.\frac{dN}{dM}\right|_{\eta\mathrm{Dal}} \propto 2M \times \frac{2}{3}\frac{\alpha}{\pi}\frac{\Gamma(\eta\to\gamma\gamma)}{M^2} \times \left(1-\frac{M^2}{m_\eta^2}\right)^3 \left(1+\frac{2m_\mu^2}{M^2}\right)\left(1-\frac{4m_\mu^2}{M^2}\right)^{1/2} \times \left|F_\eta(M^2)\right|^2 \quad (2)$$

$$\left.\frac{dN}{dM}\right|_{\omega\mathrm{Dal}} \propto 2M \times \frac{\alpha}{3\pi}\frac{\Gamma(\omega\to\gamma\pi^0)}{M^2} \times \left(1+\frac{2m_\mu^2}{M^2}\right) \times$$
$$\times\left(1-\frac{4m_\mu^2}{M^2}\right)^{1/2} \times \left[\left(1+\frac{M^2}{m_\omega^2-m_{\pi^0}^2}\right)^2 - \frac{4m_\omega^2 M^2}{(m_\omega^2-m_{\pi^0}^2)^2}\right] \times \left|F_\omega(M^2)\right|^2 \quad (3)$$

$$\left.\frac{dN}{dM}\right|_{\rho} \propto \frac{\alpha^2 m_\rho^4}{3(2\pi)^4} \times \left(1-\frac{4m_\pi^2}{M^2}\right)^{3/2} \times$$
$$\times\left(1-\frac{4m_\mu^2}{M^2}\right)^{1/2}\left(1+\frac{2m_\mu^2}{M^2}\right) \times (2\pi MT)^{3/2} \times \frac{1}{(M^2-m_\rho^2)^2+(M\Gamma_{\mathrm{tot}}^0)^2} \quad (4)$$

In the fit, the following parameters are left free:

- the three normalizations, one for each of the line shapes above
- the parameters Λ_η^{-2} and Λ_ω^{-2}, contained in the form factors $|F_\eta(M^2)|^2$ and $|F_\omega(M^2)|^2$.

Concerning the line shape for the $\rho \to \mu^+\mu^-$ decay, the parameters Γ_ρ (entering in the definition of Γ_{tot}), m_ρ and the effective temperature T_ρ are fixed respectively to $\Gamma_\rho = 151$ MeV (from the PDG), $m_\rho = 768$ MeV/c^2 (also from the PDG) and $T_\rho = 170$ MeV (consistent with the expected value obtained by statistical model fits on particle ratios in pp interactions). All the parameters involved in the analysis procedure have been varied to study their influence on the fits and determine the systematic uncertainties on the results: the weights defining the acceptance×efficiency correction, the fit range (and the choice of the free parameters), the value for the $\sigma_{\eta'}/\sigma_\omega$ ratio, the charge correlation parameter for the combinatorial background and finally the level of the open charm contribution. The analysis leads to the following values:

$$\Lambda_\eta^{-2} = 1.946 \pm 0.096 \text{ (stat.) } \pm 0.015 \text{ (syst.) } (\mathrm{GeV}/c^2)^{-2}$$
$$\Lambda_\omega^{-2} = 2.248 \pm 0.030 \text{ (stat.) } \pm 0.009 \text{ (syst.) } (\mathrm{GeV}/c^2)^{-2}$$

Once the final fit parameters and their errors are fixed, the contributions of the $\eta \to \mu^+\mu^-\gamma$ and $\omega \to \mu^+\mu^-\pi^0$ processes are disentangled, making it then possible to present the two form factors in the usual way, in order to compare them with the previous experimental points, see Figure 3. The pole parameters and their errors as obtained from the combined fit to both Dalitz decays are shown as inserts. Both figures include the NA60 data obtained in peripheral In-In, the Lepton-G data, and the expectations from VMD for comparison. Within the errors, perfect agreement between the three data sets is seen in both cases. Irrespective of the much reduced errors, the form factor of the η is still close to the expectation from VMD. The form factor of the ω, on the other hand, strongly deviates from the VMD, showing a relative increase close to the kinematic cut-off by a factor of ~ 10.

In conclusion, we have presented a new measurement of the electromagnetic transition form factors of the η and ω mesons. The analysis has been performed by studying the Dalitz decays $\eta \to \mu^+\mu^-\gamma$ and $\omega \to \mu^+\mu^-\pi^0$ in p-A collisions at 400 GeV/c with the NA60 apparatus. The new preliminary values presented here are in good agreement with those already available; improving the precision of the measure by a factor ~ 2, they confirm on more solid grounds the discrepancy between the VMD prediction for the ω and the experimental observations.

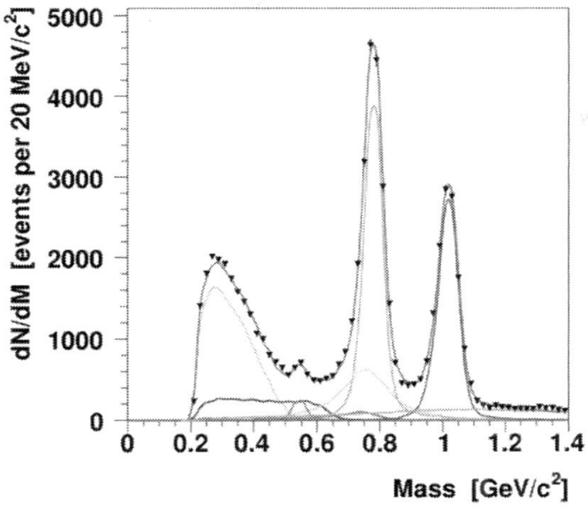

Figure 1. Fit on the target-integrated mass spectrum.

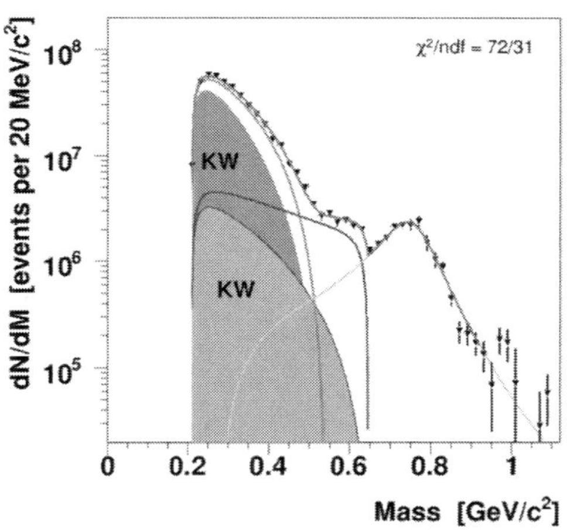

Figure 2. Fit on the acceptance-corrected mass spectrum. The shaded areas indicate the Kroll-Wada (KW) expectations for the η and ω Dalitz decays for point-like particles, defined by QED [1].

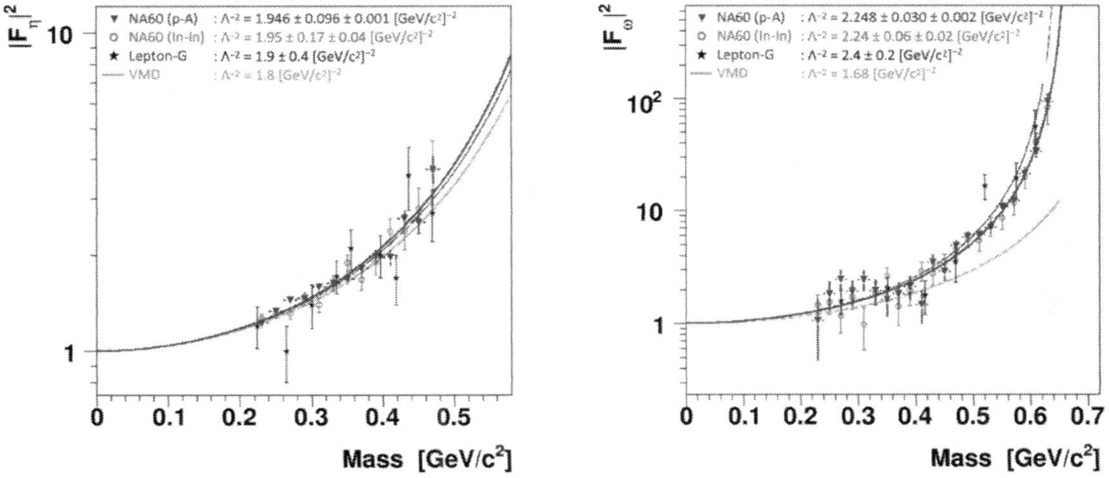

Figure 3. Electromagnetic transition form factors for the η (left) and ω (right) mesons as a function of M.

References

[1] N. M. Kroll and W. Wada, *Internal Pair Production Associated with the Emission of High-Energy Gamma Rays*, Phys. Rev. **98** (1955) 1935.

[2] C. Amsler *et al.* [Particle Data Group], *Review of Particle Physics*, Phys. Lett. B **667** (2008) 1

[3] L. G. Landsberg, *Electromagnetic Decays of Light Mesons*, Phys. Rep. **128** (1985) 301.

[4] R. Arnaldi *et al.* [NA60 Collaboration], *Study of the electromagnetic transition form-factors in $\eta \to \mu^+\mu^-\gamma$ and $\omega \to \mu^+\mu^-\pi^0$ decays with NA60*, Phys. Lett. B **677** (2009) 260 [arXiv:0902.2547].

[5] R. Arnaldi *et al.* [NA60 Collaboration], *Evidence for the production of thermal-like muon pairs with masses above 1 GeV/c^2 in 158A GeV Indium-Indium Collisions* Eur. Phys. J. C **59** (2009) 607 [arXiv:0810.3204].

[6] M. Keil *et al.* [NA60 Collaboration], *The NA60 vertex telescope for particle tracking in heavy-ion and proton-nucleus collisions* Nucl. Instrum. Meth. A **568** (2006) 263

Dilepton production in proton-proton collisions at top RHIC energy

J Manninen[1], E L Bratkovskaya[2], W Cassing[1] and O Linnyk[2]

[1] Institut für Theoretische Physik, Universität Giessen, 35392 Giessen, Germany
[2] Institut für Theoretische Physik, Universität Frankfurt, 60054 Frankfurt, Germany

E-mail: Mannisenjaakko@gmail.com

Abstract. We study dielectron production in proton-proton collisions at top RHIC beam energy within an extended statistical hadronization model. The invariant mass spectrum of correlated dielectron pairs is evaluated in the low invariant mass region and calculated results are compared with the PHENIX experiment. The model is found to be able to describe the data very well up to invariant masses of 1 GeV with few adjustable parameters.

1. Introduction

Correlated dilepton radiation has been considered for a long time to be an important tool to study the hot and dense nuclear matter created in relativistic nuclear collisions. The idea behind is that dileptons are emitted during all stages of the collision evolution and the electromagnetic cross sections with the created medium are believed to be small and so electromagnetic signals from the early stages might survive the preceding evolution of the system. Thus, by studying the dilepton emission one can extract information also on the earlier, possibly exotic phases of the collision evolution. For this to hold true, however, one has to carefully model the final (and dominant) stages of the collision evolution where dilepton radiation is predominantly governed by the decays of light mesons.

The PHENIX collaboration has recently measured the correlated dielectron invariant mass spectrum in proton-proton [1] as well as in $Au+Au$ [2] collisions at the top RHIC beam energy. A large and so far un-explained excess of dielectrons in (semi)central heavy-ion collisions is seen in the data below the mass of ϕ meson when compared with various model calculations while the spectrum in $p+p$ collisions can be fairly well understood within the hadronic freeze-out cocktail calculations [1]. We present in this paper our baseline hadronic freeze-out cocktail calculation for the dielectron production in proton-proton collisions in the low invariant mass region while the analysis of the heavy-ion collisions as well as an extension to invariant masses larger than the ϕ meson mass will be presented in a forthcoming publication [3].

2. Statistical hadronization model

The statistical hadronization model (SHM) has been used successfully in describing the hadronic yields and rapidity densities in high energy proton-proton collisions at SPS and RHIC [4-7]. In this work we employ the simplest version of the SHM, i.e. the analysis is performed in the grand-canonical ensemble, because calculations become considerably easier once we do not require exact conservation of conserved quantities. The price to pay for the easier calculations is that one has to introduce a free parameter for each of the conserved quantities that are conserved in average only.

In the SHM, the primary hadron multiplicity of hadron type i is calculated (omitting the fugacities) according to

$$N_i = V\frac{2J_i+1}{(2\pi)^3}\int \gamma_S^{n_s}\ \mathrm{e}^{-\sqrt{p^2+m_i^2}/T}\ \mathrm{d}^3p. \quad (1)$$

In Eq. (1) J_i denotes the spin, p the three-momentum and m_i the mass of the particle of the hadron species i. In principle the model has 6 free parameters: T, μ_B, μ_Q, μ_S, γ_S and the system volume V, but we know that the central rapidity region is close to net charge free[1] and, furthermore, the dilepton radiation is dominated by decays of neutral mesons, whose multiplicities do not depend explicitly on the chemical potentials, and thus to a good approximation we can set the chemical potentials to zero. In this case we are left with 3 free parameters (T, γ_S and V) that describe the yields of all relevant hadrons. The γ_S (< 1) [8] parameter takes into account the empirical fact that the SHM tends to over-estimate the strange hadron yields in the elementary particle collision systems.

The SHM has been fitted to the STAR data measured in the central rapidity region in $p+p$ collisions [7] and we use the resulting fit parameters T=170 MeV and $\gamma_S = 0.6$ in our analysis. We have re-adjusted the system volume such that the π^0 rapidity density at mid-rapidity is reproduced. All calculations are performed assuming that in each of the events there is only one large cluster produced. This picture is justified under special assumptions discussed for example in [9]. For the resonances with width larger than 2 MeV, Eq. (1) is convoluted with the relativistic Breit-Wigner distribution in order to take into account the finite widths of the resonant states and the two dimensional integration over the momentum and mass is performed in order to calculate the primary yields.

The evaluation of the dilepton emission with the SHM proceeds as follows: First, we calculate the mean primary yields of different hadrons according to Eq. (1) and then - using these values - we sample event-by-event Poisson distributions for the primary yields of all hadrons and resonances. Thus, even though the temperature, γ_S and the system volume are kept the same for all events, we have somewhat different amounts of primary hadrons produced in each of the events and so the energy, for example, is not fixed but fluctuates from event to event.

Next, unstable resonances decay into stable ones and as a final step, we let the vector mesons decay into dielectrons. The branching ratios of different hadrons and resonances into dileptons are typically very small and thus we take into account the dilepton radiation from the dominant sources only, which at RHIC energies are the low mass vector mesons. We do not consider the direct dilepton emission from most of the resonances in the cases where the contribution would be negligible. One should notice, however, that in our approach it is important to include all the known resonances (≈ 300) because eventually they decay into the light mass vector mesons[2] and thus indirectly affect the dilepton yields. The dilepton channels taken into account in our analysis are listed in Table 1.

In general, the probability of a hadron to decay into a certain channel depends both on the mass of the parent hadron as well as on the energies of the daughter particles. We have evaluated in detail the mass dependent partial widths into dileptons as well as the total widths of the vector mesons in this analysis while for the resonance decays, which do not directly produce dielectrons, we have taken into account trivial mass threshold effects only in our analysis. Formulae for the dielectron partial widths are known from the literature and we have used the standard

[1] estimates for the μ_B in the central rapidity region vary from around $20-30$ MeV in central $Au+Au$ collisions to around 10 MeV in $p+p$ collisions ; $\mu_S \approx \mu_B/4$ and $\mu_Q \approx \mu_B/40$ are thus also small

[2] for example about half of the ρ^0 mesons are produced primarily while the other half come from resonance decays

Table 1. Decay channels relevant for dielectron production in $p+p$ collisions at $\sqrt{s}$=200 GeV.

Hadron	direct	Dalitz	other
π^0		$\pi^0 \to \gamma\, e^+e^-$	
η^0		$\eta^0 \to \gamma\, e^+e^-$	$\eta^0 \to \pi^+\pi^- e^+e^-$
η'		$\eta' \to \gamma\, e^+e^-$	$\eta' \to \pi^+\pi^- e^+e^-$
ρ^0	$\rho^0 \to e^+e^-$		
ω^0	$\omega^0 \to e^+e^-$		$\omega^0 \to \pi^0\, e^+e^-$
ϕ^0	$\phi^0 \to e^+e^-$		$\phi^0 \to \eta\, e^+e^-$

expressions [10] while the form factors used as well as a discussion on other details can be found for example in [11].

The SHM is very efficient in counting the Lorentz invariant relative yields of different hadrons. This model, however, does not take into account the initial dynamical evolution of the colliding partons from the beam particles. Typically the SHM has been compared with experimental data which has been corrected for the limited acceptance in rapidity and transverse momentum, in which case the initial parton dynamics do not affect the final results. The PHENIX data has not been corrected for the acceptance effects and so in order to be able to compare our calculations with the data, we need to introduce an additional model that correctly accounts for the dynamics not taken into account in the SHM. We wish to do this as simply and transparently as possible so that we do not mask the main physics interest of this paper with complicated dynamical corrections. Thus, both in the longitudinal direction as well as in the transverse direction, we model the initial parton dynamics by simply assuming a two dimensional Gaussian distribution for the momentum distribution of the clusters. Obviously, the means both in rapidity and p_T are zero while we have fitted the widths of the clusters' p_T and y distribution such that the pion distributions become compatible with the measurements.

3. Dielectron production in the LMR in $p+p$ collisions at $\sqrt{s}$=200 GeV

We compare our calculated dielectron invariant mass spectrum with the PHENIX data [1] measured in proton-proton collisions at the top RHIC beam energy in Figure 1. The contributions from the different vector meson decays are shown separately while the thick black line denotes the sum of all relevant contributions. One can see that the model can do a fairly good job in describing the measured data in the LMR and it is clear that the aforementioned details like mass dependent partial widths must be taken carefully into account in order to describe the shape of the data at all invariant masses. We have also added the background contribution from open charm decays (denoted with cc in the figure) but one can see that this contribution is negligible at all invariant masses except between the ω and ϕ meson peaks. We point out here that we have also taken into account the limited experimental mass resolution in the figure by running our calculated curves through a Gaussian smearing filter with a resolution of 10 MeV. This way especially the ϕ meson peak and also the ω peak become broader and compatible with the experimental data.

4. Conclusions

We have demonstrated that an extended statistical hadronization model can describe the dilepton radiation in ultra-relativistic proton-proton collisions at RHIC. In this work, we have implemented the statistical hadronization model including most of the known resonances and thus we have gone beyond the standard hadronic freeze-out cocktail calculations which typically

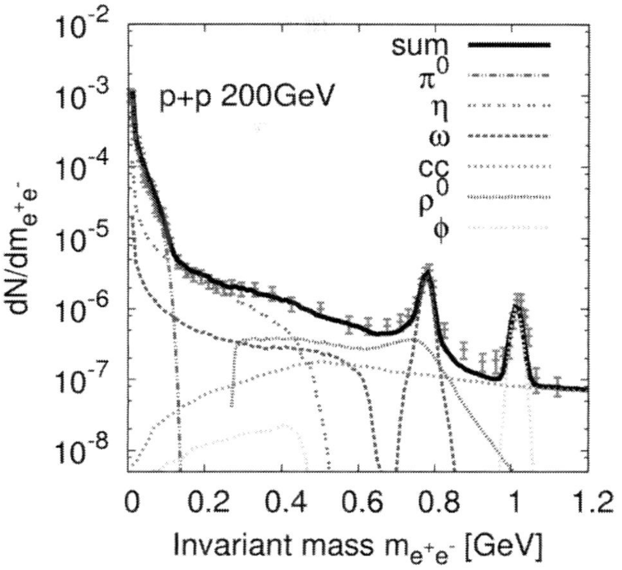

Figure 1. Invariant mass spectrum of correlated pairs of electrons and positrons in proton-proton collisions at $\sqrt{s} = 200$ GeV. The data are from the PHENIX collaboration [1] while the contribution from different dilepton emitting sources are calculated as explained in the text. The full thick black line denotes the sum from all relevant sources.

include only a few different hadron and resonances species whose relative yields are taken either from the experiment or put in by hand. In our approach, the relative yields of different hadron species and thus the relative contribution at different invariant masses of the dielectron spectra is described with only 3 free parameters in the low invariant mass region. Thus, our work underlines the usefulness of the statistical hadronization model since together with simple and intuitive supplementary corrections for the initial state dynamics, we can address not only 4π integrated data but also the differential spectra measured by the PHENIX collaboration.

Acknowledgments

The authors would like to thank A. Toia for stimulating discussions. Furthermore, E.L.B. and O.L. are grateful for financial support from the 'HIC for FAIR' center of the 'LOEWE' program and J.M. for support from DFG.

References

[1] Adare A *et al.* [PHENIX Collaboration], *Phys. Lett.* B **670** (2009) 313
[2] Adare A *et al.* [PHENIX Collaboration], *Phys. Rev. C* **81** (2010) 034911
[3] Manninen J, Bratkovskaya E L, Cassing W and Linnyk O, arXiv:1005.0500 (2010)
[4] Becattini F, Gazdzicki M, Keränen A, Manninen J and Stock R, *Phys. Rev. C* **69** (2004) 024905
[5] Kraus I, Cleymans J, Oeschler H and Redlich K, *Phys. Rev. C* **79** (2009) 014901
[6] Abelev B I *et al.* [STAR Collaboration], *Phys. Rev. C* **79** (2009) 034909
[7] Becattini F, Castorina P, Milov A and Satz H, *Eur. Phys. J. C* **66** (2010) 377
[8] Koch P, Müller B and Rafelski J, *Phys. Rept.* **142** (1986) 167
[9] Becattini F, Manninen J and Gazdzicki M, *Phys. Rev.* C **73** (2006) 044905
[10] Landsberg L G, *Phys. Rept.* **128** (1985) 301
[11] Cassing W and Bratkovskaya E L, *Phys. Rept.* **308** (1999) 65

Low mass dimuons within a hybrid approach

Elvira Santini[1], **Marcus Bleicher**[2]
[1]Institut für Theoretische Physik, Goethe-Universität, Max-von-Laue-Str. 1, D-60438 Frankfurt am Main, Germany
[2]Frankfurt Institute for Advanced Studies (FIAS), Ruth-Moufang-Str. 1,D-60438 Frankfurt am Main, Germany

E-mail: santini@th.physik.uni-frankfurt.de

Abstract. We analyse dilepton emission from hot and dense hadronic matter using a hybrid approach based on the Ultrarelativistic Quantum Molecular Dynamics (UrQMD) transport model with an intermediate hydrodynamic stage for the description of heavy-ion collisions at relativistic energies. Focusing on the enhancement with respect to the contribution from long-lived hadron decays after freeze-out observed at the SPS in the low mass region of the dilepton spectra (often referred to as "the excess"), the relative importance of the emission from the equilibrium and the non-equilibrium stages is discussed.

1. Introduction

Electromagnetic probes, such as photons and lepton pairs, are penetrating probes of the hot and dense matter. Once created these particles pass the collision zone essentially without further interaction and can therefore mediate valuable information on the electromagnetic response of the strongly interacting medium. The analysis of such response is tightly connected to the investigation of the in-medium modification of the vector meson properties. Vector mesons can directly decay into a lepton-antilepton pair. One therefore aims to infer information on the modifications induced by the medium on specific properties of the vector meson, such as its mass and/or its width, from the invariant mass dilepton spectra.

Measurements of dileptons in heavy ion collisions have indeed revealed many exciting phenomena, triggering a copious theoretical activity. A first generation of ultra-relativistic heavy ion collision experiments performed in the nineties observed an enhancement of dilepton production in heavy systems at low invariant mass as compared to conventional hadronic cocktails and models [1, 2]. The enhancement could be later explained by the inclusion of a substantial medium modification of the ρ meson properties, see e.g. Refs. [3, 4, 5] for reviews or the review on photons and dileptons by S. Bathe in this volume. However, no discrimination between a "melting" of the ρ meson spectral function as expected within many-body hadronic models [6, 7, 8, 9, 10] and a dropping of the ρ meson mass according to the Brown-Rho scaling hypothesis [11] and the Hatsuda and Lee sum rule prediction [12] could be achieved. Recently, a substantial improvement in statistics and mass resolution in low-mass dilepton spectra has been achieved by the NA60 collaboration [13], who measured dimuon production in In-In collisions at 158 A GeV. The improved experimental accuracy enabled a subtraction of final-state hadron decay contributions (the so-called cocktail), except for ρ and charm decays. The dimuon spectra of the remaining "excess" over the cocktail strongly favour the broadening over the dropping mass scenario. Typically, model calculations aimed at the interpretation of the data require a convolution of the dilepton emission rates over a realistic space-time model of the heavy-ion collision. Assuming local equilibrium, thermal fireball/hydrodynamics calculations have been performed over the last years and quite successfully applied to the interpretation of the NA60 data (see e.g. Fig.4 in Ref. [14]). Here, we present recent results on dimuon emission obtained using a hybrid approach for the description of the evolution dynamics. In such an approach, both thermal and non-thermal sources play a noticeable role and can be separately investigated. We will focus on the low invariant mass region of the dilepton spectra ($M < 1$ GeV), where the $\rho^* \to \mu\mu$ emission plays the dominant role. Extension of the model to the intermediate mass region ($1.0 < M < 1.5$ GeV) requires addition of further contributions and is currently under development.

2. Dilepton phenomenology within a hybrid model

The dynamics of the In+In collisions is simulated employing a hybrid approach based on the integration of an ideal hydrodynamic evolution into the UrQMD transport model [15]. This approach has been successfully applied to many bulk observables at SPS energies [15, 16, 17, 18, 19] and first applications to electromagnetic probes have been performed recently [20, 21]. Here, we describe it only briefly and refer the reader to the indicated references for details.

During the first stage of the evolution the particles are described as a purely hadronic cascade within UrQMD. The coupling to the hydrodynamical evolution proceeds when the two Lorentz-contracted nuclei have passed through each other. At this time, the spectators continue to propagate in the cascade and all other hadrons are mapped to the hydrodynamic grid. Subsequently, a $(3+1)$ ideal hydrodynamic evolution is performed using the SHASTA algorithm [22, 23]. The hydrodynamic evolution is gradually merged into the hadronic cascade. Transverse slices, of thickness 0.2 fm, are transformed to particles whenever in all cells of each individual slice the energy density drops below five times the ground state energy density. The employment of such a gradual transition allows us to obtain a rapidity independent transition temperature without artificial time dilatation effects. When merging, the hydrodynamic fields are transformed to particle degrees of freedom via the Cooper-Frye equation. The created particles proceed in their evolution in the hadronic cascade where final state interactions and decays of the particles occur within the UrQMD framework.

Concerning dilepton emission, during the locally equilibrated hydrodynamic stage the production of lepton pairs is described by radiation rates for a strongly interacting medium in thermal equilibrium. Invoking vector meson dominance the latter can be related, at low invariant masses, to the spectral properties of the vector mesons, with the ρ meson giving the dominant contribution. The thermal dilepton rate reads then [3]:

$$\frac{d^8 N_{ll}}{d^4x d^4q} = -\frac{\alpha^2 m_\rho^4}{\pi^3 g_\rho^2}\frac{L(M^2)}{M^2} f_B(q_0;T)\,\mathrm{Im}\,D_\rho(M,q;T,\mu_B)\,, \tag{1}$$

where α denotes the fine structure constant, $M^2 = q_0^2 - q^2$ the dilepton invariant mass squared, f_B the Bose distribution function (for a moving fluid this must be substituted with the Jüttner function), and $L(M)$ a lepton phase space factor that quickly approaches one above the lepton pair threshold. The electromagnetic response of the strongly interacting medium is then contained in $\mathrm{Im}\,D_\rho(M,q;T,\mu_B)$, the imaginary part of the in-medium ρ meson propagator,

$$D_\rho(M,q;T,\mu_B) = \frac{1}{M^2 - m_\rho^2 - \Sigma_\rho(M,q;T,\mu_B)}. \tag{2}$$

In this application, the self-energy contributions taken into account are $\Sigma_\rho = \Sigma^0 + \Sigma^{\rho\pi} + \Sigma^{\rho N}$, where Σ^0 is the vacuum self-energy and $\Sigma^{\rho\pi}$ and $\Sigma^{\rho N}$ denote the contribution to the self-energy due to the direct interactions of the ρ with, respectively, pions and nucleons of the surrounding heat bath. The self-energies have been calculated according to Ref. [24], where they were evaluated in terms of empirical scattering amplitudes from resonance dominance at low energies and Regge-type behaviour at high energy. Finally, in the evolution stage that precedes or follows the hydrodynamical phase dimuon emission from the ρ-meson is calculated employing the time integration method that has long been applied in the transport description of dilepton emission (see e.g. [25]).

In Fig. 1 hybrid model calculations are compared to recent acceptance-corrected NA60 data [14]. The calculations have been exemplary performed using a hadron gas equation of state (HG-EoS) for the hydrodynamical evolution. Since in such a scenario all the energy density is stored in hadronic degrees of freedom, for the sake of consistency dimuon emission according to Eq. (1) has been evaluated during the whole hydrodynamical phase, hottest cells included (in the central region some cells have initial temperatures up to $T \sim$240 MeV). In general, however, if a phase transition from quark-gluon to hadronic matter occurs, the hadronic thermal rate would be only a fraction of the total thermal rate. Keeping in mind that the total life-time of the fireball affects the total dilepton yield too, we can quite generally say that the inclusion of a phase in which quarks and gluons are the relevant degrees of freedom would result in a reduction of the hadronic emission with respect to the calculation presented here, provided that the life-time of the fireball is comparable for the two equations of state. Calculations with such an EoS are on-going and suggest this reduction to amount to about 30%-40% across the entire dimuon invariant mass region considered here. Anyway, this baseline calculation performed with the HG-EoS still allows to point out qualitatively the main features of the present approach. More systematic studies and discussions will be presented elsewhere [26] in the next future.

We observe that the cascade emission dominates the invariant mass region around the vector meson peak for both low and intermediate transverse pair momenta p_T. At low p_T (left panel of Fig. 1), the very low invariant masses, $M < 0.5$ GeV, are filled by the thermal radiation with in-medium spectral function. The sum of both contributions, however, leads to an overestimation of the vector meson peak region at low transverse momenta of the dilepton pair. The reason for the discrepancy might partially lie in the specific spectral function used here and/or, presumably more severely, on the eventual presence of not yet negligible residual in-medium modification of the ρ meson spectral function during the cascade stage, that are here neglected. An additional source of uncertainty is the dependence of the total thermal yield on the specific criterium adopted to perform the switch from the hydrodynamic to the transport description. Further investigations are required to clarify these model dependences and quantify their effect on the total dilepton emission. With increasing p_T (right panel of Fig. 1), dilepton emission at very low invariant masses is reduced and the total spectra are almost completely determined by the cascade emission. This is not trivial. Indeed, thermal emission had already shown discrepancies for $p_T > 1$ GeV [27], pointing to the necessity to account for non-thermal contributions. In the present approach, the latters appear quite naturally.

3. Acknowledgments

The authors thanks the organizers for the exciting and sunny conference. This work was supported by the Hessen Initiative for Excellence (LOEWE) through the Helmholtz International Center for FAIR (HIC for FAIR). We thank the Center for Scientific Computing for providing computational resources.

References

[1] Agakishiev G *et al.* (CERES) 1995 *Phys. Rev. Lett.* **75** 1272–1275
[2] Mazzoni M A (HELIOS/3.) 1994 *Nucl. Phys.* **A566** 95c–102c
[3] Rapp R and Wambach J 2000 *Adv. Nucl. Phys.* **25** 1 (*Preprint* hep-ph/9909229)

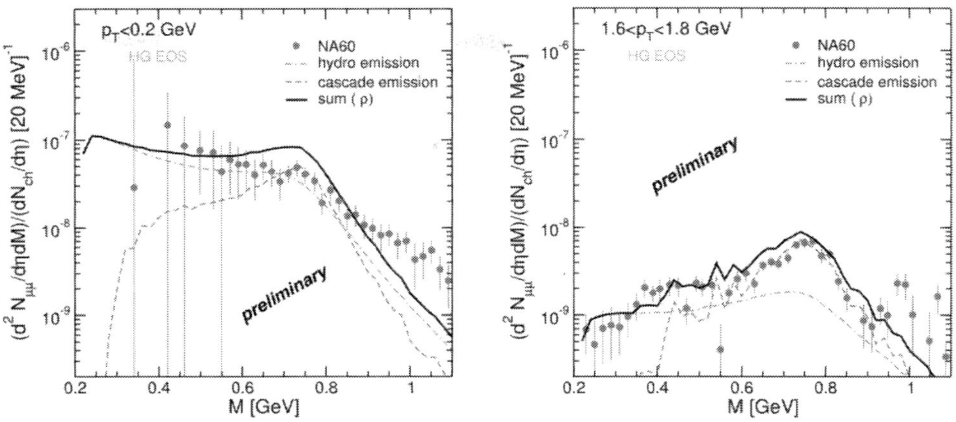

Figure 1. Left panel: Acceptance-corrected invariant mass spectra of the excess dimuons in In-In collisions at 158A GeV for transverse pair momenta $p_T < 0.2$ GeV, compared to hybrid model calculations based on thermal radiation from in-medium modified ρ meson spectral function and non-thermal cascade emission. Experimental data from Ref. [14]. Right panel: Same as in the left panel, but for the transverse momenta window $1.6 < p_T < 1.8$ GeV.

[4] Gale C and Haglin K L 2003 Electromagnetic radiation from relativistic nuclear collisions *Quark-gluon plasma. Vol. 3* ed Hwa R C and Wang X N (River Edge, USA: World Scientific) p 348 (*Preprint* hep-ph/0306098)
[5] Rapp R, Wambach J and van Hees H 2009 The Chiral Restoration Transition of QCD and Low Mass Dileptons *Relativistic Heavy Ion Physics* ed Stock R (Landolt-Bernstein series) (*Preprint* 0901.3289)
[6] Rapp R, Chanfray G and Wambach J 1997 *Nucl. Phys.* **A617** 472–495 (*Preprint* hep-ph/9702210)
[7] Friman B and Pirner H J 1997 *Nucl. Phys.* **A617** 496–509 (*Preprint* nucl-th/9701016)
[8] Peters W, Post M, Lenske H, Leupold S and Mosel U 1998 *Nucl. Phys.* **A632** 109–127 (*Preprint* nucl-th/9708004)
[9] Lutz M F M, Wolf G and Friman B 2002 *Nucl. Phys.* **A706** 431–496 (*Preprint* nucl-th/0112052)
[10] Santini E *et al.* 2008 *Phys. Rev.* **C78** 034910 (*Preprint* 0804.3702)
[11] Brown G E and Rho M 1991 *Phys. Rev. Lett.* **66** 2720–2723
[12] Hatsuda T and Lee S H 1992 *Phys. Rev.* **C46** 34–38
[13] Arnaldi R *et al.* (NA60) 2006 *Phys. Rev. Lett.* **96** 162302 (*Preprint* nucl-ex/0605007)
[14] Arnaldi R *et al.* (NA60) 2009 *Eur. Phys. J.* **C61** 711–720 (*Preprint* 0812.3053)
[15] Petersen H, Steinheimer J, Burau G, Bleicher M and Stocker H 2008 *Phys. Rev.* **C78** 044901 (*Preprint* 0806.1695)
[16] Petersen H and Bleicher M 2009 *Phys. Rev.* **C79** 054904 (*Preprint* 0901.3821)
[17] Petersen H, Steinheimer J, Bleicher M and Stocker H 2009 *J. Phys.* **G36** 055104 (*Preprint* 0902.4866)
[18] Petersen H, Mitrovski M, Schuster T and Bleicher M 2009 *Phys. Rev.* **C80** 054910 (*Preprint* 0903.0396)
[19] Graf G *et al.* 2010 *J. Phys.* **G37** 094010 (*Preprint* 1001.4937)
[20] Baeuchle B and Bleicher M 2010 *Phys. Rev.* **C81** 044904 (*Preprint* 0905.4678)
[21] Santini E, Petersen H and Bleicher M 2010 *Phys. Lett.* **B687** 320–326 (*Preprint* 0909.4657)
[22] Rischke D H, Bernard S and Maruhn J A 1995 *Nucl. Phys.* **A595** 346–382 (*Preprint* nucl-th/9504018)
[23] Rischke D H, Pursun Y and Maruhn J A 1995 *Nucl. Phys.* **A595** 383–408 (*Preprint* nucl-th/9504021)
[24] Eletsky V L, Belkacem M, Ellis P J and Kapusta J I 2001 *Phys. Rev.* **C64** 035202 (*Preprint* nucl-th/0104029)
[25] Schmidt K *et al.* 2009 *Phys. Rev.* **C79** 064908 (*Preprint* 0811.4073)
[26] Santini E *et al.* in preparation
[27] van Hees H and Rapp R 2008 *Nucl. Phys.* **A806** 339 (*Preprint* 0711.3444)

Systematic study of v_2 at 62.4 and 200 GeV in Cu + Cu and Au + Au Collisions at RHIC-PHENIX

Maya Shimomura for the PHENIX Collaboration
Graduate School of Pure and Applied Sciences, Univ. of Tsukuba, Tenno-dai 1-1-1, Tsukuba, Ibaraki, Japan
E-mail: mayap@bnl.gov

Abstract. The transverse momentum (p_{T}) and centrality dependence of the azimuthal anisotropy (v_2) are measured for charged hadron species at 62.4 and 200 GeV in Cu + Cu and Au + Au collisions by the PHENIX experiment at RHIC. The results are consistent with eccentricity scaling and with quark number (n_{q}) + transverse kinetic energy ($\mathrm{KE_T}$) scaling. Also, we found that v_2 divided by the participant eccentricity of the initial geometry proportionally increases with the number of participants to the 1/3 power except at small N_{part} in Cu+Cu at $\sqrt{s_{\mathrm{NN}}}$ = 62.4 GeV. Taking these scalings (n_{q}, $\mathrm{KE_T}$, eccentricity and $\mathrm{N}_{part}^{1/3}$) into account, there is a universal scaling for v_2 with different energies, collision sizes and particle species. The results indicate that v_2 is determined by more than just the geometrical eccentricity and also depends on the size of the collision. Moreover, using blast wave fitting, we report that both the behaviors of v_2 and p_{T} distributions can be understood from the thermal nature of produced particles based on hydro-dynamical behavior.

1. Introduction

Relativistic heavy ion collisions have been considered as a unique way to create the quark-gluon plasma (QGP), which is the phase of de-confined quarks and gluons. The Relativistic Heavy Ion Collider (RHIC) at Brookhaven National Laboratory was constructed to create and study the QGP. Azimuthal anisotropy of particles produced in relativistic heavy ion collisions is the one of the most powerful probes for investigating the characteristics of the QGP. Especially the strength of the elliptic anisotropy (v_2), which is defined by the second harmonics of Fourier expansion for the azimuthal distribution of produced particles with respect to the reaction plane, is expected to be sensitive to the early stage of heavy ion collisions. The anisotropy in the momentum phase space is transferred from the geometrical anisotropy of the initial collisional region because of the pressure gradient. Thus, the measured v_2 reflects the dynamical properties of the dense matter produced in the collisions.

2. Motivation

One of the most remarkable findings at RHIC is that the strength of v_2 can be described well by hydro-dynamical models in the low transverse momentum region ($\sim$ 1 GeV/c) [1]. In the intermediate transverse momentum region (1 $\sim$ 4 GeV/c), v_2 is consistent with n_{q} and KE_{T} ($= m_{\mathrm{T}}$ - m_0) scaling, and the result supports a quark-recombination model [2]. For a more comprehensive understanding of v_2, we have carried out systematic measurements of v_2, by measuring v_2 for identified charged hadrons in Au+Au and Cu+Cu collisions at $\sqrt{s_{\mathrm{NN}}}$ = 200

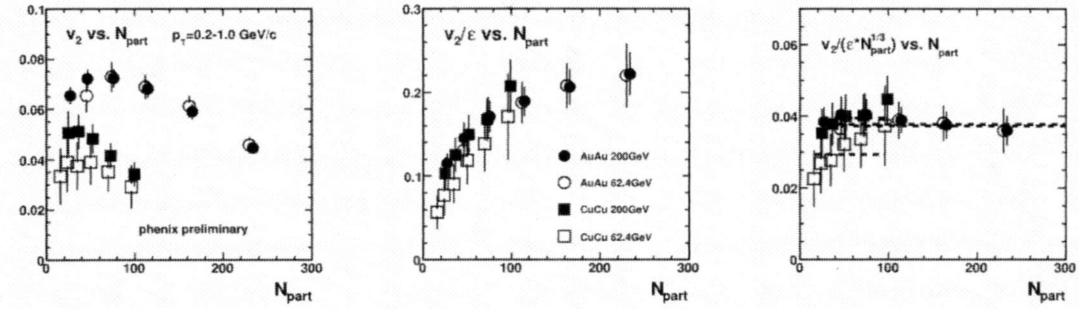

Figure 1. Comparison of integrated v_2 as a function of N_{part} for four different collision systems. Left panel shows v_2 vs. N_{part}, middle panel is v_2/ε vs. N_{part} and right panel is $v_2/(\varepsilon \cdot N_{\text{part}}^{1/3})$ vs. N_{part}. Closed symbols indicate the results of $\sqrt{s_{\text{NN}}}$ = 200 GeV, open symbols are for $\sqrt{s_{\text{NN}}}$ = 62.4 GeV. Circles indicate Au+Au collisions and squares Cu+Cu. The statistical and systematic errors are included in the bars.

and 62.4 GeV, and studied the dependence on collision energy, size and species of the produced particles. We expect that the v_2 is determined not only by the initial ellipticity, but also by the collision volume [3].

3. Results

3.1. Collision Energy, System Size and Species Dependence

The left panel in Figure 1 shows the comparison of v_2 in Au+Au and Cu+Cu at $\sqrt{s_{\text{NN}}}$ = 200 and 62.4 GeV as a function of N_{part}. The values of v_2 agree well at $\sqrt{s_{\text{NN}}}$ = 200 and 62.4 GeV in the errors, but have clear differences between the Cu+Cu and Au+Au. Since Au+Au and Cu+Cu collisions have different initial geometrical eccentricities at the same N_{part}, normalizing v_2 by eccentricity, ε, (eccentricity scaling) shows that the data follows a universal curve between Cu + Cu and Au + Au as shown in the middle panel in Figure 1. Here, we use the participant eccentricity, which includes the effect of participant fluctuations [4]. One can see that v_2/ε is not a constant and it depends on N_{part} . Therefore, v_2 can be normalized by ε at the same N_{part}, but ε is not enough to determine v_2 . We found that v_2/ε is proportional to $N_{\text{part}}^{1/3}$ as shown in the right panel of Figure 1. $v_2/(\varepsilon \cdot N_{\text{part}}^{1/3})$ is independent of the collision system except for small N_{part} in Cu+Cu at $\sqrt{s_{\text{NN}}}$ = 62.4 GeV. This exception might indicate that this is a region where the matter has not reached sufficient thermalization, although the errors are too large to discuss the difference. Therefore, a scan of collision energies and sizes would be very important for further study of this exception. Figure 1 is for p_{T} = 0.2 - 1.0 GeV/c. The results for p_{T} = 1.0 - 2.0 and 2.0 - 4.0 GeV/c have the same tendency as well.

3.2. Quark number(n_{q}) + KE_{T} scalings and Universal v_2

The left panel of Figure 2 shows that various particle species are mostly consistent with n_{q} + KE_{T} scaling in Au+Au at at $\sqrt{s_{\text{NN}}}$ = 200 GeV, From the systematic study, it is found that v_2 in Au+Au at $\sqrt{s_{\text{NN}}}$ = 62.4 GeV and in Cu+Cu at $\sqrt{s_{\text{NN}}}$ = 200 GeV are also consistent with n_{q} + KE_{T} scaling for centralities 0 - 50%. In addition to the fact that $v_2(p_{\text{T}})$ does not depend on collision energy at RHIC energies, v_2 normalized by n_{q} + KE_{T}, eccentricity, and $N_{\text{part}}^{1/3}$ scaling follows a universal curve as shown in the right panel of Figure 2. This figure includes the 45 curves for π/K/p in Au+Au at $\sqrt{s_{\text{NN}}}$ = 200 GeV, in Au+Au at $\sqrt{s_{\text{NN}}}$ = 62.4

Journal of Physics: Conference Series **270** (2011) 012041 doi:10.1088/1742-6596/270/1/012041

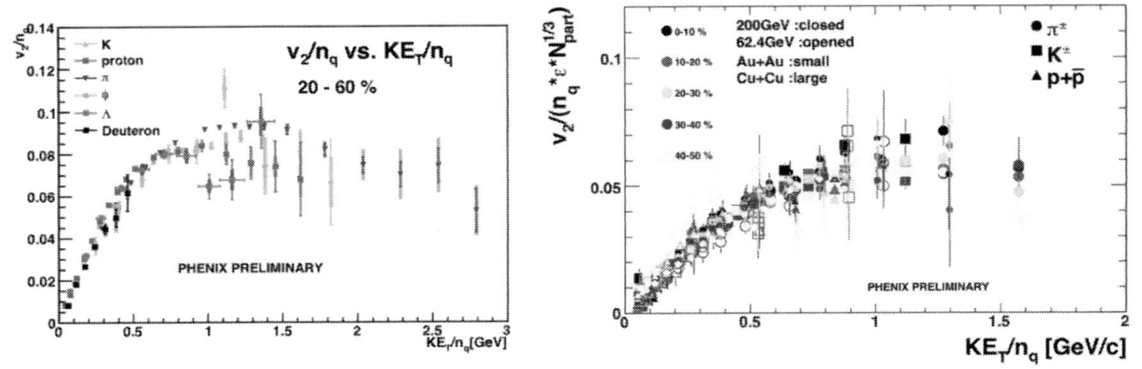

Figure 2. v_2 with the scalings. Left panel shows v_2/n_q vs. KE_T/n_q for π, K, p, Λ, ϕ and Deuteron in Au+Au at $\sqrt{s_{NN}}$ = 200 GeV at 20-60 % centrality bin. Right panel shows $v_2/(\varepsilon \cdot N_{\rm part}^{1/3} \cdot n_{\rm q})$ vs. KE_T/n_q for π/K/p in Au+Au at $\sqrt{s_{NN}}$ = 200 GeV, in Au+Au at $\sqrt{s_{NN}}$ = 62.4 GeV and in Cu+Cu at $\sqrt{s_{NN}}$ = 200 GeV at five centrality bins for 0-50 % as 10 % step for each system. There are 45 curves. Applying polynomial fitting, χ^2/NDF is 2.1.

GeV and in Cu+Cu at $\sqrt{s_{NN}}$ = 200 GeV for the five centrality bins from 0 - 50% in 10% steps. The χ^2/NDF of the polynomial fitting is 2.1.

3.3. Radial Flow Effect with Blast-Wave Fit

To understand the $N_{\rm part}$ dependence and $KE_{\rm T}$ scaling behavior for the v_2, we use a Blast-wave model to extract dynamical properties of the matter, especially at freeze-out. This model is a two-parameter model describing a boosted thermal source based on relativistic hydrodynamics [5]. The two parameters, the radial velocity ($\beta_{\rm T}$) and the freeze-out temperature ($T_{\rm fo}$), are extracted from the invariant cross section data according to the following equation:

$$\frac{dN}{m_{\rm T}dm_{\rm T}} \propto \int_0^R rdrm_{\rm T}I_0(\frac{p_{\rm T}\sinh\rho}{T_{\rm fo}})K_1(\frac{m_{\rm T}\cosh\rho}{T_{\rm fo}}), \tag{1}$$

where I_0 and K_1 represent modified Bessel functions with ρ the transverse boost $\rho(r) = \tanh^{-1}\beta_{\rm T}(r)$, $\beta_{\rm T}(r) = \beta_{\rm s}(\frac{r}{R})$, and $m_{\rm T} = \sqrt{p_{\rm T}^2 + m^2}$. Re-plotting the measured $p_{\rm T}$ spectra, weighted by measured v_2 ($p_{\rm T}$), we obtain the $p_{\rm T}$ spectra in and out-of plane separately for π/K/p. We use two data sets: Cu+Cu and Au+Au at $\sqrt{s_{NN}}$ = 200 GeV. Applying the Blast-wave fitting to the $p_{\rm T}$ spectra in and out-of plane separately, the $\beta_{\rm T}$ and $T_{\rm fo}$ in and out-of plane are obtained separately. From these values, $\beta_{\rm T2} = (\beta_{\rm Tin} - \beta_{\rm Tout})/(\beta_{\rm Tin} + \beta_{\rm Tout})/2$ is calculated. $\beta_{\rm T2}$ indicates the amplitude of the second harmonic of the $\beta_{\rm T}$ azimuthal distribution.

As results, we have observed that the blast wave model can reproduce the KET scaling. Moreover, it was found that $\beta_{\rm T2}/\varepsilon$ agrees between Au+Au and Cu+Cu and it is flat at $N_{\rm part} \geq 40$ when scaled by eccentricity as shown in Figure 3. The left panel of Figure 4 shows the $\beta_{\rm T}$ vs. $N_{\rm part}$ for in and out-of plane in Cu+Cu and Au+Au collisions. The magnitude of $\beta_{\rm T}$ is clearly different between in and out-of plane, and it agrees well between Au+Au and Cu+Cu, especially for the in-plane data. Since v_2 is proportional to $\beta_{\rm T2}$ in this model, v_2 should be scaled by participant eccentricity. However, as shown in Figure 1, this is not what is seen in the measured v_2 results. Therefore, this implies that v_2 is not determined by only $\beta_{\rm T2}$ but also other effects such as the freeze-out temperature ($T_{\rm fo}$) of radial flow. The right panel of Figure 4 shows the $T_{\rm fo}$ vs. $N_{\rm part}$ for both in and out-of plane. The magnitude of $T_{\rm fo}$ agrees well between Au+Au and Cu+Cu collision at the same $N_{\rm part}$. It can be seen that $T_{\rm fo}$ depends on $N_{\rm part}$, and it can influence v_2 since the larger $T_{\rm fo}$ makes $p_{\rm T}$ spectra flatter. Thus, the size dependence of v_2

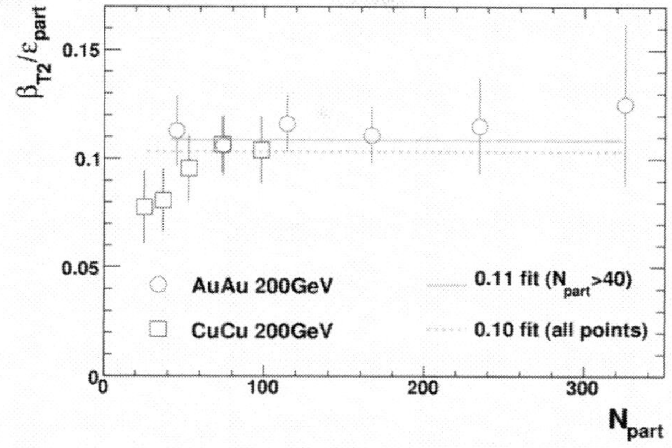

Figure 3. β_{T2}/ε vs. N_{part} in Cu+Cu and Au+Au at 200 GeV. This result is obtained by fitting to PHENIX PRELIMINARY results.

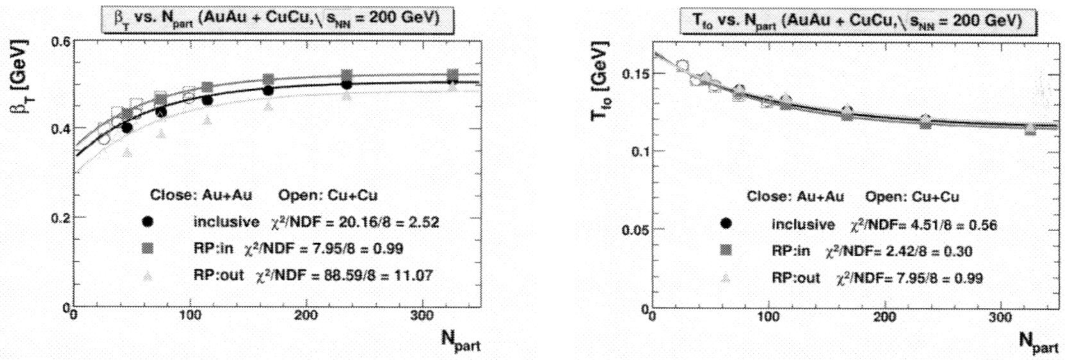

Figure 4. Left panel shows that β_T vs. N_{part} for in and out-of plane in Cu+Cu and Au+Au at 200 GeV. This result is obtained by fitting to PHENIX PRELIMINARY results. Right panel shows that T_{fo} vs. N_{part} for in and out-of plane in Cu+Cu and Au+Au at 200 GeV. This result is obtained by fitting to PHENIX PRELIMINARY results.

can be understood as thermal freeze-out nature of produced particles based on hydrodynamical behavior, which is different from that of chemical freeze-out.

4. Acknowledgements

I would like to express my great thanks to the organizers of this conference, Hot Quarks 2010, for the opportunity to present these results.

References

[1] S. S. Adler et al., *Phys. Rev. Lett.* **91**, 182301 (2003)
[2] A. Adare et al., *Phys. Rev. Lett.* **98**, 162301 (2007)
[3] S. A. Voloshin, A. M. Poskanzer and R. Snellings (2008) nucl-ex/08092949
[4] B. Alver et al., *Phys. Rev. Lett.* **98**, 242302 (2007)
[5] S. S. Adler et al., *Phys. Rev.* **C72**, 014903 (2005)

Journal of Physics: Conference Series **270** (2011) 012042 doi:10.1088/1742-6596/270/1/012042

Relativistic Viscous Hydrodynamics for Multi-Component Systems with Multiple Conserved Currents

Akihiko Monnai, Tetsufumi Hirano

Department of Physics, The University of Tokyo, Tokyo 113-0033, Japan

E-mail: monnai@nt.phys.s.u-tokyo.ac.jp

Abstract. We would like to formulate relativistic dissipative hydrodynamics for multi-component systems with multiple conserved currents. This is important for analyses of the hot matter created in relativistic heavy ion collisions because particle creations and annihilations of various particle species are frequently taking place there. We show that consistent formulation in such systems involves many non-trivialities, and derive constitutive equations that satisfy Onsager reciprocal relations and describe the systems without ambiguity.

1. Introduction

One of the most important discoveries of Relativistic Heavy Ion Collider experiments is that relativistic ideal hydrodynamic models quantitatively describe the collective behavior of the quark-gluon plasma (QGP) created in the collisions [1]. The success is a great motivation to take non-equilibrium processes into account to explain small discrepancies between theoretical predictions and experimental data. Large Hadron Collider experiments also expect heavy ion collisions of higher energies, and viscous hydrodynamic models would become more important as one might observe *less* strongly-coupled QGP because of asymptotic freedom in QCD.

The ambiguities of viscous hydrodynamic models lie in the formalism of relativistic dissipative hydrodynamics itself. If one considers only the first order dissipative correction to the entropy current, theories are known to become acausal and unstable [2]. Thus varieties of theories with the second order correction have been proposed. However, the resulting constitutive equations for the dissipative currents differ among theories. Also, most viscous hydrodynamic models are based on the theories for single component systems with binary collisions. Considering that the QGP is a multi-component system with particle creation and annihilation, it is important that one establishes a multi-component formalism which takes account of inelastic scatterings. This naturally motivates us to consider multiple conserved currents because one needs to solve continuity equations for conserved charges, rather than the ones for particle numbers.

2. Formulation of Relativistic Dissipative Hydrodynamics

We would like to formulate relativistic dissipative hydrodynamics for multi-component and multi-conserved current systems [3]. Relativistic hydrodynamics is a general theoretical framework which describes collective motion of strongly-coupled systems with the energy-momentum conservation $\partial_\mu T^{\mu\nu} = 0$, the charge conservations $\partial_\mu N_J^\mu = 0$ and the law of increasing entropy $\partial_\mu s^\mu \geq 0$. Here the index J denotes different types of conserved quantum number. Typically, one can consider the baryon number B or the strangeness S for such quantities in relativistic heavy ion collisions.

We introduce tensor decompositions of the energy-momentum tensor $T^{\mu\nu}$ and the conserved currents N_J^μ in terms of the flow field u^μ to define thermodynamic quantities

$$T^{\mu\nu} = (e_0 + \delta e)u^\mu u^\nu - (P_0 + \Pi)\Delta^{\mu\nu} + W^\mu u^\nu + W^\nu u^\mu + \pi^{\mu\nu}, \tag{1}$$

$$N_J^\mu = (n_{J0} + \delta n_J)u^\mu + V_J^\mu, \tag{2}$$

where $\Delta^{\mu\nu} = g^{\mu\nu} - u^\mu u^\nu$ is the projection operator. There are $2 + N$ equilibrium quantities: energy density e_0, hydrostatic pressure P_0 and type-J charge density n_{J0}. On the other hand, one has $10 + 4N$ dissipative currents: bulk pressure Π, energy density distortion δe, energy current W^μ, shear stress tensor $\pi^{\mu\nu}$, type-J charge density distortion δn_J and type-J charge current V_J^μ. Although thermodynamic stability conditions [4, 5] require δe and δn_J to be zero, we have to keep them for the moment to correctly count the number of the dissipative currents. In ideal hydrodynamics, the conservation laws and the equation of state $P_0 = P_0(e_0, \{n_{J0}\})$ give sufficient number of equations to describe systems because the number of unknowns is $5 + N$ in such systems. In dissipative hydrodynamics, one needs *constitutive equations* because there are the $10 + 4N$ additional unknowns Π, δe, W^μ, $\pi^{\mu\nu}$, δn_J and V_J^μ. We would like to derive them from the law of increasing entropy, because it embodies irreversible processes.

The energy-momentum tensor and the conserved currents are expressed in relativistic kinetic theory with the phase space distribution f^i as

$$T^{\mu\nu} = \sum_i \int \frac{g_i d^3p}{(2\pi)^3 E_i} p_i^\mu p_i^\nu f^i, \quad N_J^\mu = \sum_i \int \frac{q_i^J g_i d^3p}{(2\pi)^3 E_i} p_i^\mu f^i, \tag{3}$$

where g_i and q_i^J are degeneracy and type-J quantum number of i-th particle species, respectively. We introduce the higher moment equations for the energy-momentum conservation and for the charge conservations

$$\partial_\alpha I^{\mu\nu\alpha} = \sum_i \int \frac{g_i d^3p}{(2\pi)^3 E_i} p_i^\mu p_i^\nu p_i^\alpha \partial_\alpha f^i = Y^{\mu\nu}, \tag{4}$$

$$\partial_\alpha I_J^{\mu\alpha} = \sum_i \int \frac{q_i^J g_i d^3p}{(2\pi)^3 E_i} p_i^\mu p_i^\alpha \partial_\alpha f^i = Y_J^\mu, \tag{5}$$

to derive second order constitutive equations. $Y^{\mu\nu}$ and Y_J^μ are to be determined from the law of increasing entropy. The conventional Israel-Stewart theory [5] has only Eq. (4) because single component systems with binary collisions are considered. However, one has to introduce the new sets of equations (5) to consistently match the numbers of equations with that of dissipative currents in multi-component and/or multiple conserved current systems.

The entropy production can be written in kinetic theory as

$$\partial_\mu s^\mu = -\partial_\mu \sum_i \int \frac{g_i d^3p}{(2\pi)^3 E_i} p_i^\mu \left[f^i \ln f^i - \frac{1}{\epsilon}(1 + \epsilon f^i)\ln(1 + \epsilon f^i) \right] = \sum_i \int \frac{g_i d^3p_i}{(2\pi)^3 E_i} p_i^\mu y^i \partial_\mu f^i, \tag{6}$$

where y^i is defined in $f^i = [\exp(y^i) - \epsilon]^{-1}$. The sign factor ϵ is $+1$ for fermions and -1 for bosons. We estimate the distortion of distribution $\delta f^i = f^i - f_0^i$ in terms of dissipative currents by generalizing the Grad's moment method [5] for the systems with multiple conserved currents. If we require that the resulting constitutive equations satisfy Onsager reciprocal relations [6, 7], the only possible moment expansion would be

$$\delta f^i = -f_0^i(1 + \epsilon f_0^i)(p_i^\mu \sum_J q_i^J \varepsilon_\mu^J + p_i^\mu p_i^\nu \varepsilon_{\mu\nu}), \tag{7}$$

where ε_μ^J and $\varepsilon_{\mu\nu}$ are self-consistently determined so that the distribution reproduces the energy-momentum tensor and the conserved currents (3). Then the ε's are expressed as linear combinations of the dissipative currents. On the other hand, the entropy production (6) is expressed in terms of ε's and Y's as

$$\partial_\mu s^\mu = \sum_J \varepsilon_\nu^J Y_J^\nu + \varepsilon_{\nu\rho} Y^{\nu\rho} \geq 0. \tag{8}$$

The law of increasing entropy requires that Y's are linear combinations of ε's. Then the second order constitutive equations for multi-component systems with multiple conserved currents are obtained by writing the moment equations (4) and (5) with the dissipative current:

$$\begin{aligned}
\Pi &= -\zeta \nabla_\mu u^\mu - \zeta_{\Pi\delta e} D\frac{1}{T} + \sum_J \zeta_{\Pi\delta n_J} D\frac{\mu_J}{T} \\
&- \tau_\Pi D\Pi + \sum_J \chi_{\Pi\Pi\Pi}^{aJ} \Pi D\frac{\mu_J}{T} + \chi_{\Pi\Pi\Pi}^{b} \Pi D\frac{1}{T} + \chi_{\Pi\Pi\Pi}^{c} \Pi \nabla_\mu u^\mu \\
&+ \sum_J \chi_{\Pi W}^{aJ} W_\mu \nabla^\mu \frac{\mu_J}{T} + \chi_{\Pi W}^{b} W_\mu \nabla^\mu \frac{1}{T} + \chi_{\Pi W}^{c} W_\mu D u^\mu + \chi_{\Pi W}^{d} \nabla^\mu W_\mu \\
&+ \sum_{J,K} \chi_{\Pi V_J}^{aK} V_\mu^J \nabla^\mu \frac{\mu_K}{T} + \sum_J \chi_{\Pi V_J}^{b} V_\mu^J \nabla^\mu \frac{1}{T} + \sum_J \chi_{\Pi V_J}^{c} V_\mu^J D u^\mu + \sum_J \chi_{\Pi V_J}^{d} \nabla^\mu V_\mu^J \\
&+ \chi_{\Pi\pi} \pi_{\mu\nu} \nabla^{\langle\mu} u^{\nu\rangle},
\end{aligned} \tag{9}$$

$$\begin{aligned}
W^\mu &= -\kappa_W \left(\frac{1}{T} D u^\mu + \nabla^\mu \frac{1}{T} \right) + \sum_J \kappa_{W V_J} \nabla^\mu \frac{\mu_J}{T} \\
&- \tau_W \Delta^{\mu\nu} D W_\nu + \sum_J \chi_{WW}^{aJ} W^\mu D\frac{\mu_J}{T} + \chi_{WW}^{b} W^\mu D\frac{1}{T} \\
&+ \chi_{WW}^{c} W^\mu \nabla_\nu u^\nu + \chi_{WW}^{d} W^\nu \nabla_\nu u^\mu + \chi_{WW}^{e} W^\nu \nabla^\mu u_\nu \\
&- \sum_J \tau_{W V_J} \Delta^{\mu\nu} D V_\nu^J + \sum_{J,K} \chi_{W V_J}^{aK} V_J^\mu D\frac{\mu_K}{T} + \sum_J \chi_{W V_J}^{b} V_J^\mu D\frac{1}{T} \\
&+ \sum_J \chi_{W V_J}^{c} V_J^\mu \nabla^\nu u_\nu + \sum_J \chi_{W V_J}^{d} V_J^\nu \nabla_\nu u^\mu + \sum_J \chi_{W V_J}^{e} V_J^\nu \nabla^\mu u_\nu \\
&+ \sum_J \chi_{W\pi}^{aJ} \pi^{\mu\nu} \nabla_\nu \frac{\mu_J}{T} + \chi_{W\pi}^{b} \pi^{\mu\nu} \nabla_\nu \frac{1}{T} + \chi_{W\pi}^{c} \pi^{\mu\nu} D u_\nu + \chi_{W\pi}^{d} \Delta^{\mu\nu} \nabla^\rho \pi_{\nu\rho} \\
&+ \sum_J \chi_{W\Pi}^{aJ} \Pi \nabla^\mu \frac{\mu_J}{T} + \chi_{W\Pi}^{b} \Pi \nabla^\mu \frac{1}{T} + \chi_{W\Pi}^{c} \Pi D u^\mu + \chi_{W\Pi}^{d} \nabla^\mu \Pi,
\end{aligned} \tag{10}$$

$$\begin{aligned}
V_J^\mu &= \sum_K \kappa_{V_J V_K} \nabla^\mu \frac{\mu_K}{T} - \kappa_{V_J W} \left(\frac{1}{T} D u^\mu + \nabla^\mu \frac{1}{T} \right) \\
&- \sum_K \tau_{V_J V_K} \Delta^{\mu\nu} D V_\nu^K + \sum_{K,L} \chi_{V_J V_K}^{aL} V_K^\mu D\frac{\mu_L}{T} + \sum_K \chi_{V_J V_K}^{b} V_K^\mu D\frac{1}{T} \\
&+ \sum_K \chi_{V_J V_K}^{c} V_K^\mu \nabla_\nu u^\nu + \sum_K \chi_{V_J V_K}^{d} V_K^\nu \nabla_\nu u^\mu + \sum_K \chi_{V_J V_K}^{e} V_K^\nu \nabla^\mu u_\nu \\
&- \tau_{V_J W} \Delta^{\mu\nu} D W_\nu + \sum_K \chi_{V_J W}^{a} W^\mu D\frac{\mu_K}{T} + \chi_{V_J W}^{b} W^\mu D\frac{1}{T} \\
&+ \chi_{V_J W}^{c} W^\mu \nabla^\nu u_\nu + \chi_{V_J W}^{d} W^\nu \nabla_\nu u^\mu + \chi_{V_J W}^{e} W^\nu \nabla^\mu u_\nu \\
&+ \sum_K \chi_{V_J\pi}^{aK} \pi^{\mu\nu} \nabla_\nu \frac{\mu_K}{T} + \chi_{V_J\pi}^{b} \pi^{\mu\nu} \nabla_\nu \frac{1}{T} + \chi_{V_J\pi}^{c} \pi^{\mu\nu} D u_\nu + \chi_{V_J\pi}^{d} \Delta^{\mu\nu} \nabla^\rho \pi_{\nu\rho}
\end{aligned}$$

$$+ \sum_K \chi^{aK}_{V_J\Pi}\Pi\nabla^\mu\frac{\mu_K}{T} + \chi^b_{V_J\Pi}\Pi\nabla^\mu\frac{1}{T} + \chi^c_{V_J\Pi}\Pi Du^\mu + \chi^d_{V_J\Pi}\nabla^\mu\Pi, \quad (11)$$

$$\begin{aligned}
\pi^{\mu\nu} &= 2\eta\nabla^{\langle\mu}u^{\nu\rangle} \\
&- \tau_\pi D\pi^{\langle\mu\nu\rangle} + \sum_J \chi^{aJ}_{\pi\pi}\pi^{\mu\nu}D\frac{\mu_J}{T} + \chi^b_{\pi\pi}\pi^{\mu\nu}D\frac{1}{T} + \chi^c_{\pi\pi}\pi^{\mu\nu}\nabla_\rho u^\rho + \chi^d_{\pi\pi}\pi^{\rho\langle\mu}\nabla_\rho u^{\nu\rangle} \\
&+ \sum_J \chi^{aJ}_{\pi W}W^{\langle\mu}\nabla^{\nu\rangle}\frac{\mu_J}{T} + \chi^b_{\pi W}W^{\langle\mu}\nabla^{\nu\rangle}\frac{1}{T} + \chi^c_{\pi W}W^{\langle\mu}Du^{\nu\rangle} + \chi^d_{\pi W}\nabla^{\langle\mu}W^{\nu\rangle} \\
&+ \sum_{J,K} \chi^{aJ}_{\pi V_J}V_J^{\langle\mu}\nabla^{\nu\rangle}\frac{\mu_K}{T} + \sum_J \chi^b_{\pi V_J}V_J^{\langle\mu}\nabla^{\nu\rangle}\frac{1}{T} + \sum_J \chi^c_{\pi V_J}V_J^{\langle\mu}Du^{\nu\rangle} + \sum_J \chi^d_{\pi V_J}\nabla^{\langle\mu}V_J^{\nu\rangle} \\
&+ \chi_{\pi\Pi}\Pi\nabla^{\langle\mu}u^{\nu\rangle}. \quad (12)
\end{aligned}$$

Here $D = u^\mu\partial_\mu$ and $\nabla^\mu = \Delta^{\mu\nu}\partial_\nu$ are time- and space-like derivatives. The angle brackets denote the operation $A^{\langle\mu}B^{\nu\rangle} = [\frac{1}{2}(\Delta^\mu{}_\alpha\Delta^\nu{}_\beta + \Delta^\mu{}_\beta\Delta^\nu{}_\alpha) - \frac{1}{3}\Delta^{\mu\nu}\Delta_{\alpha\beta}]A^\alpha B^\beta$. ζ's, κ's and η are first order transport coefficients; ζ is bulk viscosity, κ_{WW} energy conductivity, $\kappa_{V_JV_J}$ charge conductivity and η shear viscosity. τ's are relaxation times, and χ's are second order transport coefficients. The ratios of the first and the second order coefficients can be determined in kinetic theory.

3. Discussion and Conclusions

We have derived constitutive equations for multi-component systems with multiple conserved currents in arbitrary frame. We would like to emphasize four points. Firstly, the equations have Onsager cross terms, although most of the conventional formalisms do not consider them carefully. These terms contain important physics such as Soret and Dufour effects, which are chemical diffusion caused by thermal gradient and its counter-effect respectively. The reciprocal relations are satisfied. Secondly, there are relaxation terms – terms with time-like derivatives of the dissipative currents themselves – in the constitutive equations. These terms play important roles in preserving causality because they cause exponential damping of the dissipative currents. Thirdly, we have independent equations for energy and charge currents. This means that our formalism can describe systems in an arbitrary frame unlike the conventional formalisms in which one has to choose either energy or particle frames. Fourthly, we have several second order terms which do not appear in Ref. [5]. For further detailed discussion including comparison with other approaches [8, 9, 10], see Ref. [3].

Acknowledgments

The authors acknowledge fruitful discussions with T. Hatsuda, T. Kodama, T. Koide, T. Kunihiro and S. Muroya. The work of A.M. is supported by JSPS Research Fellowships for Young Scientists. The work of T.H. was partly supported by Grant-in-Aid for Scientific Research No. 19740130 and by Sumitomo Foundation No. 080734.

References

[1] Hirano T, van der Kolk N and Bilandzic A 2010 *Lect. Notes Phys.* **785** 139
[2] Hiscock W A and Lindblom L 1985 *Phys. Rev.* D **31** 725
[3] Monnai A and Hirano T 2010 *arXiv:1003.3087* [nucl-th] (to appear in Nucl. Phys. A)
[4] Monnai A and Hirano T 2009 *Phys. Rev.* C **80** 054906
[5] Israel W and Stewart J M 1979 *Annals Phys.* **118** 341
[6] Onsager L 1931 *Phys. Rev.* **37** 405
[7] Onsager L 1931 *Phys. Rev.* **38** 2265
[8] Baier R, Romatschke P, Son D T, Starinets A O and Stephanov M A 2008 *JHEP* **0804** 100
[9] Tsumura K and Kunihiro T 2010 *Phys. Lett.* B **690** 255
[10] Betz B, Henkel D and Rischke D H 2009 *J. Phys.* G **36** 064029

Journal of Physics: Conference Series **270** (2011) 012043 doi:10.1088/1742-6596/270/1/012043

Elliptic flow studies in heavy-ion collisions using the CMS detector

Sergey Petrushanko (for the CMS collaboration)
Skobeltsyn Institute of Nuclear Physics, Moscow State University, Leninskiye Gory, 119991 Moscow, Russia

E-mail: Serguei.Petrouchanko@cern.ch

Abstract. The azimuthal anisotropy of charged particle emission in heavy ion collisions is a sensitive probe of the properties and the dynamical evolution of the produced matter. The systematic study of elliptic flow (v_2), and higher order harmonics can yield information about the equilibration timescale, the nuclear equation of state and the viscosity in the different stages of the system evolution, as well as the initial state conditions of the heavy ion collisions. Experimentally, it is important to be able to reconstruct the flow using different methods, since they have different sensitivity to non-flow correlations and fluctuations which affect the extracted signal. In CMS, the nuclear reaction plane can be determined independently using several different detector subsystems. The reconstruction of elliptic flow using calorimetry and the tracking system is presented.

1. Introduction

In non-central collisions between two nuclei the beam direction and the impact parameter vector define a reaction plane for each event. A measurement of the azimuthal anisotropy of particle production with respect to the reaction plane is one of the important tools for studying the properties of the dense matter created in ultra relativistic heavy-ion collisions [1, 2]. There is a number of predictions for elliptic flow in the heavy-ion programme at the LHC [3, 4, 5]; compared with the RHIC energies a decrease, increase or saturation is possible.

This report is dedicated to studying the capability of the CMS detector [6] at the LHC to reconstruct the reaction plane and to measure elliptic flow, using calorimetry and the tracking system. The high tracking efficiency and low rate of fake tracks at CMS, together with a large calorimetric coverage, provide a precise measurement of global event characteristics, event by event [7, 8].

The first measurements of the proton-proton collisions with the CMS detector at the LHC was begun in November 2009 [9]. The start of the heavy-ion programme at the LHC with lead-lead collisions at the energy in c.m.s. $\sqrt{s} = 2.76$ TeV per nucleon pair is planned for the end of 2010.

2. Methods

The elliptic flow parameter, v_2, is defined as the second harmonic coefficient in the Fourier expansion of the particle azimuthal distribution with respect to the reaction plane:

$$\frac{dN}{d\varphi} = \frac{N_0}{2\pi}[1 + 2v_1 \cos(\varphi - \Psi_R) + 2v_2 \cos 2(\varphi - \Psi_R) + ...], \tag{1}$$

where Ψ_R is the true reaction plane angle and N_0 stands for full multiplicity. Then v_2 is the average over particles of $\cos(2(\varphi - \Psi_R))$.

There exists a wealth of anisotropic flow measurement methods, each of which has its advantages and limitations. Here we have used a v_2 determination method based on plane angle measurement. Usually the true elliptic flow coefficient in the event plane (EP) method is evaluated by dividing the observed v_2 value by a factor, R [10], which accounts for the event plane resolution:

$$v_2\{EP\} = \frac{v_2^{\mathrm{obs}}\{EP\}}{R} = \frac{\langle\cos 2(\varphi - \Psi_2)\rangle}{\langle\cos 2(\Psi_2 - \Psi_R)\rangle}. \tag{2}$$

Here the event plane angle Ψ_2 is the estimate of the true reaction plane angle Ψ_R. The mean was taken over all charged particles in a given event and then over all events. In order to avoid the trivial autocorrelation of particles, the event plane angle Ψ_2 and, hence, R are calculated from the angular distribution of a sample of events, and v_2 from another event sample with the same multiplicity. The samples can be selected, for instance, in two distinct regions of pseudorapidity, such as $\eta < 0$ and $\eta > 0$.

The event plane angle, Ψ_n, can be determined from the measured n-th harmonics via the standard method [10, 11]:

$$\tan n\Psi_n = \frac{\sum_i w_i \sin(n\varphi_i)}{\sum_i w_i \cos(n\varphi_i)}, \qquad n \geq 1, \quad 0 \leq \Psi_n < 2\pi/n, \tag{3}$$

where φ_i is the azimuthal angle of the i-th particle and w_i is a weight. The sum runs over all particles in each given event.

Other techniques like cumulant [12] and Lee-Yang zero's [13] methods for elliptic flow measurement with CMS detector at the LHC also will be useful.

3. Analysis and discussion

For the estimation of the azimuthal anisotropy of particles in heavy ion collisions, the HYDJET event generator [14] was used with a full GEANT simulation of the CMS detector responses.

Study of event plane resolution with CMS calorimeters. It was found that the CMS calorimetric system can be used for the determination of the event plane, using the energy responses of the calorimeter towers $\omega_i = E_i$ in Eq. (3) [7]. Although the anisotropic flow has to be maximal at midrapidity, the much larger total energy deposition in the CMS calorimeter endcaps ($1.5 < |\eta| < 3.$) results in reduced relative fluctuations and, accordingly, in a much better event plane resolution. Moreover, energy flow measurements in the endcaps are less sensitive to the magnetic field than in the barrel region ($|\eta| < 1.5$).

Study of v_2 reconstruction with the CMS tracker. A sample of 10^5 Pb+Pb events at impact parameter $b = 9$ fm at $\sqrt{s} = 5.5$ TeV per nucleon pair within the pseudorapidity window $|\eta| < 2.4$ (the CMS tracker acceptance) was utilized. The standard settings were used to reconstruct tracks (i.e. more than 12 hits per track and a track fit probability above 1 %). A cut on $p_T > 0.9$ GeV/c was set in both simulated and reconstructed events. The number of reconstructed tracks per event is about 170 at this centrality.

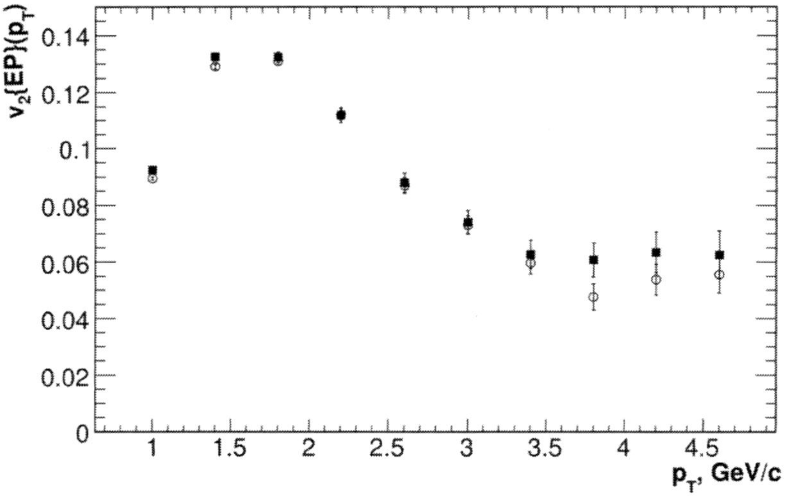

Figure 1. The p_T dependence of $v_2\{EP\}$ in Pb+Pb collisions for impact parameter $b = 9$ fm, calculated with the simulated (open circles) and reconstructed events (closed squares). Statistical errors are shown for 10^5 events. Non-flow systematic uncertainties not included.

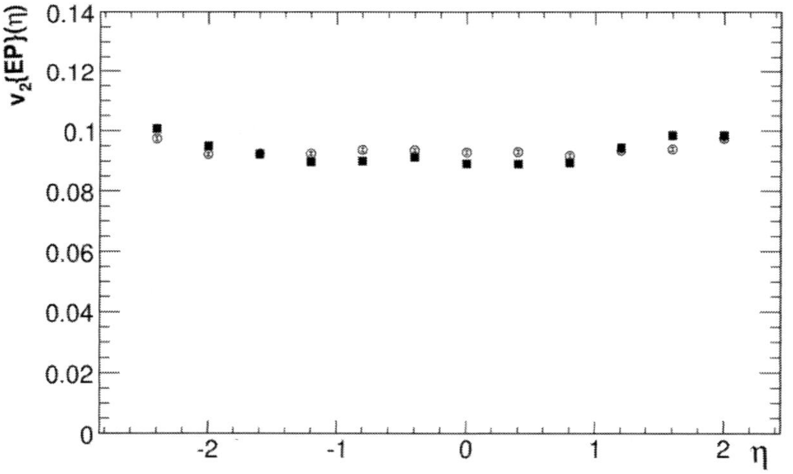

Figure 2. The η dependence of $v_2\{EP\}$ in Pb+Pb collisions for impact parameter $b = 9$ fm, calculated with the simulated (open circles) and reconstructed events (closed squares). Statistical errors are shown for 10^5 events. Non-flow systematic uncertainties not included.

The differential p_T and η dependencies of the elliptic flow in Pb+Pb collisions for impact parameter $b = 9$ fm are shown in Fig. 1 and Fig. 2, respectively. The dependencies were calculated using the event plane angle determination from Eq. (3). For the p_T dependence, two sub-event sets were used, with $\eta > 0$ and $\eta < 0$. For the η dependence, the factor R in each histogram bin was calculated using particles from other bins (excluding neighboring bins).

4. Summary

We have shown that the CMS detector at the LHC will be able to determine the reaction plane using the calorimeters and the tracker. The transverse momentum dependence of the elliptic flow coefficient v_2 can be reconstructed in the CMS tracker with high accuracy. The CMS track reconstruction performance induces a systematic error estimated to be about 3% on the v_2 determination.

Acknowledgments

I would like to thank the members of the CMS collaboration for providing the materials. I wish to express the gratitude to the organizers of the Hot Quarks Workshop 2010 for the possibility to give my talk. This work was partly supported by Russian Foundation for Basic Research (grants No 08-02-91001 and No 08-02-92496).

References

[1] J.-Y.Ollitrault, *Phys.Rev.* **D** 46, 229 (1992).
[2] H.Sorge, *Phys. Rev.Lett.* 82, 2048 (1999).
[3] N.Armesto, arXiv:0903.1330 (2009).
[4] G.Eyyubova et al., *Phys.Rev.* **C** 80, 064907 (2009).
[5] E. Zabrodin et al., arXiv:1001.3602 (2010)
[6] CMS Collaboration, *JINST* 3, S08004 (2008).
[7] D.d'Enterria et al., *J.Phys.* **G** 34, 2307 (2007).
[8] V.L.Korotkikh et al., *Phys.Atom.Nucl.* 71, 2142 (2008).
[9] CMS Collaboration, *J.High Energy Phys.* 02, 041 (2010).
[10] A.M.Poskanzer, S.A.Voloshin, *Phys.Rev.* **C** 58, 1671 (1998).
[11] S.A.Voloshin, Y.Zhang, *Z.Phys.* **C** 70, 665, (1996).
[12] N.Borghini, P.M.Dinh, J.-Y.Ollitrault, *Phys.Rev.* **C** 64, 054901 (2001).
[13] R.S.Bhalerao, N.Borghini, J.-Y.Ollitrault, *Nucl.Phys* **A** 727, 373 (2003).
[14] I.P.Lokhtin, A.M.Snigirev, *Eur.Phys.J.* **C** 46, 211 (2006).

Journal of Physics: Conference Series **270** (2011) 012044 doi:10.1088/1742-6596/270/1/012044

v_4 for identified particles at RHIC from viscous hydrodynamics

Matthew Luzum
Institut de physique théorique, CEA Saclay, 91191 Gif-sur-Yvette, France
E-mail: matthew.luzum@cea.fr

Abstract. Using ideal and viscous hydrodynamics, the ratio of azimuthal moments $v_4/(v_2)^2$ is calculated for pions, protons, and kaons in $\sqrt{s} = 200A$ GeV Au+Au collisions. For any value of viscosity here is little dependence on particle species. Ideal hydrodynamics and data show a flat curve as a function of p_t. Adding viscosity in the standard way destroys this flatness. However, it can be restored by replacing the standard quadratic ansatz for δf (the viscous correction to the distribution function at freeze-out) with a weaker momentum dependence.

1. Introduction

The azimuthal distribution of detected particles in a relativistic heavy ion collision contains valuable information about the medium created in such a collision. This angular distribution is typically characterized by its Fourier components

$$E\frac{d^3N}{d^3\mathbf{p}} = v_0\left[1 + \sum_{n=1}^{\infty} 2v_n \cos(n\,\phi)\right], \tag{1}$$

where ϕ is the angle of the outgoing particle momentum with respect to the collision plane for each event.

The first non-vanishing coefficient is the elliptic flow coefficient v_2, which is an important observable that is sensitive to bulk properties of the collision fireball and has been the object of much study [1]. Much less studied is the next coefficient, v_4. In these proceedings, results are presented from ideal and viscous hydrodynamic simulations of Au+Au collisions at the Relativistic Heavy Ion Collider (RHIC). The main results of these calculations can be found in Ref. [2], to which this can be considered a companion article. The details of the viscous hydrodynamic model can be found in Ref. [3].

2. Analytic prediction

A simple argument by Borghini and Ollitrault [4] indicates that v_4 is expected, at large p_t, to be largely determined by v_2. It is then most useful to construct the ratio $v_4/(v_2)^2$ and investigate deviations from this "ideal hydrodynamic" scaling.

To summarize, one can define a function $u_{\max}(\phi)$, which is the magnitude of maximum fluid velocity on the freeze-out surface which is flowing in the azimuthal direction ϕ. For symmetric collision systems, this fluid velocity on the freeze-out surface will have only even Fourier components:

$$u_{\max}(\phi) = U(1 + 2V_2\cos(2\phi) + 2V_4\cos(4\phi) + \ldots). \tag{2}$$

If one plugs this expression into the ideal (zero viscosity) Cooper-Frye formula (neglecting quantum statistics)

$$E\frac{d^3N}{d^3\mathbf{p}} \propto \int p_\mu d\Sigma^\mu \exp\left(-\frac{p_\mu u^\mu}{T}\right), \tag{3}$$

performs a saddle-point integration, and expands to leading order in these "intrinsic" elliptic and quadrangular flow coefficients $V_2 \ll 1$ and $V_4 \ll 1$, the result is

$$v_2(p_t) = \frac{V_2 U}{T}(p_t - m_t v) \tag{4}$$

$$\begin{aligned} v_4(p_t) &= \frac{1}{2}\frac{(V_2U)^2}{T^2}(p_t - m_t v)^2 + \frac{V_4 U}{T}(p_t - m_t v) \\ &= \frac{1}{2}v_2(p_t)^2 + \frac{V_4}{V_2}v_2(p_t), \end{aligned} \tag{5}$$

with $m_t = \sqrt{p_t^2 + m^2}$ and $v \equiv U/\sqrt{1+U^2}$. This suggests that it is useful to construct the ratio $v_4/(v_2)^2$:

$$\frac{v_4}{(v_2)^2} = \frac{1}{2} + \frac{V_4 T}{V_2^2 U (p_t - m_t v)} = \frac{1}{2} + \frac{V_4}{V_2}\frac{1}{v_2(p_t)}. \tag{6}$$

So, according to this prediction, the ratio should approach 1/2 at large p_t, with $1/p_t$ corrections at smaller transverse momenta. Note that this prediction does not depend on any details of

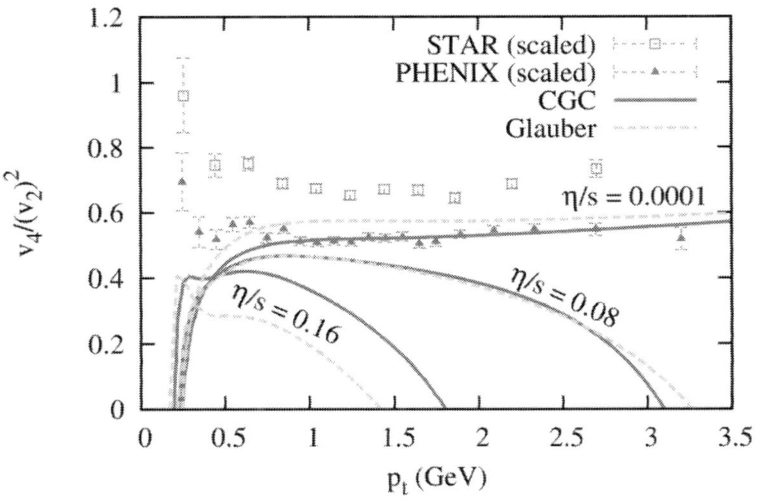

Figure 1. Charged hadron $v_4/(v_2)^2$ as a function of transverse momentum, using Glauber and CGC initial conditions with impact parameter $b = 10$ fm, and using the standard quadratic ansatz for δf. Data are from STAR (40-50% centrality) [5] and PHENIX (45-50% centrality) [6], scaled down by a factor of 1.38 to account for fluctuations [7, 2].

Journal of Physics: Conference Series **270** (2011) 012044 doi:10.1088/1742-6596/270/1/012044

the collision, as long as the assumptions are satisfied (that V_4 and V_2 are small and that one is considering large enough p_t compared to the freeze out temperature). Also note that, when plotted as a function of v_2 instead of p_t, there should be no dependence on the particle species.

3. Results

First, one should note that although hydrodynamics is expected in general to provide a description of data that is more and more reliable with decreasing p_t, the results of these calculations are of most interest at intermediate values and larger. This is true for two reasons: v_4 and $(v_2)^2$ are both very small at low p_t, and the calculated ratio is insensitive to numerical error only for p_t above $\sim 0.2-0.3$ GeV. Even in the range where the calculations are numerically reliable, the results of hydrodynamic simulations become more universal with increasing p_t. At smaller p_t, there is more and more sensitivity to details of the model. For example, the extreme sensitivity at low p_t to the freeze-out temperature seen in Ref. [2] indicates that one should use caution when interpreting these low p_t results, as an improved treatment of freeze-out (such as a hybrid approach with a hadronic afterburner) could change them significantly. Nevertheless, one can gain much insight by looking at the more robust results at intermediate p_t.

Results from ideal hydrodynamics confirm the simple predictions of Section 2. As detailed in Ref. [2], when the viscosity is set to zero, the ratio $v_4/(v_2)^2$ approaches roughly 1/2 at large p_t with a correction that behaves as $1/p_t$, as predicted from Eq. (6). Both the magnitude and sign of this correction depend strongly on the freeze-out temperature. A freeze-out temperature close to $T_f = 140 MeV$ (the temperature that gives the best fit to other data [3]) gives a flat dependence on p_t, which is what is seen experimentally (see, e.g., Fig. 1).

Using different choices of initial conditions for the hydrodynamic evolution [Glauber-type or Color-Glass-Condensate (CGC)-type] has a small effect, indicating little sensitivity to the initial eccentricity. This is in contrast to v_2 itself, which is largely driven by the initial eccentricity.

As predicted, and as seen experimentally [8, 9], there is no dependence on particle species at

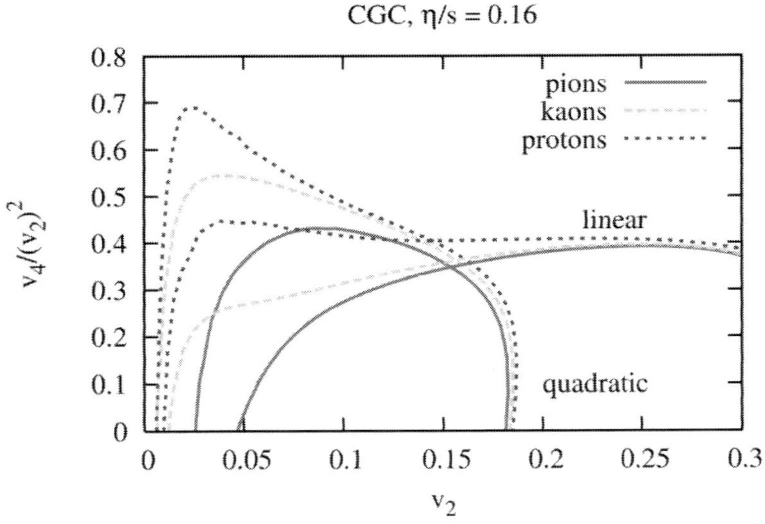

Figure 2. Identified particle $v_4/(v_2)^2$ versus v_2 from viscous hydrodynamics with CGC initial conditions and $\eta/s = 0.16$ at $b = 8$ fm using the standard quadratic ansatz as well as a linear ansatz for the viscous correction, δf, to the equilibrium distribution function at freeze-out [10].

high p_t. Since there is little dependence on transverse momentum, this is true whether $v_4/(v_2)^2$ is plotted versus p_t or v_2.

Data, however, show a value that is closer to 1, for a large range of centrality and p_t. Most of the discrepancy can be understood by noting that v_4 and v_2 are obtained separately when averaged over events. Event-by-event fluctuations in v_4 and v_2, which are not included in these hydrodynamic calculations, then increase the ratio $v_4/(v_2)^2$ [7].

To add viscosity to the calculation, one must first make a choice for δf, the viscous correction to the equilibrium distribution function at freeze-out. Until recently, all groups used a quadratic ansatz for the momentum dependence of δf, but the correct form is not known, and it can have a significant effect [10].

Adding viscosity and using the standard quadratic ansatz results in a significant change to the p_t dependence of $v_4/(v_2)^2$, as shown in Fig. 1. There is again little dependence on the initial conditions. Although there is some dependence of the overall size of $v_4/(v_2)^2$ on initial conditions, the shape of the curve is largely determined by only the viscosity. In addition, a large centrality dependence is introduced (not shown). Figure 2 shows that, as with ideal hydrodynamics results, there is little dependence on particle species at high p_t when plotted as a function of v_2, even when the viscous corrections are large, and regardless of the choice for δf.

Using a weaker momentum dependence for δf can make the p_t dependence of $v_4/(v_2)^2$ more flat (see Fig. 2), and at the same time reduce the dependence on impact parameter. This may indicate that the standard quadratic ansatz is not correct and could give insight into the behavior of the hadron gas present at freeze-out (see Ref. [10] for more details).

4. Conclusions

Viscous hydrodynamic simulations of heavy ion collisions were performed, and the ratio of azimuthal moments $v_4/(v_2)^2$ was computed. It was found that, as in ideal hydrodynamics, identified particles in viscous hydrodynamic simulations have the same value, except at low p_t, when plotted as a function of v_2. In addition, the shape as a function of p_t depends little on the initial conditions, but depends strongly on the shear viscosity to entropy density ratio η/s, when implemented using the standard quadratic ansatz for the viscous correction to the equilibrium distribution function at freeze-out. In contrast, a weaker momentum dependence for δf can restore the flat dependence on transverse momentum that is seen in ideal hydrodynamics and data, and reduce the strong centrality dependence, which is not seen in experiment.

Acknowledgments

The author would like to thank collaborators Jean-Yves Ollitrault and Clément Gombeaud, as well as Paul Romatschke for the original version of the viscous hydro code. This work was funded by "Agence Nationale de la Recherche" under grant ANR-08-BLAN-0093-01.

References

[1] Voloshin S A, Poskanzer A M and Snellings R 2008 (*Preprint* 0809.2949)
[2] Luzum M, Gombeaud C and Ollitrault J Y 2010 *Phys. Rev.* C **81** 054910 (*Preprint* 1004.2024)
[3] Luzum M and Romatschke P 2008 *Phys. Rev.* C **78** 034915 (*Preprint* 0804.4015)
[4] Borghini N and Ollitrault J Y 2006 *Phys. Lett.* B **642** 227–231 (*Preprint* nucl-th/0506045)
[5] Bai Y 2007 Ph.D. thesis University of Utrecht
[6] Adare A *et al.* (PHENIX) 2010 (*Preprint* 1003.5586)
[7] Gombeaud C and Ollitrault J Y 2010 *Phys. Rev.* C **81** 014901 (*Preprint* 0907.4664)
[8] Huang S (PHENIX) 2008 *J. Phys.* G **35** 104105 (*Preprint* 0804.4864)
[9] Huang S (PHENIX) 2009 *J. Phys.* G **36** 064061
[10] Luzum M and Ollitrault J Y 2010 *Phys. Rev.* C **82** 014906 (*Preprint* 1004.2023)

Hot Quarks 2010 IOP Publishing
Journal of Physics: Conference Series **270** (2011) 012045 doi:10.1088/1742-6596/270/1/012045

Directed flow measurements from STAR

Jiayun Chen (for the STAR Collaboration)

Institution of Particle Physics, Huazhong Normal University (CCNU), Wuhan 430079, Peoples Republic of China

The Key Laboratory of Quark and Lepton Physics (Huazhong Normal University), Ministry of Education, Wuhan 430079, Peoples Republic of China

E-mail: chenjy@iopp.ccnu.edu.cn

Abstract. STAR measurements of v_1 for identified particles for Au+Au collisions at $\sqrt{s_{NN}}$ = 200 GeV are presented. A negative v_1 slope is observed for pions, protons, antiprotons and kaons (K_S^0) in 10-70% central collisions. For pion v_1 slope, the comparison between the current result and available models (RQMD, UrQMD, AMPT, QGSM and hydro model with a tilted source) is made. In 5-30% central collisions, surprisingly sizable difference between v_1 of protons and antiprotons is observed. The centrality and energy dependence of proton v_1 are also presented. The charged hadron v_1 for Cu+Cu collisions at $\sqrt{s_{NN}}$ = 22.4 GeV is measured and compared with results obtained previously.

1. Introduction and Motivation

At RHIC, it is expected that the Quark-Gluon Plasma (QGP) can be created in the laboratory by colliding heavy ions at high energies. In a relativistic heavy ion collision, two nuclei are smashed together at near light-speed, and the collision produces thousands of particles due to the tremendous amount of energy deposited. In order to characterize the collision, anisotropic flow is used to describe the emission pattern, with respect to the reaction plane, of out-going particles. Anisotropic flow is quantified by Fourier coefficients of particle's distribution in azimuthal angle measured with respect to the reaction plane [1]. The first coefficient of such Fourier expansion is regarded as directed flow, and is evaluated by

$$v_1 = \langle \cos(\phi - \Psi_1) \rangle / Res(\Psi_1), \quad (1)$$

where ϕ denotes the azimuthal angle of outgoing particles, and Ψ_1 is the azimuthal angle of the reconstructed first-order event plane. A correction factor, namely the first-order event plane resolution ($Res(\Psi_1)$), is applied here to take into account the smearing between the reconstructed event plane and the true reaction plane [1]. To calculate the event plane resolution, we first calculate the correlation between two sub-event planes reconstructed in different η windows (details in Ref. [2, 3]), then calculate the full event plane resolution according to Eq. 11 in Ref. [1].

Directed flow (v_1) describes the side-splash of particles away from the mid-rapidity [4] and it probes the early dynamics of the system in relativistic heavy-ion collisions. The shape of the directed flow as function of rapidity for identified particles is sensitive to the equation of state (EOS) and may carry a phase transition signal [5]. It is argued that directed flow, as a function of rapidity, may exhibit a flatness at mid-rapidity due to a strong, tilted expansion of the source.

Such tilted expansion gives rise to anti-flow [6] (or a 3rd flow component [7]). The anti-flow is perpendicular to the source surface, and is in the opposite direction to the bouncing-off motion of nucleons. If the tilted expansion is strong enough, it can even overcome the bouncing-off motion and results in a negative $v_1(y)$ slope at mid-rapidity, potentially producing a wiggle-like structure in $v_1(y)$. Although these calculations are made for collisions at SPS energies, yet the origination of the negative slope is the strong expansion of a tilted source, which is also relevant at top energies at RHIC. Indeed the hydrodynamic calculation with a tilted source as the initial condition [8] can give the similar negative $v_1(y)$ slope as that found in data. A wiggle structure for nucleons is also found in the Relativistic Quantum Molecular Dynamics (RQMD) model [9], and it is attributed to the baryon stopping together with the positive space-momentum correlation. In this picture, no phase transition is needed, and pions also exhibit a wiggle because that the shadowing effect causes pions and nucleons to flow in opposite directions. To distinguish between the baryon stopping and the anti-flow associated with a phase transition, it is desirable to measure the $v_1(y)$ for identified particles and their slope around mid-rapidity.

2. Dataset and cuts

In total 54 million events for Au + Au collisions at $\sqrt{s_{NN}}$ = 200 GeV are used in this study, all taken with a minimum-bias trigger by the STAR experiment during the RHIC run 2007. Event cuts and track selections can be found in Ref. [10]. In this analysis, the event plane angle ψ_1 is determined from the sideward deflection of spectator neutrons measured by STAR's Shower Maximum Detectors at Zero Degree Calorimeters (ZDCSMD), which are located close to beam rapidity and have the minimum contribution from non-flow. The description of measuring v_1 using the ZDCSMD event plane can be found in [2, 3].

There are about 420k events used in the analysis of v_1 for Cu + Cu collisions at $\sqrt{s_{NN}} = 22.4$ GeV. STAR's Beam Beam Counters (BBCs) [11] are used to reconstruct the first order event plane at this energy. BBCs are scintillator annuli mounted around the beam pipe beyond the east and west pole-tips of the STAR magnet at about 375 cm from the center of the nominal interaction region (IR). They have a η coverage of $3.8 < |\eta| < 5.2$ and a full azimuthal (2π) coverage. For v_1 study in the Time Projection Chamber (TPC) [12] region ($|\eta| < 1.3$), the full event plane is combined from the east BBC and west BBC event planes. However, for v_1 study in the FTPC region ($2.5 < |\eta| < 4.2$), only the east (west) BBC event plane, which is located at the backward (forward) pseudorapidity, is used to study v_1 in the forward (backward) FTPC. This is to avoid the possible self-correlation due to the overlap in η between the FTPC and the BBC coverage. The self-correlation arises when using particles from the same η region that is used to reconstruct the event plane to study flow, and it should be avoided in any flow studies. Errors presented in figures are statistical.

3. Directed flow results

In the left panel of Fig. 1, $v_1(y)$ of pions, kaons (K_S^0), protons and antiprotons are presented for centrality 10-70%. Following convention, the sign of spectator v_1 in the forward region is chosen to be positive, to which the measured sign of v_1 for particles of interest is only relative. Fitting with a linear function, slopes are $-0.15 \pm 0.05(\%)$, $-0.46 \pm 0.06(\%)$, $-0.27 \pm 0.01(\%)$, $-0.02 \pm 0.11(\%)$ and $-0.17 \pm 0.02(\%)$ for protons, antiprotons, pions, charged kaons and K_S^0, respectively. The $v_1(y)$ slope for the three produced particle species (pions, kaons (K_S^0) and antiprotons) are found negative at mid-rapidity, which is consistent with the anti-flow picture. In particular K_S^0 are less sensitive to the nucleon shadowing effect due to the small kaon/nucleon cross section, yet they show negative slope. This is again in favor of the anti-flow picture. However, surprisingly it is observed that v_1(y) of protons exhibits a clearly flatter shape than that of antiprotons. This feature cannot be explained by the anti-flow alone and will be discussed later in this paper. In the right panel of Fig. 1, pions $v_1(y)$ at mid-rapidity is plotted

Journal of Physics: Conference Series **270** (2011) 012045 doi:10.1088/1742-6596/270/1/012045

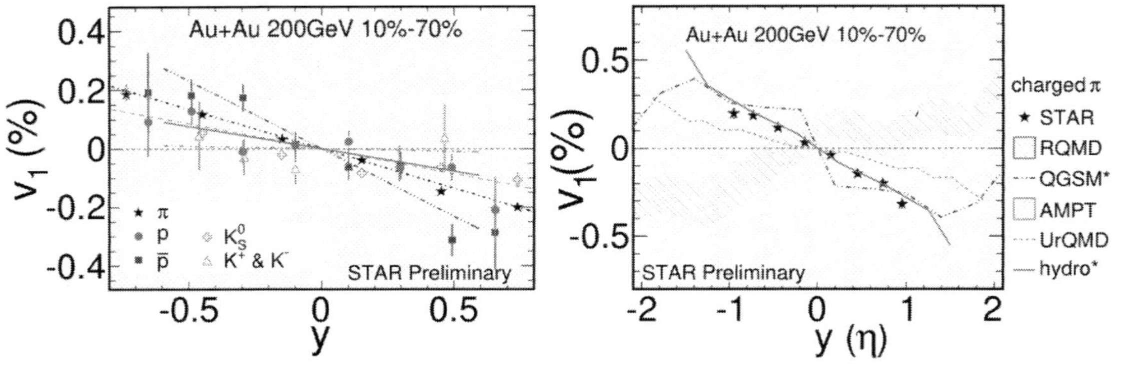

Figure 1. Left panel: Pions, kaons (K^0_S), protons and antiprotons v_1 as a function of rapidity for 10-70% Au + Au collisions at $\sqrt{s_{NN}} = 200$ GeV. The dashed lines presented the linear fit (see text for detail). Right panel: model calculations of pion $v_1(y)$ (or $v_1(\eta)$, for the QGSM model only) for minimum-bias events for Au + Au collisions at $\sqrt{s_{NN}} = 200$ GeV. Pions $v_1(y)$ measurement from the left panel is re-plotted for comparison. QGSM* model presents the basic Quark-Gluon String model with parton recombination [15]. Hydro* model presents the hydrodynamic expansion from a tiled source [8].

together with five model calculations, namely, RQMD [9], UrQMD [13], AMPT [14], QGSM with parton recombination [15], and hydrodynamic calculation with a tilted source [8]. Many model calculations either predict wrong sign of pions v_1 or wrong magnitude. The calculation that assumes a hydrodynamic expansion starting from a tilted source, which is a characteristic of anti-flow, describes the data best.

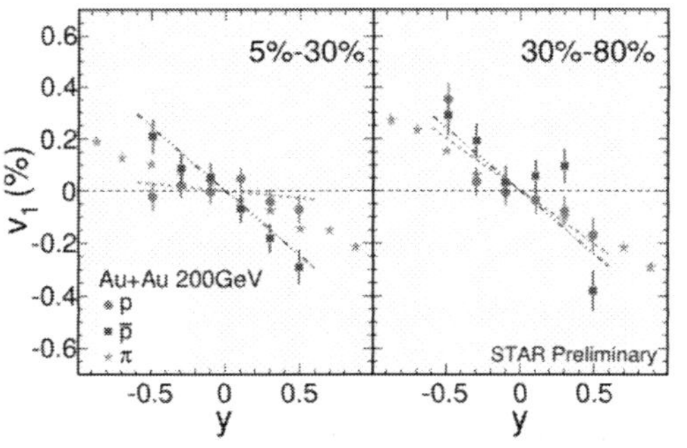

Figure 2. Directed flow *vs.* rapidity for protons (circles), antiprotons (squares) and charged pions (stars) for Au+Au collisions at $\sqrt{s_{NN}} = 200$ GeV, for centrality 5%-30%(left panel) and 30%-80%(right panel). Dashed and dotted lines are linear fits of $v_1(y)$ for protons and antiprotons, respectively.

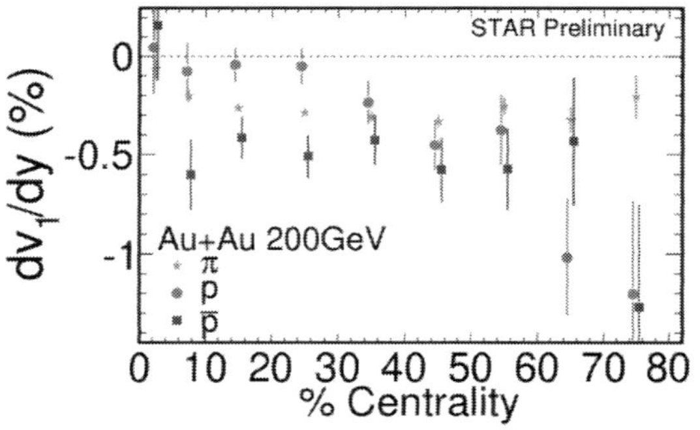

Figure 3. Charged pions (stars), protons (circles) and antiprotons (squares) $v_1(y)$ slope (dv_1/dy) as a function of centrality for Au + Au collisions at $\sqrt{s_{NN}} = 200$ GeV.

Fig. 2 shows $v_1(y)$ for protons, antiprotons and charged pions for different collision centralities. Fitting with a linear function, in 5-30% central collisions, the slope is -0.06 ± 0.06(%) and -0.50 ± 0.08(%) for protons and antiprotons, respectively; while in 30-80% central collisions, it is -0.40 ± 0.07(%) and -0.48 ± 0.09(%), respectively. Negative v_1 slope for protons and

Journal of Physics: Conference Series **270** (2011) 012045 doi:10.1088/1742-6596/270/1/012045

antiprotons is observed at mid-rapidity in 30-80% centrality. A sizable difference in $v_1(y)$ slope is seen in 5-30% central collisions but not much is seen in 30-80% central collisions.

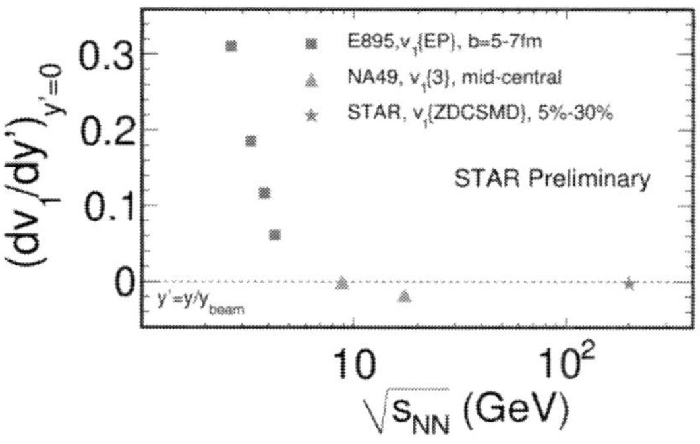

Figure 4. Protons $v_1(y')$ slope (dv_1/dy') at mid-rapidity as a function of center of mass collision energy, where $y' = y/y_{beam}$.

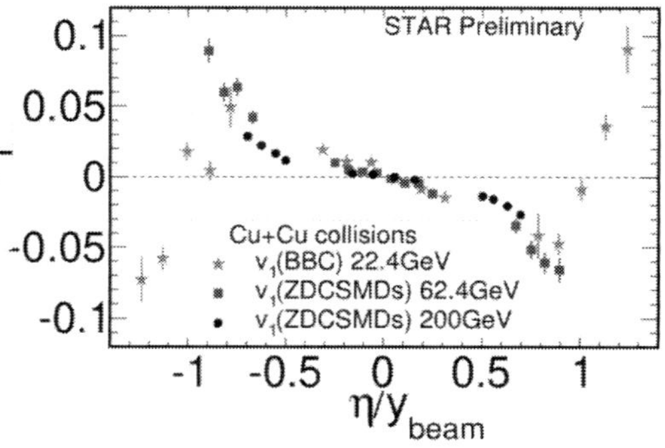

Figure 5. Directed flow *vs.* η/y_{beam} of charged hadrons from Cu+Cu collisions at 22.4 GeV(stars), 62.4 GeV(squares) and 200 GeV(circles) [2].

The centrality dependence of v_1 at mid-rapidity is studied further in Fig. 3, in which $v_1(y)$ slope is plotted as a function of centrality for protons, antiprotons and charged pions. Two observations are noteworthy to be mentioned: i) If the difference between v_1 of protons and antiprotons is caused by the anti-flow alone, then such difference is expected to be accompanied by strongly negative v_1. In data, the large difference between protons and antiprotons v_1 slopes is seen in 5-30% centrality, while strongly negative v_1 slope is found for protons, antiprotons and charged pions in a different centrality (30-80%), and ii) the hydrodynamics model with tilted source (which is a characteristic of anti-flow), while correctly predicts the pions $v_1(y)$ slope, does not predict the difference in $v_1(y)$ between particle species [16]. Both observations suggest that additional mechanism is needed to explain the centrality dependence of the difference between the $v_1(y)$ slopes of protons and antiprotons.

In Fig. 4, protons $v_1(y')$ slope F ($= dv_1/dy'$) around mid-rapidity is plotted as a function of collision energy, where $y' = y/y_{beam}$. Values for the $v_1(y')$ slope are extracted via a polynomial fit of the form $Fy' + Cy'^3$. At low energies, a different slope, that is, the slope of $\langle p_x(y')\rangle$ in stead of $v_1(y)$, has been measured for protons by the E877 [18] and the E895 [17] Collaboration. The $d\langle p_x\rangle/dy'$ decreases steadily with increasing beam energy over the E877 and the E895 energy range. The similar trend is shown in Fig. 4 with $v_1(y)$ slope for protons. The slope decreases rapidly with increasing energy, reaching zero around $\sqrt{s_{NN}} = 9$ GeV. It changes its sign to negative around $\sqrt{s_{NN}} = 17$ GeV, as shown by the NA49 measurement. The E877 result is not included in this plot because no dv_1/dy' from the E877 experiment is available. With previous measurements, which include only one point above $\sqrt{s_{NN}} = 9$ GeV, one cannot conclude if protons v_1 slope continues to decrease or stays close to zero when energy increases. The addition of the data point from RHIC indicates that protons v_1 slope remains close to zero at $\sqrt{s_{NN}} = 200$ GeV. Judging over a broad energy range, the transition of the protons v_1 slope sign from positive to negative happens around $\sqrt{s_{NN}} = 9$ GeV, interestingly, it coincides with the energy vicinity where $\langle k^+\rangle/\langle\pi^+\rangle$ exhibits a horn [19].

Fig. 5 shows the charged hadron v_1 result for Cu+Cu collisions for the centrality 0-60% at $\sqrt{s_{NN}} = 22.4$ GeV. The result is compared to published results for $\sqrt{s_{NN}} = 62.4$ and 200 GeV in centrality 30-60% [2]. When viewed in η/y_{beam} frame, a comment trend is observed for all energies in the range of $|\eta|/y_{beam} < 1$.

4. Summary
In summary, STAR's measurement of directed flow of pions, kaons (K_S^0), protons and antiprotons for Au + Au collisions at $\sqrt{s_{NN}} = 200$ GeV are presented. In 10-70% central collisions, $v_1(y)$ slopes of pions, kaons (K_S^0) and antiprotons are found mostly negative around mid-rapidity. Sizable difference is seen between v_1 of protons and antiprotons in 5-30% central collisions. Comparison to models (RQMD, UrQMD, AMPT, QGSM with parton recombination, hydrodynamics with a tilted source) are made. It is found that hydrodynamic calculation with a tilted source gives the most reasonable description of $v_1(y)$ slope for pions. However, it does not predict the difference between protons and antiprotons v_1 [16].) Anti-flow can explain the negative $v_1(y)$ slope while it alone has difficulties in explaining the centrality dependence of the difference between the $v_1(y)$ slopes of protons and antiprotons. The v_1 results at 22.4 GeV from Cu + Cu collisions are presented, it is consistent with the published results for Cu + Cu collisions at $\sqrt{s_{NN}} = 62.4$ GeV and $\sqrt{s_{NN}} = 200$ GeV.

Acknowledgments
This work was supported in part by the National Natural Science Foundation of China under Grants 10775058 & 11075060, the MOST of China under Grant 2008CB817707.

References
[1] Poskanzer A M and Voloshin S A 1998 *Phys. Rev.* C **58** 1671
[2] Abelev B I *et al.* (STAR Collaboration) 2008 *Phys. Rev. Lett.* **101** 252301
[3] Chen J (for the STAR Collaboration) 2008 *J. Phys.* G **35** 044072
[4] Sorge H 1997 *Phys. Rev. Lett.* **78** 2309
[5] Stöcker H 2005 *Nucl. Phys.* A **750** 121
[6] Brachmann J, Soff S, Dumitru A, Stöcker H, Mahruhn J A, Greiner W, Bravina L V and Rischke D H 2000 *Phys. Rev.* C **61** 024909
[7] Csernai L P and Röhrich D 1999 *Phy. Lett.* B **458** 454
[8] Bożek P and Wyskiel I 2010 *Phys. Rev.* C **81** 054902
[9] Snellings R J, Sorge H , Voloshin S, Wang F Q and Xu N 2000 *Phys. Rev. Lett* **84** 2803
[10] Tang A (for the STAR Collaboration) 2010 *J. Phys.: Conference Series* **230** 012018
[11] Bieser F S *et al.* 2003 *Nucl. Instrum. Method* A **499** 766
[12] Anderson M *et al.* 2003 *Nucl. Instrum. Methods* A **499** 659
[13] Bleicher M and Stöcker H 2002 *Phys. Rev.* B **526** 309
[14] Chen J Y, Zuo J X, Cai X Z, Liu F, Ma Y G and Tang A H 2010 *Phys. Rev.* C **81** 014904
[15] Bleible J, Burau G, Faessler A and Fuchs C 2007 *Phys. Rev.* C **81** 024912
[16] Piotr Bożek 2010 *private communication*
[17] Liu H *et al.* (E895 Collaboration) 2000 *Phys. Rev. Lett.* **84** 5488
[18] Barrette J *et al.* (E877 Collaboration) 1997 *Phys. Rev.* C **56** 3254; Barrette J *et al.* (E877 Collaboration) 1997 *Phys. Rev.* C **55** 1420
[19] Afanasiev S V *et al.* (NA49 Collaboration) 2002 *Phys. Rev.* C **66** 054902; Alt C *et al.* (NA49 Collaboration) 2008 *Phys. Rev.* C **77** 024903; Alt C *et al.* (NA49 Collaboration) 2006 *Phys. Rev.* C **73** 044910; Anticic T *et al.* (NA49 Collaboration) 2004 *Phys. Rev.* C **69** 024902

Search for collective expansion in pp collisions at the LHC

Michele Floris For the ALICE Collaboration
CERN, Geneva, Switzerland
E-mail: michele.floris@cern.ch

Abstract. Proton-proton collisions at LHC energies reach a multiplicity density comparable to nuclear collisions at lower energy. It is therefore natural to ask whether the collective bulk behavior observed in heavy-ion collisions develops already in p-p collisions. In previous experiments, the study of p_t distributions of identified particles in the framework of blast wave models provided considerable insight on the collective behavior and on the freeze-out parameters of the fireball created in heavy-ion collisions. These ideas have recently been applied also to p-p collisions at RHIC. The ALICE experiment, thanks to its excellent PID capabilities and p_t coverage, offers an ideal test-bench for these studies at the LHC. In this work, we discuss the performance and analysis strategy of ALICE for blast wave studies and present some preliminary results on identified particle spectra, based on the data collected in the late 2009.

1. Introduction

Collective behaviour is one of the defining features of the matter produced in relativistic heavy ion collisions at RHIC ($\sqrt{s} = 200$ GeV) and SPS ($\sqrt{s} = 17$ GeV) energies, where several different observables are understood as due to a common velocity flow in the expanding fireball [1, 2, 3]. With the term "collectivity", it is usually meant that the particles in a given space cell move with the same mean velocity. That is, one observes space-momentum correlations which may indicate significant final-state interactions. Collectivity could be a consequence of the pressure gradients developed in a thermalized medium, but this is not a necessary condition [2]. Proton-proton collisions at LHC energy reach multiplicity densities comparable to those of nuclear collisions at lower energies. At $\sqrt{s} = 7$ TeV, for instance, dN/dy easily reaches values of several tens of particles, comparable to semi-central Cu-Cu collisions at RHIC [4, 5, 6]. It is therefore natural to ask whether collective behaviour already develops in p-p collisions at these energies.

There is no single "smoking gun" of collectivity: the understanding and discovery of collective behaviour will only come from the simultaneous study of several different observables: radial flow, HBT radii as a function of k_T, correlations between non-identical particles, elliptic flow, etc. The study of these observables will follow previous analyses of heavy ion collisions, but in some cases there will be important differences. In the case of elliptic flow (v_2), for instance the determination of the reaction plane and the subtraction of non-flow effects will be highly non-trivial [7]. In this paper, we focus on the study of radial flow through blast wave models and discuss the implications for p-p collisions.

A crucial part of this programme will be the study of the evolution of the different observables with multiplicity and with event shape: collective effects are more likely to be observed in high multiplicity and "soft" (see below) events. In the ALICE experiment, the number of pixel chips hit in a collision is available on-line. It can thus be used in the trigger logic, to produce a "high multiplicity" trigger. This is a unique feature, which allows the experiment to enrich the sample of collected high multiplicity events (the luminosity of the LHC is high enough that it is not possible to collect all the minimum bias events produced). Hadron interactions are usually classified as either "hard" or "soft", where by hard one means that there was an associated high E_t parton-parton interaction, leading to two collimated high p_t jets of particles. Collectivity is an inherently soft effect, so one would aim for a sample depleted of hard events. Studies to separate hard and soft events were already done in the past, for instance in Ref. [9] by the CDF collaboration, where two qualitatively different samples were indeed isolated. An issue with the strategy depicted in this reference, however, is that the selection was based on transverse energy clusters, so it could bias the transverse momentum distribution needed for this study. ALICE is pursuing several different approaches to separate hard and soft events and to characterize the shape of events: study of clusters (*a la* CDF or high density clusters), event shape variables, $\Delta\eta - \Delta\phi$ and multiplicity correlations [10]. As none of the selections outlined above will be able to isolate purely soft events, similar to [9] a data driven approach will be adopted, looking for samples with different properties. The comparison of different event selection strategies will provide an estimate of the bias related to the selection itself.

2. Blast Wave models and conservation of energy and momentum

The term "blast wave" identifies a family of models inspired by hydrodynamics, which provide an analytical parametrization of the freezeout conditions. In its original implementation, the spectra of identified particles were described using only 2 parameters: a freezeout temperature and a velocity of radial expansion [11]. Over the years, this model was improved by relaxing some of its assumptions or by adding more parameters, to the purpose of describing more observables [3, 12]. A recent development, which could be particularly relevant for p-p collisions, is the so-called "Tsallis blast wave". In this model, the Boltzmann-Gibbs statistics is replaced by Tsallis statistics, which adds a further parameter, q, supposed to quantify the degree of non-equilibrium of the system. Tsallis statistics is reduced to the Boltzmann-Gibbs one for $q \to 1$. This model was applied to STAR p-p and Au-Au data [13], and it was found to give a very satisfactory description of the spectra up to $p_t \simeq 3$ GeV/c. In Au-Au collision the parameter q was found to increase with decreasing centrality, as expected from its interpretation. In p-p collisions, however, it was not possible to fit both mesons and baryons with a single set of parameters. It is worth to mention that the interpretation of the parameter q is not straightforward nor uncontroversial (see for instance Ref. [14]) and one should be cautious before drawing physics conclusions. Blast wave models provide a useful tool to quantify the evolution of the spectra with event shape and multiplicity in a few simple parameters, and as such can contribute to the overall understanding of the system. Moreover, being able to describe different particles with the same set of parameters implies a specific mass ordering of the low p_t part of the spectrum, which could hint at the presence of radial flow. In p-p collisions one cannot ignore the power-law tail, which sets in at LHC energies already at moderately low p_t. The Tsallis blast wave represents a possible approach to address this issue.

Another important aspect are the correlations and constraints induced by energy and momentum conservation on particle production, which cannot be ignored in small systems, including p-p collisions. These "trivial" phase-space effects introduce distortions to the p_t spectra, which weaken with increasing multiplicity and will need to be separated from collectivity/flow. These effects were investigated in Ref. [8], and it was found that p-p and peripheral Au-Au collisions are much more similar to central Au-Au once they are taken into account. The formalism developed in that reference, however, is still being debated, in particular because it depends on four unobservable parameters, which must be fitted to the data. It should be also stressed that those effects are complementary, not alternative, to collectivity: energy and momentum conservation are there whether the system is flowing or not.

Journal of Physics: Conference Series **270** (2011) 012046 doi:10.1088/1742-6596/270/1/012046

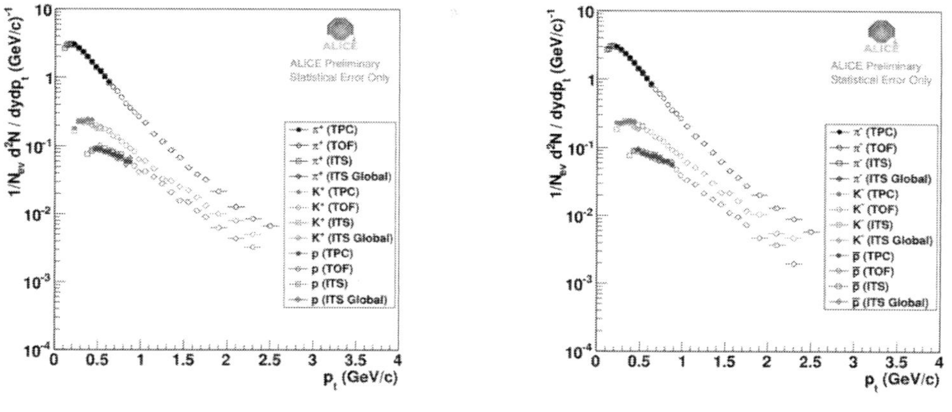

Figure 1. Results from the different PID analyses; positive (left) and negative (right) particles.

3. Identified particle spectra with the ALICE detector

Identified particle spectra over a wide p_t range are the basic ingredient for this kind of analysis. The ALICE detector is ideally suited for this, since it was designed to cope with the high multiplicities expected in Pb-Pb collisions at $\sqrt{s} = 5.5$ TeV, and to identify particles over a wide range of momenta. A detailed description of the apparatus can be found in [15]. In this section, we focus on the first ALICE measurements of π, K, p at $\sqrt{s} = 900$ GeV, based on a sample of about 300 000 events, collected at the end 2009 during the commissioning of the LHC. These particles are identified using the specific energy loss measurement in the Time Proportional Chamber (TPC) and in the Inner Tracking System (ITS) or using the time of flight information from the TOF detector at higher p_t. In the present work, the PID detectors are used independently, and the results (after checking for consistency) are combined at a later stage in the analysis, weighting the spectra for the not common part of the systematic error. The basic strategy is the same for all the detectors: the distribution of the response function is sliced in bins of p_t and fitted with a superposition of gaussian-like functions, to extract the yield of the different species. The ITS can be used both in conjunction with the TPC or as a standalone tracker. In conjunction with the TPC, it allows to reduce the contamination from secondaries, by constraining the measurement close to the interaction vertex. As a standalone tracker, despite the poorer p_t resolution, it allows to reconstruct tracks below $p = 200$ MeV/c, which are deflected or decay before they can reach the TPC. This is very important for blast wave studies, where the possible presence of flow mainly influences the low p_t part of the spectrum. The ITS dE/dx information is used in two independent analyzes, based on either ITS standalone or TPC/ITS tracks (see [16] for details). The TOF analysis uses the sub-sample of tracks reconstructed in the TPC which have a matching hit in this detector.

Figure 1 shows the very good agreement between the results of the different analyses. The spectra are normalized to inelastic collisions with a technique similar to that employed in the first ALICE publications [4, 5]. The combined spectra are fitted in Fig. 2 with the Lévy function $\frac{d^2N}{dp_t dy} \propto p_t \left(1 + \frac{m_t - m_0}{nT}\right)^{-n}$, which provides a very good description of the data and can be used to extrapolate in order to extract yields and mean p_t. The Hagedorn function was also found to give a satisfactory description of the spectra, and yields results within 5%. Our results were also compared to several commonly used Monte Carlo models (Phojet and Pythia with tunes D6T, CSC and Perugia0), but none was found to give a satisfactory description of the data.

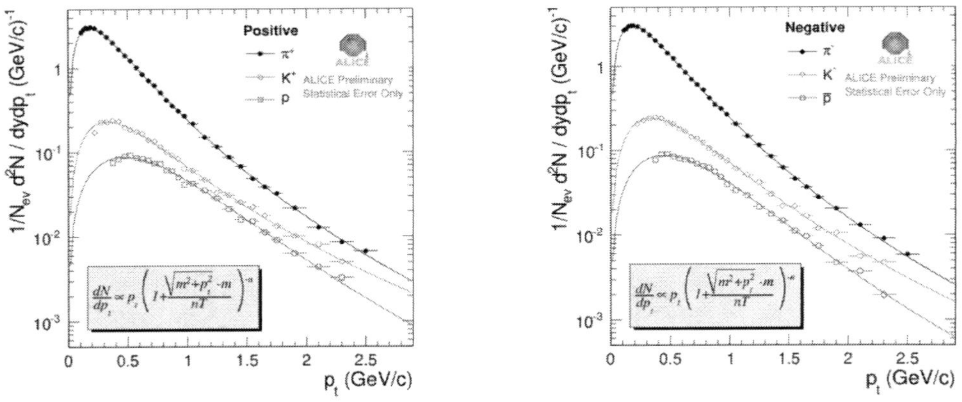

Figure 2. Combined spectra; positive (left) and negative (right) particles.

4. Conclusions

Proton collisions at LHC energies reach rather high multiplicity densities, and it becomes important to establish whether the signatures of collective behaviour observed in nuclear collisions at lower energies with similar multiplicity are already observed in p-p collisions. In this paper we argue that a conclusion on this topic will only come from the simultaneous understanding of different observables and we discussed in some detail one of those: radial flow trough blast wave fits. The basic ingredient for this kind of analysis are spectra of identified particles. ALICE has excellent experimental capabilities for these studies and we present some preliminary results for identified particle spectra integrated in centrality and event shape. These results show that the capabilities of the detector in terms of PID and tracking match the expectations, but the 900 GeV statistics does not allow for a more differential analysis. Concerning the specific question of flow, one can expect first results to come from the detailed analysis of the much larger 7 TeV sample (several hundred millions events collected in 2010), which is currently being analyzed.

References

[1] Kolb P F and Heinz U W 2003 Invited review for *Quark Gluon Plasma 3* ed R C Hwa and X N Wang (Singapore: World Scientific) p634 (*Preprint* nucl-th/0305084 [arXiv])
[2] Heinz U W *Proc. 2nd CERN-CLAF School of High-energy Physics* p165 (*Preprint* hep-ph/0407360 [arXiv])
[3] Retiére F and Lisa M 2004 Phys. Rev. C **70** 044907
[4] Aamodt K*et al.* [ALICE Collaboration] 2010 Eur. Phys. J. C **68** 89
[5] Aamodt K *et al.* [ALICE Collaboration] 2010 *Preprint* hep-ex/004.3514 [arXiv]
[6] Hollis R S *et al.* [PHOBOS Collaboration] 2006 Rom. Rep. Phys. **58** 37
[7] Eyyubova G *Can we observe the elliptic flow in p-p collisions at LHC?* These proceedings
[8] Chajkecki Z and Lisa M 2009 Phys. Rev. C **79** 034908
[9] Acosta D 2002 Phys. Rev. D **65** 072005
[10] Paic G 2009 PoS High-pTphysics09:007
[11] Schnedermann E, Sollfrank J and Heinz U 1993 Phys. Rev. C **48** 2462
[12] Wiedemann U A and Heinz U 1997 Phys. Rev. C **56** 3265
[13] Tang Z *et al.* 2009 Phys. Rev. C **79** 051901
[14] G. Barnaföldi *On the origin of Tsallis-Pareto like distributions in hadron-hadron collisions*, These proceedings
[15] Aamodt K *et al.* [ALICE Collaboration] 2008 JINST **3** S08002
[16] M. Chojnacki *Measurement of identified charged hadron spectra with ALICE using the ITS in proton-proton collisions at LHC*, These proceedings

Journal of Physics: Conference Series **270** (2011) 012047 doi:10.1088/1742-6596/270/1/012047

Can we observe the elliptic flow in *pp* collisions at LHC?

Gyulnara Eyyubova
for the ALICE collaboration
Department of Physics, University of Oslo, Norway
E-mail: gyulnare@student.matnat.uio.no

Abstract. The azimuthal anisotropy of charged particles in heavy ion collisions is an important probe of quark-gluon plasma evolution at early stages of reaction. Now, with the start of Large Hadron Collider (LHC), the collective effects in proton-proton collisions are also of great interest. In this work we analyze the ability of different methods to extract elliptic flow in *pp* collisions at LHC. The analysis is based on Monte Carlo simulations of proton-proton collisions with PYTHIA, PHOJET and EPOS event generators at energies $\sqrt{s} = 900$ GeV and 7 TeV.

1. Introduction

In non-central nucleus-nucleus collisions an initial overlap geometry is anisotropic. Unequal pressure gradients then produce an anisotropic momentum distribution of particles with respect to the event-by-event reaction plane. The observed particle yield versus azimuthal angle gives information on the early collision dynamics and is described by means of Fourier series:

$$\frac{dN}{d\varphi} = \frac{N_0}{2\pi}[1 + 2\sum_{n=1}^{\infty} v_n \cos(n(\varphi - \Psi_R))], \tag{1}$$

where Ψ_R is the reaction plane angle. The second harmonic v_2 in Eq.(1) is called the elliptic flow. The measurements of v_2 at RHIC are commonly interpreted in terms of hydrodynamic expansion of a thermalized relativistic fluid. The STAR experiment at RHIC has measured azimuthal correlations in proton-proton collisions as a reference for nucleus-nucleus collisions in order to understand systematic errors of v_2 calculation, resulting from non-flow effect, such as resonance decays and inter- and intra-jet correlations [1]. It was assumed that there are no correlations due to elliptic flow in *pp* collisions.

At LHC energy, the expected large energy densities suggest the possibility of collective behavior even in *pp* collisions. Here the multiplicity tail contains more than 100 charged particles in central pseudorapidity region $|\eta| < 1$ at $\sqrt{s} = 7$ TeV[2]. Assuming that the matter formed in *pp* collisions has the same properties as in $A + A$ collisions, one also expects the formation of elliptic flow in *pp* collisions. There exist various attempts to apply hydrodynamical calculation to the *pp* collision system (see, e.g., [3, 4, 5]). Besides hydrodynamics, the mechanism of asymmetry can also arise due to initial conditions in Regge field theory [6] and in string interaction models [7]. The predicted values of the v_2 vary from 0.03 till 0.15 in different models. The aim of the

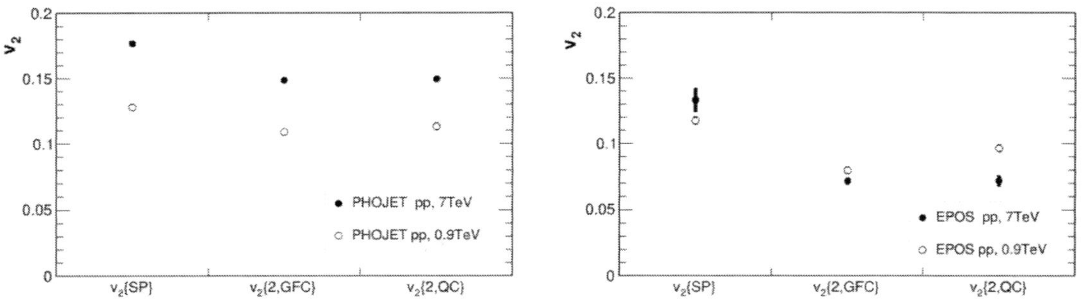

Figure 1. v_2-coefficient with the two-particle correlation methods for the pp collisions at $\sqrt{s} = 900$ and 7000 GeV generated by PHOJET (left) and EPOS (right).

present work is to study the possibility of extraction of elliptic flow in pp collisions at energies between $\sqrt{s} = 900$ GeV and 7 TeV using various MC event generators.

2. Elliptic flow analysis

Methods for the flow extraction. The main obstacle for v_2 determination is that it cannot be extracted directly as a Fourier coefficient in Eq.(1) and should be calculated via particle correlations since a position of the reaction plane Ψ_R is unknown. Apart from the collective correlations w.r.t. the reaction plane, there exist the so-called non-flow correlations. Systematic uncertainties resulting from the non-flow correlations are much affected by multiplicity. Several anisotropic flow reconstruction methods, with their own advantages and limitations, are usually employed for the analysis of data, e.g., the Reaction Plane method, the Scalar Product, and the 2nd order cumulant method [8, 9]. These methods exhibit different sensitivity to two- or more-particle azimuthal correlations. Due to non-flow correlations, v_2 can be reliably extracted in two-particle correlation methods if $v_2 > 1/\sqrt{M}$ [10], where M is the event multiplicity. There are also higher order correlation methods based on cumulants and many-particle correlation methods, such as Lee-Yang Zeroes method [10]. They allow for measurements of azimuthal anisotropies down to values of order of $1/M^{3/4}$ for 4-particle correlation methods and $1/M$ for genuine multiparticle methods. The multiplicity and the number of events are relevant to multiparticle correlation methods as well. Unfortunately, these methods can hardly be employed with the currently available number of recorded pp interactions of ALICE at the LHC.

Monte Carlo event generators at our disposal. None of available microscopic Monte Carlo (MC) models describes the development of anisotropic flow in elementary hadron–hadron interactions yet. Three popular MC event generators, namely, PYTHIA6.4 [11], PHOJET [12] and EPOS [13], were applied in these proceedings.

PYTHIA treats each hadron as a set of partons characterized by parton distributions. Initial parton shower is started by a parton from each set. One parton from each of the initial showers participates in a hard collision, where a number of outgoing partons is produced. Other partons from the showers may experience semihard interactions. The fragmentation process into the colorless hadrons takes place via the string fragmentation (so-called Lund model is used), cluster decays and decays of unstable particles and resonances. Both PHOJET and EPOS event generators belong to the family of microscopic models based on dual topological unitarization and utilized the color-exchange mechanism of string excitation. Here the hadronic interactions are treated in terms of Reggeon and Pomeron exchanges. In PHOJET the Pomeron exchange is subdivided into soft processes with small momentum transfer and at least one hard process with

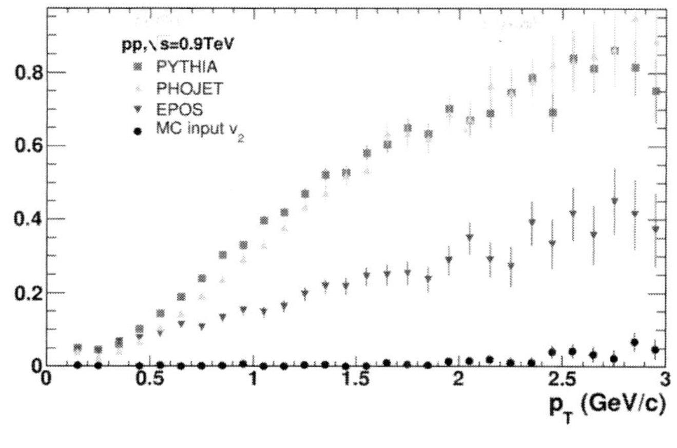

Figure 2. $v_2(p_T)$-dependence of charged particles for pp collisions at $\sqrt{s} = 900\,\mathrm{GeV}$ generated by the MC models.

large momentum transfer. In EPOS the complete picture of hadron-hadron collision consists of the three colorless objects, namely, the two off-shell remnants of the target and projectile, and the parton ladders between the two active partons on either side. As one can see below, jets in EPOS are playing a minor role compared to jets in PYTHIA or in PHOJET.

Results. The present analysis is based on the minimum bias pp events at $\sqrt{s} = 900$ GeV and 7 TeV with the mean multiplicity for MC models in the region $|\eta| < 1$ about 6 and 10 respectively. The v_2-coefficient was reconstructed by the Scalar Product Method $v_2\{SP\}$, and by the cumulants of the second order with Generating function $v_2\{2, GFC\}$ and with Q-vector $v_2\{2, QC\}$ [14]. Fig.1 shows the p_T-integrated flow extracted for PHOJET- and EPOS-simulated pp collisions. Analysis of PHOJET data favors increase of the v_2 with rising energy, whereas for EPOS no strong energy dependence is observed. The p_T dependencies of obtained coefficients for $v_2\{2, QC\}$ method are presented in Fig.2. Note that these coefficients are below 10% at $p_t \leq 0.5\,\mathrm{GeV/c}$, but quickly rise up to 40% (EPOS) or even 90% (PYTHIA) at $p_T \approx 3\,\mathrm{GeV/c}$. Since the intrinsic elliptic flow in the MC models is essentially zero, the reconstructed v_2 values are solely due to non-flow correlations. The integrated v_2 values also differ by a factor of 1.5 ($v_2^{PYT/PHO} \approx 0.16$ versus $v_2^{EPOS} \approx 0.096$). As the multiplicities are practically the same, the stronger non-flow correlations in PYTHIA and PHOJET are attributed to jet contributions. It is worth noting that the presence of real flow can diminish the non-flow effects, as seen in Fig.3. Here the EPOS events are modified to carry certain amount of v_2. The contribution of non-flow decreases with increasing true v_2. However, for the methods used, even for the input $v_2 = 15\%$ at the particle level, the non-flow contribution to the extracted v_2 coefficient is about 20%.

How can we disentangle flow and non-flow terms? If the correlation is dominated by elliptic flow, it should be approximately constant as a function of pseudorapidity. Flow correlates particles at all $\Delta\eta$, while non-flow effects from jets dominates short range correlations (small $\Delta\eta$) and can be reduced by the η-gap technique [15]. The η-gap analysis was performed within the Scalar Product method [9]. The general formula of the method is:

$$\langle\langle v_2(\eta, p_T)\rangle\rangle = \left\langle\left\langle \frac{Qu_i(\eta, p_T)}{M^q} \right\rangle\right\rangle \Bigg/ \sqrt{\left\langle \frac{Q^a Q^{b*}}{M^a M^b} \right\rangle}\,, \tag{2}$$

where $u_i = e^{i2\varphi_i}$ is a unit vector associated with the i-th particle and $Q = \sum u_i$ is the event flow vector. Subevents a and b were separated by $\Delta\eta$. Fig.4 shows the dependence of calculated elliptic flow coefficient on $\Delta\eta$ in PYTHIA-simulated pp data at $\sqrt{s} = 7\,\mathrm{TeV}$. The observed correlations decrease with increase of $\Delta\eta$. Still the obtained non-flow value of the v_2 is quite large even for $\Delta\eta \geq 0.9$. Larger $\Delta\eta$ can further reduce the correlations. Another way of non-flow

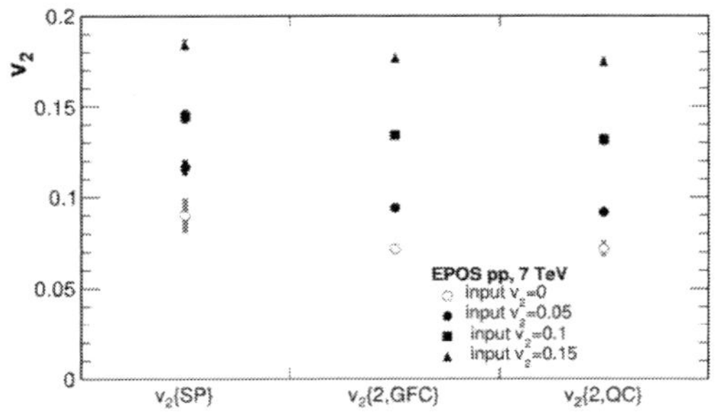

Figure 3. v_2-coefficient with the two-particle correlation methods for the EPOS-simulated *pp* collisions with different values of input flow.

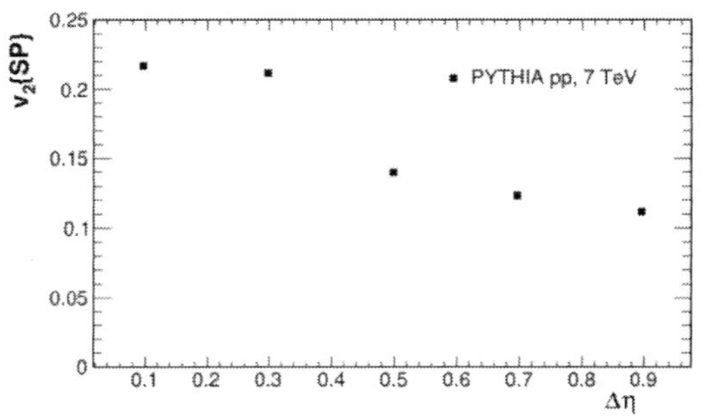

Figure 4. η-gap dependence of the azimuthal correlations restored from PYTHIA-simulated *pp* collisions.

separation may be the application of cuts on $\Delta\eta \times \Delta\varphi$ region around leading particle in order to decrease jet correlations.

3. Conclusions

Proton-proton collisions simulated by PYTHIA, PHOJET and EPOS at $\sqrt{s} = 900$ and 7000 GeV are analyzed by two-particle correlation methods. The p_T-integrated v_2 coefficients reconstructed by the methods are found to vary from 10% to 15%. These values are attributed solely to the non-flow correlations.

To weaken the non-flow correlations η-gap separation of two subevents is applied. This reduces non-flow value for the PYTHIA events by a factor 1/3. Using an even larger $\Delta\eta$ will allow for a further reduction of the non-flow contributions.

I want to express deep gratitude to R. Snellings, C. Loizides, N. van der Kolk, L. Bravina, L. Malinina, E. Zabrodin for help and useful comments. This work was supported by the QUOTA Program and by the Norwegian Research Council (NFR).

References

[1] Adams J *et al.* [STAR Collaboration] 2003 *Phys. Rev. Lett.* **93** 252301 (STAR)
[2] Aamodt K *et al.* [ALICE Collaboration] 2010 *Eur. Phys. J.* C**68** 345
[3] d'Enterria D *et al.* 2010 *Eur. Phys. J* C**66** 173
[4] Luzum M and Romatschke P 2009 *Phys. Rev. Lett.***103** 262302
[5] Chaudhuri A 2010 *Phys. Lett.* B**692** 15
[6] Boreskov K, Kaidalov A and Kancheli O 2008 *Eur.Phys.J.* C**58** 445
[7] Bautista I *et al.* 2010 *J.Phys* G**37** 015103;
[8] Poskanzer A M and Voloshin S A 1998 *Phys. Rev.* C**58** 1671
[9] Adler C *et al.* [STAR Collaboration] 2002 *Phys. Rev.* C**66** 034904
[10] Borghini N, Dinh P M and Ollitrault J Y 2001 *Phys. Rev.* C**64** 054901
[11] Sjostrand T, Mrenna S and Skands P Z 2008 *Comput. Phys. Commun.* **178** 852
[12] Engel R and Ranft J 1996 *Phys. Rev.* D**54** 4244
[13] Werner K, Liu F M and Pierog T 2006 *Phys. Rev* C **74** 044902
[14] Bilandzic A, Snellings R and Voloshin S 2010 Flow analysis with cumulants: direct calculations *Preprint* nucl-ex/1010.0233
[15] Alver B *et al.* 2010 *Phys. Rev.* C**81** 034915

NA61/SHINE ion program

Maja Mackowiak for the NA61 Collaboration
Warsaw University of Technology, Faculty of Physics, Koszykowa 75, 00-662 Warszawa, Poland
E-mail: `majam@if.pw.edu.pl`

Abstract. The Super Proton Synchrotron (SPS) at CERN covers one of the most interesting regions of the phase diagram (T - μ_B) of strongly interacting matter. The study of central Pb+Pb collisions by NA49 indicate that the threshold for deconfinement is reached already at the low SPS energies. Theoretical considerations predict a critical point of strongly interacting matter at energies accessible at the SPS. The NA61/SHINE experiment, a successor of the NA49 project, will study hadron production in p+p, p+A, h+A, and A+A reactions at various energies. The broad physics program includes the investigation of the properties of strongly interacting matter, as well as precision measurements of hadron spectra for the T2K neutrino experiment and for the Pierre Auger Observatory and KASCADE cosmic-ray projects. The main physics goals of the NA61/SHINE ion program are to study the properties of the onset of deconfinement at low SPS energies and to find signatures of the critical point of strongly interacting matter. To achieve these goals a broad range in the (T - μ_B) phase diagram will be covered by performing an energy (10*A*-158*A* GeV/c) and system size (p+p, B+C, Ar+Ca, Xe+La) scan. The first data for this 2-D scan were taken in 2009, i.e. p+p interactions at 20, 30, 40, 80, 158 GeV/c beam energy. This contribution will summarize physics arguments for the NA61/SHINE ion program, show the detector performance and present the current status of the experiment and plans for the next years.

1. Introduction

The NA61/SHINE experiment (Fig. 1) [1] is located at the SPS accelerator at CERN. It is the continuation of the fixed target NA49 experiment and will study hadron+hadron, hadron+nucleus and nucleus+nucleus collisions. The NA61/SHINE collaboration consists of 130 physicists from 24 institutions. SHINE stands for *SPS Heavy Ion and Neutrino Experiment*. The NA61/SHINE physics program includes:

- hadron production measurements for neutrino and cosmic-ray experiments (not discussed here (for details see [1])
- ion collision program studying the onset of deconfinement, searching for the critical point of strongly interacting matter and investigating high p_T physics

2. Detector

The main component of the NA61/SHINE detector (Fig. 1) are four large volume Time Projections Chambers. The first two (VTPC-1/2) are located in the magnetic field of two superconducting dipole magnets. The other two (MTPC-L/R) are positioned downstream of the magnets. An additional small gap TPC (GTPC) is installed on the beam axis between VTPC-1 and VTPC-2. Behind the TPCs are three ToF-walls. Two of them (ToF-L/R) were inherited from the NA49 experiment. A forward ToF-wall was added in 2007 to extend particle identification in the low momentum range.

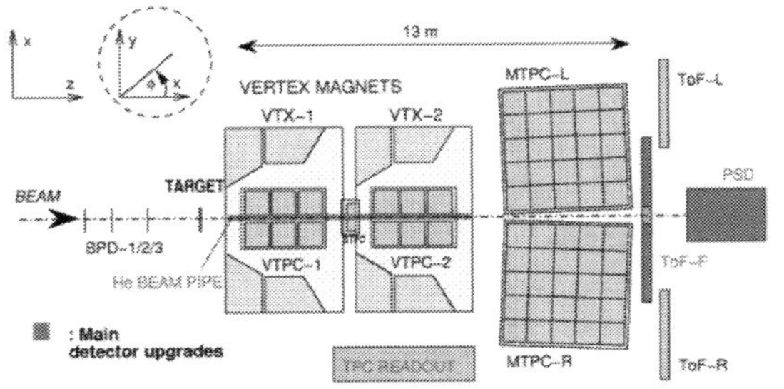

Figure 1. NA61/SHINE detector layout. Upgrades are shown in red color.

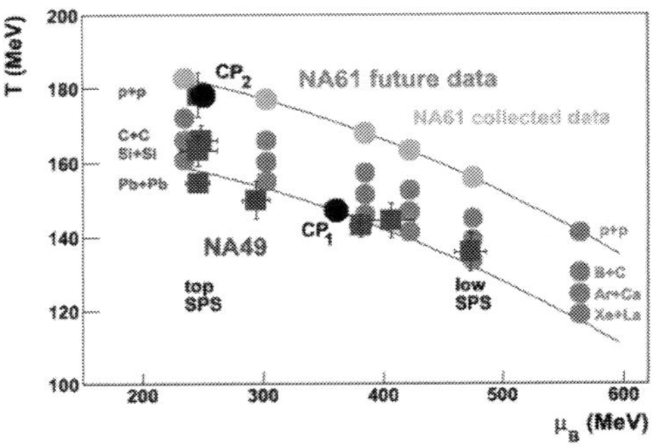

Figure 2. System size and energy scan of the phase diagram. Estimated (squares [NA49]) and extrapolated (circles) chemical freeze-out points ([2]). Green circles indicate data collected in 2009. Black circles - two locations of critical point analyzed by NA49.

In 2008 a new TPC read-out system was constructed which increased the event recording rate by a factor of 10 compared to NA49. Two other upgrades are planned mainly for the ion beam program. The first one (partially installed in 2010) is the Projectile Spectator Detector (PSD), which will allow precise measurement of the energy of projectile spectators and a reconstruction of the reaction plane. The determination of the number of projectile spectators is crucial for measurements of event-by-event multiplicity fluctuations. The PSD will allow single nucleon discrimination. The second upgrade is a He-beam pipe which will reduce δ-electron production by the beam by a factor of 10. It will be installed in 2011.

3. NA61/SHINE ion program

One of the most important tasks of heavy ion collision physics is to explore the phase diagram of strongly interacting matter. Theorists expect that for large values of baryon chemical potential (μ_B) there is a first order phase transition between hadron matter and quark gluon plasma. In contrast, for small μ_B values the transition is expected to be a smooth crossover. The 1st-order phase transition line ends with a critical point of a 2nd-order phase transition. One of the main goals of NA61/SHINE is to study the transition line and search for the critical point. To achieve these goals NA61/SHINE will perform a 2-D scan (energy and system size) of the phase diagram (Fig. 2). The first part of the scan, i.e. collection of p+p collisions was/is done in 2009/2010. Also in 2010 a test of a B beam from fragmentation of Pb ions will be performed. B+C interactions will be collected in 2011(13). Ar+Ca and Xe+La are planned for 2012 and 2014. p+Pb interactions will be recorded in 2011/2012.

3.1. Onset of deconfinement

When the early stage of matter created in an ion collision hits the transition line three signals should be observed (so-called: kink, horn, step) according to the SMES model [3]. Such signals have been observed in the NA49 experiment in central Pb+Pb collisions near 30*A* GeV energy [4] (Fig. 3). In contrast to the structures seen in Pb+Pb (Au+Au) reactions a smooth behavior is seen for elementary interactions. NA61/SHINE will check where the evidence of the onset of deconfinement appears for light systems (B+C, Ar+Ca, Xe+La).

3.2. Critical Point

The critical point (CP) of the phase transition is predicted to lie in a region of the phase diagram accessible at SPS energies [5]. Enhanced fluctuations in multiplicity and transverse momentum are expected if hadronization and freeze-out happen near the critical point [6]. The NA49

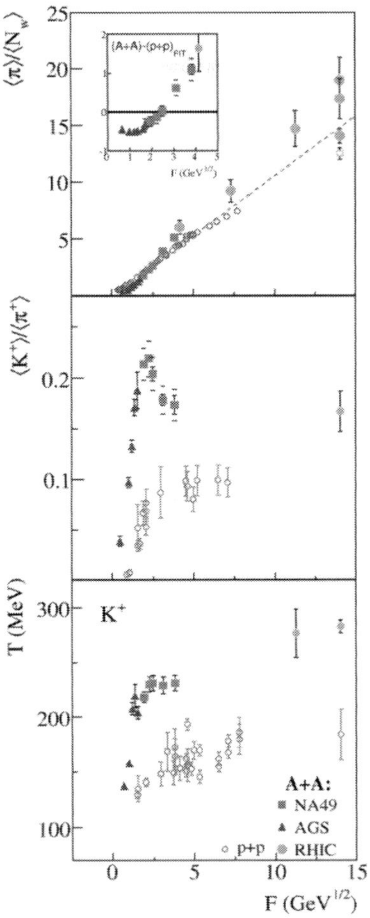

Figure 3. Signals of the onset of deconfinement: kink, horn, step [4]. Measurements are plotted as a function of the Fermi measure $F \approx \sqrt{\sqrt{s_{NN}}}$.

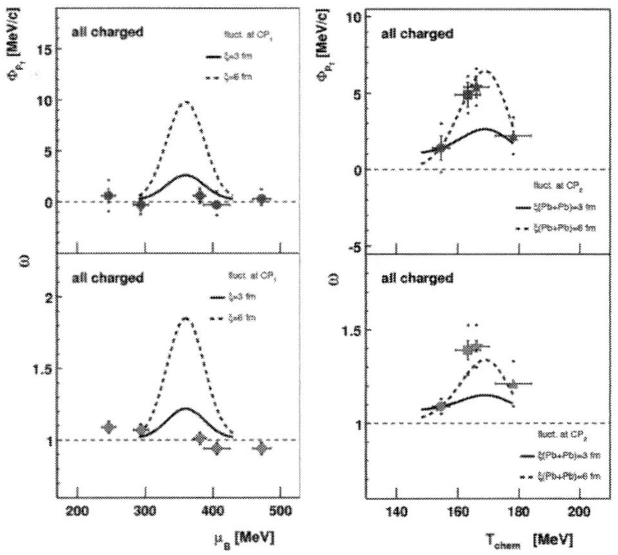

Figure 4. Transverse momentum and multiplicity fluctuations measured by NA49. Left: energy scan for central Pb+Pb collisions. Right: system size dependence (p+p, C+C, Si+Si and Pb+Pb) at 158A GeV. Lines indicate predictions for the critical point calculated for its two locations with two values of the correlation length parameter (dashed and solid lines). For details see Ref. [7, 8], and references therein.

experiment has already looked for indications of the critical point by studying average p_T (ϕ_{p_T} measure) and multiplicity fluctuations (ω) ([7, 8], and references therein). NA49 considered two locations of the CP (shown in Fig. 4). The second location of the CP (μ_B for A+A collisions at 158A GeV, T_{chem} for p+p collisions at 158 GeV) is consistent with the data. NA61/SHINE will search for a maximum of fluctuation by a 2-D scan (Fig. 2).

3.3. High p_T physics

There are two very interesting phenomena in high p_T physics which will be studied in NA61/SHINE. The first one is suppression of high p_T particles observed by RHIC experiments at $\sqrt{s_{NN}} = 62 - 200$ GeV collision energy [9]. It would be interesting to check whether suppression appears already at SPS energies. However, there is not sufficient statistics for p+p and p+Pb collisions in the NA49 experiment. NA61/SHINE will increase these statistics by an order of magnitude. Moreover, the system size and energy dependence of the nuclear modification factor R_{AA} will be investigated (for details see [10]).

The second interesting phenomenon, presented in [11, 12], is a change of the shape of the near-side peak in the two-particle azimuthal angle correlation function with decreasing collision energy in the SPS regime. It is the so-called Jet-Hole transition presented in Fig. 5. The origin of this

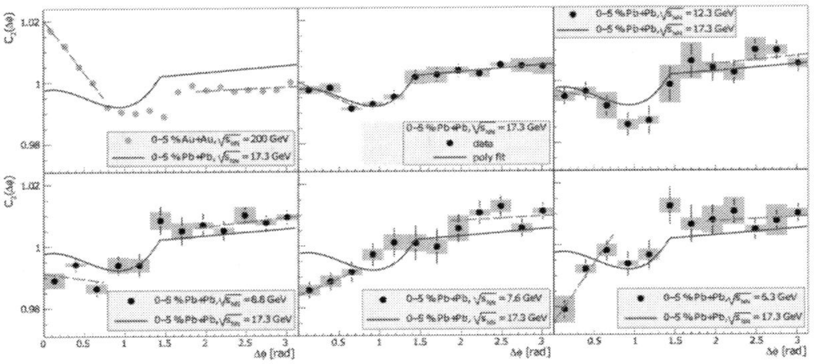

Figure 5. Two-particle azimuthal angle correlation function for central Pb+Pb (Au+Au) events at $\sqrt{s_{NN}}$= 200, 17.3, 12.3, 8.8, 7.6 and 6.3 GeV (full symbols) compared to results for central Pb+Pb collisions at 17.3 GeV. Curves represent the correlation function at $\sqrt{s_{NN}}$= 17.3 GeV and is shown for comparison at the other energies (for details see [11, 12]).

phenomenon is not known yet and further studies are required.

4. Summary

The NA61/SHINE program gives the unique opportunity to reach exciting physics in a very efficient and cost effective way. There are several other complementary projects such as the RHIC low energy scan (starting in 2010), NICA (starting in 2014) and SIS-100(SIS-300) (starting in 2015/17). The advantages of the NA61/SHINE ion program over the RHIC energy scan are:

- 2-D scan (energy and system size) of the phase diagram
- measurements of identified hadron spectra in a broad rapidity range, which in particular allows to obtain mean hadron multiplicities in full phase space
- measurements of the total number of projectile spectators including free nucleons and nucleons in nuclear fragments
- high event rate in the full SPS energy range including the lowest energies
- low p_T region accessible (signatures of the critical point should be visible mostly in the low p_T region)

On the other hand, the advantage of collider experiments is the complete azimuthal acceptance which does not depend on energy (canceling of many systematic uncertainties).

References

[1] Antoniou N *et al.* [NA61 Collaboration] 2006 CERN-SPSC-2006-034 SPSC-P-330; `http://na61.web.cern.ch/`
[2] Beccatini F, Manninen J and Gazdzicki M 2006 *Phys. Rev.* C**73** 044905
[3] Gazdzicki M and Gorenstein M 1999 *Acta Phys. Polon.* B**30** 2705
[4] Alt C *et al.* [NA49 Collaboration] 2008 *Phys. Rev.* C**77** 024903
[5] Fodor Z and Katz S D 2004 *JHEP* **0404** 050
[6] Stephanov M A Rajagopal K Shuryak E 1999 *Phys. Rev.* D**60** 114028
[7] Grebieszkow K for the NA49 Collaboration 2009 *Nucl. Phys.* A**830** 547c-550c `arXiv:0907.4101`
[8] Grebieszkow K for the NA49 and the NA61 Collaborations 2010 *Acta Phys. Polon.* B**41** 427-440 `arXiv:0911.1902`
[9] Adler S S *et al.* [PHENIX Collaboration] 2003 *Phys. Rev. Lett.* **91** 072303
[10] Laszlo A for the NA61/SHINE Collaboration 2009 *Nucl. Phys.* A**830** 559c-562c `arXiv:0907.4493`
[11] Szuba M for the NA49 Collaboration 2008 *Quark Matter* `arXiv:0805.4637`
[12] Szuba M for the NA49 Collaboration 2009 *Nucl. Phys.* A**830** 159c-162c `arXiv:0907.4403`

Quarkyonic Matter and Chiral Spirals

Toru Kojo
RIKEN BNL Research Center, Brookhaven National Laboratory, Upton, NY-11973, USA
E-mail: `torujj@quark.phy.bnl.gov`

Abstract. The nuclear matter, deconfined quark matter, and Quarkyonic matter in low temperature region are classified based on the $1/N_c$ expansion. The chiral symmetry in the Quarkyonic matter is investigated by taking into account condensations of chiral particle-hole pairs. It is argued that the chiral symmetry and parity are locally violated by the formation of chiral spirals, $\langle \bar{\psi} \exp(2i\mu_q z\gamma^0\gamma^z)\psi \rangle$. An extension to multiple chiral spirals is also briefly discussed.

1. Quarkyonic Matter

Quantum Chromodynamics (QCD) in high baryon density and low temperature has attracted continuous interests [1]. Recently large N_c [2] arguments raised conceptual questions in dense QCD [3]. It turns out that a transition scale from the nuclear to the quark matter ($\mu_q \sim \Lambda_{\rm QCD}$) and that of deconfinement ($\mu_q \sim \sqrt{N_c}\Lambda_{\rm QCD}$) are, *at least conceptually*, different. The phase where a quark density is sufficiently high to form the Fermi sea nevertheless excitations are confined, is called *Quarkyonic* phase.

To characterize the Quarkyonic matter, below we will first distinguish the nuclear matter and quark matter, and then classify the quark matter into the deconfined quark matter and the Quarkyonic matter. The $1/N_c$ expansion will be used for theoretical classifications. The following arguments based on the $1/N_c$ expansion are only approximate, nevertheless may be useful, in a similar way as the chiral limit for the chiral transition, or the heavy quark limit for the confinement-deconfinement transition.

The nuclear matter starts to appear slightly below the threshold of nucleons, $\mu_q = M_N/N_c \sim \Lambda_{\rm QCD}$, and changes into the quark matter within a small variation $\sim 1/N_c^2$ in a quark chemical potential. This picture is based on the following arguments. A change in μ_q requires a large change of the nucleon Fermi momentum k_F, since k_F is divided by a large mass in kinetic energy $\sim k_F^2/M$. Accordingly a number density of nucleons $\sim k_F^3$ increases rapidly, and nucleons start to overlap, then strong nuclear forces in short distance becomes relevant. In such a region, it is natural to change effective degrees of freedom from nucleons to quarks for which interactions are weaker in shorter distance. Here quarks need not belong to particular nucleons. A 1st order like rapid growth of a quark density characterizes transitions from the hadronic-nuclear to the quark matter. The large N_c limit provides a clear distinction between nuclear and quark matters.

Now let us discuss properties of the quark Fermi sea. Quarks deep inside of the Fermi sea are perturbative, since Pauli blocking prevents them from being affected by small momentum transfer processes. Such quarks share a large fraction of the Fermi sea, so perturbative estimates should reasonably work for bulk quantities such as pressure to which all quarks contribute. On the other hand, small momentum transfer processes can affect quarks near the Fermi surface.

Although such quarks have large momenta $\sim \mu_q$, they find comoving quarks with which they exchange small momenta, generating nonperturbative phenomena. They play deterministic roles for the phase structure (since perturbative contributions are common to different phases), transport processes, which are sensitive to excitation properties near the Fermi surface.

Now a question is how such a soft interaction looks like in finite density. In the QCD vacuum, quantum fluctuations of gluons provide the confinement, while those of quarks screen gluons reducing confining effects. The strength of fluctuations can be roughly characterized by a number of degrees of freedom: $O(N_c^2)$ for gluons, and $O(N_c)$ for quarks. In finite density, an allowed phase space for low momentum $q\bar{q}$ excitations increases, so do screening effects. The perturbative estimate of the screening scale is $\sim g_s^2 \mu_q^2 \sim \mu_q^2/N_c$, which becomes comparable with $\Lambda_{\rm QCD}$ when $\mu_q \sim \sqrt{N_c}\Lambda_{\rm QCD}$. This is the scale where the deconfinement takes place. The estimate provided here may be minimal, since any energy gaps of quarks are not taken into account: Mass gaps from Lorentz vector self-energy (chiral symmetric) and possible chiral symmetry breaking would suppress $q\bar{q}$ excitations with reducing screening effects.

It might sound strange to speak about the confinement in the region where quarks are already released from nucleons. To draw a qualitative picture, let us consider quarks in terms of quantum wavefunctions. Quarks occupy states in a color singlet way, forming a color white background. A quark excitation in such a background inevitablly accompanies a colored quark-hole, and they are confined forming a mesonic state. Glueballs are also confined, and their spectra are discrete without multi-gluon continuum. The dominant baryonic excitations are presumably baryon number solitons which are coherent excitations of quarks and quark-holes which keep the color singletness together with the background – there is no clear separation of baryons and the background. A correction to this simple argument comes from colored fluctuations in a color white background, which are already addressed in terms of screening effects.

These arguments can be tested by studying 1+1 dimensional QCD which is a confining model. In finite density, the strength of the confinement is unchanged irrespective of how closely quarks are packed within one spatial dimensional line. This is because screening effects in finite density are always same as those in vacuum, due to the same allowed phase space for $q\bar{q}$ fluctuations [4].

Finally it should be emphasized that we have not used a notion of chiral symmetry breaking/restoration to define the Quarkyonic matter. The Quarkyonic matter is solely defined by the presence of the quark Fermi sea and confined excitations. Their consequences on the chiral symmetry are discussed in the next section.

2. Local violation of the Chiral Symmetry: Quarkyonic Chiral Spirals

As μ_q increases, the conventional particle-antiparticle type chiral condensate disappears since generations of antiparticles need large energies $\sim \mu_q$ to transfer particles in the Dirac sea to the above of the Fermi sea. Yet the chiral symmetry can be broken by chiral pairs of particles and holes near the Fermi surface without costing much energy. Chiral condensates of this sort only employ particle-holes near the Fermi surface, modifying only particle dispersions around there. This is consistent with arguments on perturbative quarks deep inside of the Fermi sea.

For simplicity, only typical two cases, exciton and density wave types, will be considered below (for detail, see [4]). In the exciton case, we pick up a particle and a hole from the same spatial momentum region. Since a hole momentum must be flipped, the pair has a total momentum ~ 0. For the density wave, we pick up a particle and a hole from the opposite momentum region, so the pair has a total momentum $\sim 2\mu_q$, forming nonuniform condensates. Such situations have been first discussed in [5], in the context of the deconfined quark matter.

Since we pick up a particle and a hole with similar kinetic energies in both cases, main differences in pairing energies come from potentials. The confining potential should provide a big difference. We expect that exciton wavefunctions made of pairs with opposite large momenta are widely spread in space, costing large energies due to long strings. For density wave pairs, particles and holes comove together maintaining small sizes of $\sim 1/\Lambda_{\rm QCD}$. Therefore we naively expect that confining forces favor density wave pairings.

We first consider the one-pair problem picking up particle-hole from $p_z \sim \pm\mu_q$, $p_T \sim 0$. This setup will turn out to be a good starting point to analyze multi-pair problems [6]

For concrete arguments, we introduce a simple model of the linear confinement in which gluon propagators take the following form (screening effects are suppressed in large N_c),

$$D_{44}^{AB}(k) = -\frac{8\pi}{C_F} \times \frac{\sigma}{(\vec{k}^2)^2}\,\delta^{AB} \quad ; \quad D^{4i} = D^{ij} = 0\ , \quad (C_F = \frac{N_c^2-1}{2N_c},\ \sigma \sim \Lambda_{\rm QCD}^2) \tag{1}$$

which shows a linear potential in coordinate space. This model is inspired by Coulomb gauge analyses [7]. We have omitted perturbative parts for the sake of simplicity [8].

To investigate condensations, we analyze self-consistent equations by reducing 3+1 dimensional equations to 1+1 dimensional ones. (The essence was already discussed in [9]). Below we illustrate this for the Schwinger-Dyson equation for the quark self-energy. It looks like

$$\not{\Sigma}(p) + \Sigma_m(p) = -\int \frac{d^4k}{(2\pi)^4}\,(\gamma_4 t_A)\,S(k;\Sigma)\,(\gamma_4 t_B)\,D_{44}^{AB}(p-k)\ , \tag{2}$$

where the quark propagator is dressed and depends on $\Sigma(p)$. We will investigate the self-energy for quarks near the Fermi surface, with $p_z \sim \mu_q$ and $p_T \sim 0$.

Three points are relevant for the dimensional reduction: (I) Our gluon propagator behaves as $1/(\vec{q}^2)^2$, so dominant contributions of the integral (3) sharply concentrates on the small domain. (II) Different Dirac structures $S(k) = \gamma_4 S_4(k) + \gamma_z S_z(k) + \vec{\gamma}_T \cdot \vec{S}_T(k) + S_m(k)$, give the ratio of $|S_T|/S_z \sim k_T/k_z \sim \Lambda_{\rm QCD}/\mu_q$, so we can drop off S_T and Σ_T in the leading order of $\Lambda_{\rm QCD}/\mu_q$. (III) The quark energy is sensitive to the change of k_z while not to that of k_T. Changes in k_T is along the constant energy surface which looks flat for $\Lambda_{\rm QCD}/\mu_q \ll 1$, without changing the quark energy a lot.

Because of the restricted integral region and the insensitivity of the quark propagator to k_T variable, we can set $\vec{k}_T \simeq \vec{0}_T$ and factorize the integral equation:

$$\int dk_4 dk_z d^2\vec{k}_T\ S(k_4,k_z,\vec{k}_T)\ D_{44}(p-k) \ \rightarrow\ \int dk_4 dk_z\ S(k_4,k_z,\vec{0}_T) \int d\vec{k}_T\ D_{44}(p-k), \tag{3}$$

for which we can carry out the $\vec{k}_T$ integral. The smearing of gluon propagator yields a 1+1 dimensional confining propagator, $\sim 1/(p_z-k_z)^2$. The reduced equation then becomes

$$\bar{\not{\Sigma}}(p_4,p_z,\vec{0}_\perp) + \Sigma_m(p_4,p_z,\vec{0}_\perp) \simeq \frac{N_c g_{\rm 2D}^2}{2}\int \frac{dk_4\,dk_z}{(2\pi)^2}\ \gamma_4\ \bar{S}(k_4,k_z,\vec{0}_\perp)\ \gamma_4\ \frac{1}{(k_z-p_z)^2}\ , \tag{4}$$

where $N_c g_{\rm 2D}^2 = 4\sigma$ and we denote $\bar{\Sigma}, \bar{S}$ for components other than transverse ones. The resulting equation is nothing but the Schwinger-Dyson equation of 't Hooft model in axial gauge $A_z = 0$. The same approximations are applicable to the Bethe-Salpeter equation.

Since self-consistent equations for our one-pair problem take the same form as those of 't Hooft model [10], we can borrow results in the existing literatures [11]. Only nontrivial issues are relationships between 3+1 and 1+1 dimensional operators and condensation channels. In particular, concepts of spins are absent in one spatial dimension. In our dimensional reduction, the tranverse part of the quark propagators are suppressed. This implies that $\vec{\gamma}_T$ terms are dropped off, so that there is no spin mixing term once we quantize spins along the directions of moving particles. Introducing projection operators for moving directions, we can write spin multiplets [9],

$$\psi_\pm = \frac{1 \pm \gamma^0\gamma^z}{2}\psi\ , \qquad \varphi_\uparrow = \begin{bmatrix} \varphi_{\uparrow+} \\ \varphi_{\uparrow-} \end{bmatrix} = \begin{bmatrix} \psi_{R+} \\ \psi_{L-} \end{bmatrix} \quad ; \quad \varphi_\downarrow = \begin{bmatrix} \varphi_{\downarrow+} \\ \varphi_{\downarrow-} \end{bmatrix} = \begin{bmatrix} \psi_{L+} \\ \psi_{R-} \end{bmatrix}\ , \tag{5}$$

where indices $(+,-)$ for $+z,-z$ moving particles. Let us introduce $1+1$ dimensional "flavor", $\Phi^T \equiv (\varphi_\uparrow, \varphi_\downarrow)$, and Dirac matrices, $\Gamma^0 = \sigma^1$, $\Gamma^z = -i\sigma^2$, $\Gamma^5 = \sigma^3$. Then 3+1 dimensional quark bilinears without spin mixings are mapped onto 1+1 dimensional "flavor" *singlet* operators [4, 9],

$$\bar{\psi}\psi \to \bar{\Phi}\Phi, \quad \bar{\psi}\gamma^0\psi \to \bar{\Phi}\Gamma^0\Phi, \quad \bar{\psi}\gamma^z\psi \to \bar{\Phi}\Gamma^z\Phi, \quad \bar{\psi}\gamma^0\gamma^z\psi \to \bar{\Phi}\Gamma^5\Phi. \tag{6}$$

The last relation reflects the fact that 1+1 dimensional Γ^5 characterizes moving directions. All other quark bilinears include spin mixings, so they are "flavor" *non-singlet* in 1+1 dimensions. They do not show condensations in the 't Hooft model.

The 't Hooft model in finite density is known to show the chiral spirals rotating in chiral space as $\langle \bar{\Phi} e^{2i\mu_q z \Gamma^5} \Phi \rangle$. Mapping this, we get the chiral spiral in 3+1 dimensions,

$$\langle \bar{\psi}\psi \rangle = \Delta \cos(2\mu_q z), \quad \langle \bar{\psi}\gamma^0\gamma^z\psi \rangle = \Delta \sin(2\mu_q z). \tag{7}$$

An origin of this chiral rotation is that $\bar{\psi}_+\psi_-$ and $\bar{\psi}_-\psi_+$ moving to opposite directions each other, thus have opposite phase $e^{\pm 2i\mu_q z}$; this phase mismatch makes $\langle \bar{\psi}\gamma^0\gamma^z\psi \rangle = \langle \bar{\psi}_+\psi_- \rangle - \langle \bar{\psi}_-\psi_+ \rangle$ nonvanishing. So the chiral density wave accompanies its partner, forming spiral structures.

It is clear that a spatial average of chiral condensates vanishes because of spatial modulations. So the symmetry apparently looks restored for probes with large wavelength. Note also that our chiral tensor type condensate locally breaks parity.

Finally we briefly mention an extension of the present results to those of multi-pairs. This problem was investigated in [6]. A key observation is that in confining models, chiral spirals evolved in different spatial directions interact only weakly. Therefore our results for one-patch problem may be applied with slight modifications. Such a picture indicates that a number of directions of chiral spiral formations increases as μ_q, with a series of phase transitions.

3. Summary and Outlook

In the Quarkyonic phase, excitations are confined despite of a large quark number density. They drive the formation of the chiral spirals with large mass gaps $\sim \Lambda_{\rm QCD}$. This is in sharp contrast to the situation in the deconfined quark matter where gaps are too small to overtake screening effects [9]. So chiral spirals can be regarded as good signatures of the Quarkyonic phase. Transport processes in stars are good candidates to see consequences of chiral spirals through characteristic Nambu-Goldstone modes. A local parity violation may also provide important implications through neutrino physics. These issues deserve further investigations.

Acknowledgments

The author acknowledges his collaborators, Y. Hidaka, L. McLerran, R.D. Pisarski, and A.M. Tsvelik with whom the present ideas have been developed. This research is supported under DOE Contract No. DE-AC02-98CH10886 and Special Posdoctoral Research Program of RIKEN. He also thanks organizers and participants of the workshop Hot Quark 10.

References

[1] For a recent review, K. Fukushima and T. Hatsuda, arXiv:1005.4814 [hep-ph].
[2] G. 't Hooft, Nucl. Phys. B 72 (1974) 461.
[3] L. McLerran and R. D. Pisarski, Nucl. Phys. A 796 (2007) 83.
[4] T. Kojo, Y. Hidaka, L. McLerran and R. D. Pisarski, Nucl. Phys. A 843 (2010) 37, and references therein.
[5] D. V. Deryagin, D. Y. Grigoriev and V. A. Rubakov, Intl. Jour. Mod. Phys. A 7 (1992) 659.
[6] T. Kojo, R. D. Pisarski, and A. M. Tsvelik, arXiv:1007.0248 [hep-ph], to be published in PRD.
[7] D. Zwanziger, Phys. Rev. Lett. 90 (2003) 102001; Phys. Rev. D 69 (2004) 016002.
[8] L. Y. Glozman and R. F. Wagenbrunn, Phys. Rev. D 69 (2004) 016002.
[9] E. Shuster and D. T. Son, Nucl. Phys. B 573 (2000) 434.
[10] G. 't Hooft, Nucl. Phys. B 75 (1974) 461.
[11] For instance, B. Bringoltz, Phys. Rev. D 79 (2009) 125006.

Hot Quarks 2010 IOP Publishing
Journal of Physics: Conference Series **270** (2011) 012050 doi:10.1088/1742-6596/270/1/012050

Computation of transport coefficients around critical point based on novel diagrammatic method

Y Hidaka and T Kunihiro

Department of Physics, Kyoto University, Sakyo-ku, Kyoto 606-8502, Japan

E-mail: `hidaka@ruby.scphys.kyoto-u.ac.jp`

Abstract. We compute and discuss the temperature-dependence of the shear viscosity, focusing on that around the critical point, using the Nambu–Jona-Lasinio model in the leading order of $1/N_c$ expansion and the chiral limit. Our calculation is based on a systematic resummation method of the singular diagrams, developed recently by the present authors. In the leading order of the theory, $2 \to 2$ scattering between quark-quark, antiquark-antiquark and quark-antiquark contributes to the shear viscosity. The ratio of the shear viscosity to the entropy, η/s, decreases as temperature increases, only with a tiny cusp at $T = T_c$.

1. Introduction

Transport coefficients reflect dynamical properties of the medium slightly away from thermal equilibrium. This is also a parameter of hydrodynamics, which cannot be calculated by hydrodynamics itself. To determine the transport coefficient from a microscopic theory with accuracy is important to understand the property of the medium, and also necessary to determine the parameter of hydrodynamics such as viscosity. A diagrammatic method is a useful tool for computing the transport coefficient; there is, however, a technical difficulty to compute it due to the so called *pinch singularity*, which turns out to be circumvented by a resummation of an infinite number of diagrams. Very recently, the present authors [1] have developed a novel resummation technique for that use: Not only the leading order in the rearranged expansion, which is the linearized Boltzmann equation known in the literature [2], but also higher orders can be taken into account in this theory.

In this talk, we apply the method to a fermionic system with the chiral phase transition at finite temperature. For this purpose, we employ the Nambu–Jona-Lasinio model in the chiral limit and calculate the shear viscosity in the leading order of large-N_c expansion, although our method can systematically take into account higher order corrections.

2. Diagrammatic Method for computing transport coefficients

2.1. Nambu–Jona-Lasinio model

The Lagrangian of NJL model has a form,

$$\mathcal{L} = \overline{\psi} i \partial\!\!\!/ \psi + \frac{G_S}{2}\left[(\overline{\psi}\psi)^2 + (\overline{\psi} i\gamma_5 \tau^a \psi)^2\right]. \tag{1}$$

where ψ is the quark field with color N_c and flavor $N_f = 2$, $\partial\!\!\!/ = \gamma^\mu \partial_\mu$, and τ^a's are the Pauli matrices for the flavors. G_S is the four point coupling constant. In the large-N_c counting, G_S is

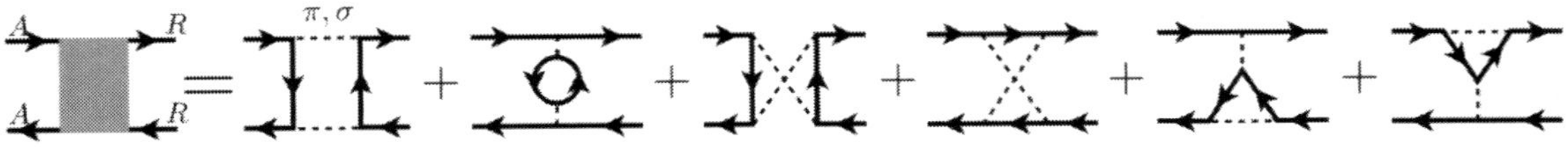

Figure 1. The four point function contributing to the shear viscosity in the leading order of $1/N_c$ expansion. The dotted line denotes the propagators of the π and σ modes.

of order $1/N_c$. In the leading order of N_c, the thermodynamic potential density is given by

$$\Omega(\sigma, T, \mu) = \frac{\sigma^2}{2G_S} - 4N_c \int \frac{d^3k}{(2\pi)^3}\left[E_k + 2T\ln\left(1 + e^{-\beta E_k}\right)\right], \tag{2}$$

where $\sigma = G_S\langle\bar{\psi}\psi\rangle$ is the mean-field of the sigma channel, and $E_k = \sqrt{\boldsymbol{k}^2 + m^2}$, with $m = -\sigma$. σ can be determined by the gap equation $\partial\Omega/\partial\sigma = 0$. The pressure and entropy density are obtained from Ω as $p = -\Omega$ and $s = -\partial\Omega/\partial T$, respectively.

2.2. Diagrammatic method

Here, we briefly review a diagrammatic method for computing transport coefficients [1]. In the linear response theory [3], the shear viscosity is given by

$$\eta = \frac{1}{10}\lim_{\omega\to 0}\frac{1}{\omega}\mathrm{Im}\int d^4x\, e^{i\omega t} i\theta(t)\left\langle[\pi_{ij}(x), \pi^{ij}(0)]\right\rangle, \tag{3}$$

where $\pi_{ij}(x) = T_{ij}(x) - g_{ij}T^k{}_k(x)/3$ is the traceless stress tensor, and $T_{\mu\nu}(x) = \bar{\psi}(x)(\gamma_\mu\partial_\nu + \gamma_\nu\partial_\mu)\psi(x)/2 - g_{\mu\nu}\mathcal{L}$ is the energy-momentum tensor. We work in Minkowski space-time with a metric, $g_{\mu\nu} = \mathrm{diag}\,(1, -1, -1, -1)$. In our diagrammatic method [1], the shear viscosity can be shown to have the following form with the quark spectral function $\rho(k)$:

$$\eta = \frac{1}{10T}\int \frac{d^4k}{(2\pi)^4}\frac{\rho(k)}{-2\mathrm{Im}\,\Sigma_R(k)} n(k^0)(1 - n(k^0))\bar{\Gamma}_{ij}(k, s, s')\Gamma^{ij}(k, s, s'), \tag{4}$$

where s denotes spin, flavor and color indices, $n(k^0) = 1/(\exp(k^0/T) + 1)$ is the Fermi-Dirac distribution function, $\bar{\Gamma}_{ij}(k, s, s')$ the vertex function including thermal and quantum corrections and $\Sigma_R(k)$ the retarded self-energy taken the spin, flavor and color average. We remark that $\bar{\Gamma}_{ij}(k, s, s')$ does not includes collision effects unlike $\Gamma(k, s, s')$, as will be elucidated below; see Eq. (6). In the leading order, $\bar{\Gamma}_{ij}(k, s, s')$ becomes

$$\bar{\Gamma}^{ij}(k, s, s') = \frac{1}{2}\bar{u}(k, s)\left(k^i\gamma^j + \gamma^i k^j - 2\delta^{ij}\frac{\boldsymbol{k}\cdot\boldsymbol{\gamma}}{3}\right)u(k, s') = 2\delta_{s,s'}B_0^{ij}(k), \tag{5}$$

with $B_0^{ij}(k) = k^ik^j - \delta^{ij}\boldsymbol{k}^2/3$ and $u(k, s')$ being a wave function of a quark. In the quasiparticle approximation, $\rho(k) = (2\pi)\delta(k^2 - m^2)$. Equation (4) shows that η diverges if $\mathrm{Im}\,\Sigma_R(k) = 0$. This is the pinch singularity mentioned above. In the $1/N_c$ expansion, $\mathrm{Im}\,\Sigma_R(k)$ is of order $1/N_c$, implying that η/s is of order N_c. $\Gamma^{ij}(k, s, s') \equiv 2\delta_{s,s'}B^{ij}(k)$ is the resummed vertex satisfying the following self-consistent equation:

$$B^{ij}(k) = B_0^{ij}(k) - \int \frac{d^4k'}{(2\pi)^4}\frac{1}{4N_c}\sum_{s,s'}\mathrm{Im}\,\mathcal{T}_{AARR}(k, s, s; k', s', s')\frac{\rho(k')}{-2\mathrm{Im}\,\Sigma_R(k')}B^{ij}(k'), \tag{6}$$

where $\mathcal{T}_{AARR}$ is a four point function that does not include diagrams with the pinch singularity. The diagrams contributing to the leading order in $1/N_c$ is shown in Fig. 1, where the indices A and R stand for advanced and retarded, respectively. In the leading order approximation, $\mathcal{T}_{AARR}$ is of oder $1/N_c$, which cancels with the contribution by the inverse of $\mathrm{Im}\,\Sigma_R(k)$ in RHS of Eq. (6), so the kernel in that equation is found to be of order 1. This means that the ladder diagram shown in Fig. 1 must be summed over. For later use, we mention that the self-energy and the four-point function are related through Ward-Takahashi identity [1, 5],

$$-2\mathrm{Im}\,\Sigma_R(k) + \int \frac{d^4k'}{(2\pi)^4}\rho(k')\frac{1}{4N_c}\sum_{s,s'}\mathrm{Im}\,\mathcal{T}_{AARR}(k,s,s;k',s',s') = 0\,. \tag{7}$$

2.3. Reduction of the self-consistent equation to a Boltzmann equation

It is useful to introduce the following collision operator to clarify the physical process in Eq. (6);

$$\begin{aligned}\mathcal{L}\varphi^{ij}(k) \equiv &-\frac{1}{k^0}\,\mathrm{Im}\,\Sigma_R(k)\varphi^{ij}(k) + \frac{1}{2k^0}\int\frac{d^4k'}{(2\pi)^4}\frac{1}{4N_c}\sum_{s,s'}\mathrm{Im}\,\mathcal{T}_{AARR}(k,k';s,s)\rho(k')\varphi^{ij}(k')\\ =&\frac{1}{16N_ck^0}\int\frac{d^4k'}{(2\pi)^4}n(k'^0)\rho(k')\int\frac{d^4q}{(2\pi)^4}(1-n(q^0))\rho(q)\int\frac{d^4q'}{(2\pi)^4}n(q'^0)\rho(q')\\ &\times|\mathcal{M}|^2(2\pi)^4\delta^{(4)}(k+q-k'-q')(\varphi^{ij}(k)-\varphi^{ij}(k')+\varphi^{ij}(q)-\varphi^{ij}(q'))\,,\end{aligned} \tag{8}$$

where $\varphi^{ij}(k) \equiv B^{ij}(k)/(-2\mathrm{Im}\,\Sigma_R(k))$. Using the collision operator, we can rewrite Eq. (8) as

$$\frac{1}{2k^0}B_0^{ij}(k) = \mathcal{L}\varphi^{ij}(k), \tag{9}$$

It can be shown [1] that Eq. (9) is equivalent to a linearized Boltzmann equation. Equation (8) contains $1\to3$ and $3\to1$ processes in addition to $2\to2$ process; however, in the leading order, only $2\to2$ process contributes to the shear viscosity, because $1\to3$ and $3\to1$ can not satisfy the on-shell condition of the quarks. $2\to2$ processes contain quark-quark, antiquark-antiquark, and quark-antiquark scattering. Scatterings of the π and σ modes are higher oder in the large-N_c expansion. The $\mathcal{M}$ of quark-quark scattering part is written as

$$|\mathcal{M}_{qqqq}|^2 = \sum_{\text{spin,flavor,color}}|\mathcal{M}_{qqqq}|^2 = |\mathcal{M}_\sigma|^2 + |\mathcal{M}_\pi|^2 + 2\,\mathrm{Re}\,(\mathcal{M}_\sigma\mathcal{M}_\pi^*)\,. \tag{10}$$

The squared amplitudes for each channel are given by

$$\begin{aligned}|\mathcal{M}_\sigma|^2 =& 16N_c^2\Big[(t-4m^2)^2|D_\sigma(k-k')|^2 + (u-4m^2)^2|D_\sigma(k-q')|^2\\ &-\frac{1}{2N_c}(4m^2s-tu)\mathrm{Re}\,D_\sigma(k-k')D_\sigma^*(k-q')\Big]\,,\\ |\mathcal{M}_\pi|^2 =& 48N_c^2\Big[t^2|D_\pi(k-k')|^2 + u^2|D_\pi(k-q')|^2 - \frac{1}{2N_c}tu\,\mathrm{Re}\,D_\pi(k-k')D_\pi^*(k-q')\Big]\,,\\ \mathrm{Re}\,(\mathcal{M}_\sigma\mathcal{M}_\pi^*) =& 12N_c\Big[u(4m^2-t)\mathrm{Re}\,D_\sigma(k-k')D_\pi^*(k-q')\\ &+t(4m^2-u)\mathrm{Re}\,D_\sigma(k-q')D_\pi^*(k-k')\Big]\,,\end{aligned} \tag{11}$$

where $t=(k-k')^2$ and $u=(k-q')^2$ are Mandelstam variables and D_σ and D_π denote propagators of the σ and π modes, respectively. We remark that the interference terms are necessary to satisfy the unitarity of S matrix, although they are higher order in the $1/N_c$ expansion. The antiquark-antiquark scattering and the quark-antiquark scattering are obtained by using crossing symmetry: $|\mathcal{M}_{\bar{q}\bar{q}\bar{q}\bar{q}}|^2(k,q,k',q') = |\mathcal{M}_{qqqq}|^2(-k',-q',-k,-q)$, and $|\mathcal{M}_{q\bar{q}q\bar{q}}|^2(k,q,k',q') = |\mathcal{M}_{qqqq}|^2(k,-q',k',-q)$.

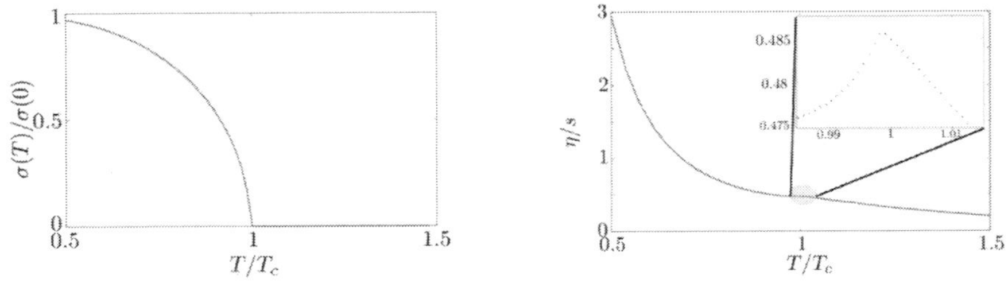

Figure 2. The preliminary results of the chiral condensate (left) and the ratio of the shear viscosity to the entropy (right) as a function of temperature.

3. Numerical results

Here we show the numerical result for the chiral condensate and the ratio of the shear viscosity to the entropy, η/s, as a function of temperature in Fig. 2. We fixed the parameters as cutoff $\Lambda = 650\text{MeV}$, $N_c = 3$, $N_f = 2$, $G_S\Lambda^2 = 4.233$. The ratio η/s decreases as temperature increases, and shows apparently no singular behavior even at $T = T_c$, except for a tiny cusp. This cusp is attributed to the vanishing of the on-shell coupling between the quarks and the mesons at $T = T_c$. One might anticipate a much more sigular behavior of the viscosity owing to the softening of the σ mode at $T = T_c$ in the chiral limit. The σ-exchange interaction in the t-channel reads $|\mathcal{M}_\sigma|^2 \sim (t - 4m^2)^2/(t - m_\sigma^2)^2$ around $T = T_c$. Since $m_\sigma = 0$ at $T = T_c$, the denominator vanishes as $t \to 0$ as expected, but so does the numerator in the chiral limit, and hence $|\mathcal{M}_\sigma|^2$ is not divergent at $T = T_c$ and $t \to 0$. This situation may change with finite (current) quark mass and chemical potential, in which case there may exist a critical point. In this case, the coupling of quarks to the critical mode that fluctuation diverges might make the shear viscosity singular. This is beyond the scope of the present talk.

4. Summary and outlook

We have calculated the temperature dependence of the shear viscosity in the leading order of $1/N_c$ expansion in the NJL model, using a systematic resummation method. The large fluctuation of the order parameter near T_c affects the shear viscosity through the σ meson exchange in the t-channel of quark-quark scattering. However, the shear viscosity shows apparently no singular behavior because the dynamical quark mass vanishes at the same time. In reality, beyond the leading order, the fluctuation of the order parameter also couples to hydrodynamics modes, which become important near T_c. It is interesting to develop a diagrammatic method taking into account such effects of hydrodynamic mode. We leave this problem in our future work.

Acknowledgments

This work was supported by the Grant-in-Aid for Scientific Research (No. 20540265) and for the Global COE Program "The Next Generation of Physics, Spun from Universality and Emergence" by the Ministry of Education, Culture, Sports, Science and Technology (MEXT) of Japan.

References

[1] Hidaka Y and Kunihiro T arXiv:1009.5154 [hep-ph]
[2] Jeon S 1995 *Phys. Rev.* D **52** 3591
[3] Zubarev D N 1974 *Nonequilibrium Statistical Thermodynamics* (New York: Plenum)
[4] Rehberg P, Klevansky S P and Hufner J 1996 *Nucl. Phys.* A **608** 356
[5] Gagnon J S and Jeon S 2007 *Phys. Rev.* D **75** 025014 [Erratum-ibid. 2007 D **76** 089902]

Hot Quarks 2010 IOP Publishing
Journal of Physics: Conference Series **270** (2011) 012051 doi:10.1088/1742-6596/270/1/012051

Ratios of Fluctuation Observables in the Search for the QCD Critical Point

Christiana Athanasiou[a], Krishna Rajagopal[a] and Misha Stephanov[b]
[a]Department of Physics, Massachusetts Institute of Technology, Cambridge, MA 02139, USA
[b]Department of Physics, University of Illinois, Chicago, Illinois 60607, USA

Abstract. The QCD critical point can be found in heavy ion collision experiments via the non-monotonic behavior of many fluctuation observables as a function of the collision energy. The event-by-event fluctuations of various particle multiplicities are enhanced in those collisions that freeze out near the critical point. Higher, non-Gaussian, moments of the event-by-event distributions of such observables are particularly sensitive to critical fluctuations, since their magnitude depends on the critical correlation length to a high power. We present quantitative estimates of the contribution of critical fluctuations to the third and fourth moments of the pion and proton, as well as estimates of various measures of pion-proton correlations, all as a function of the same five non-universal parameters. We show how to use nontrivial but parameter independent ratios among these more than a dozen fluctuation observables to discover the critical point. We also construct ratios that, if the critical point is found, can be used to overconstrain the values of the non-universal parameters.

1. Introduction

One of the distinctive features of the phase diagram is the critical point at which the first-order transition between hadron matter and QGP ends. We currently do not have a systematic way of locating this point from first principles as model and lattice calculations face many challenges and much work still needs to be done in order to overcome them. (For review, see e.g., [1].) In the meantime, if the critical point is located in a region accessible to heavy-ion collision experiments, it can be discovered experimentally in current and future experiments at RHIC, BNL and SPS. It is therefore important to define and select experimental observables that will allow us to locate the critical point, if it is located in an experimentally accessible region.

Upon scanning in the center of mass energy $\sqrt{s}$ and thus in μ_B, one should then be able to locate (or rule out the presence of) the critical point by using observables that are sensitive to the proximity of the freeze-out point to the critical point [2, 3], such as pion and proton multiplicity fluctuations. As we vary $\sqrt{s}$, if the freeze-out point approaches the critical point, we would see an increase and then a decrease, as we move away from it, in the fluctuations in the number of these particles. In this paper we describe how to use this non-monotonic behavior in the fluctuations of particle numbers near the critical point as a probe to determine its location.

2. Results

2.1. Cumulant calculations

The critical mode, σ, is the mode which develops large long wavelength fluctuations at the critical point. Its screening mass is related to the correlation length ξ by $m_\sigma \equiv \xi^{-1}$. In the case of an infinite system, ξ diverges at the critical point. In reality, ξ reaches a maximum value at the critical point but does not diverge because as it cools the system spends only a finite time in the vicinity of the critical point. Estimates of the rate of growth of ξ as the collision cools past the critical point suggest that the maximal value of ξ that can be reached is around 1.5 – 3 fm [4, 5, 6], compared to the natural $\sim$ 0.5 fm away from the critical point. Near the critical point, the σ^3 and σ^4 interaction couplings are given by $\lambda_3 = \tilde{\lambda}_3 T(T\xi)^{-3/2}$, and $\lambda_4 = \tilde{\lambda}_4(T\xi)^{-1}$, where the dimensionless couplings $\tilde{\lambda}_3$ and $\tilde{\lambda}_4$ are universal and they have been determined for the Ising universality class [7]. Throughout this paper we shall use $\tilde{\lambda}_3 = 4$ and $\tilde{\lambda}_4 = 12$ as benchmark values, because these are the midpoints of the ranges of values known for these constants [7, 8].

The long wavelength fluctuations in the σ-field manifest themselves in observable quantities in so far as they affect the fluctuations of the occupation numbers of particles that couple to the σ-field, like protons and pions, which interact with σ through the effective Lagrangian: $2G\sigma\pi^+\pi^- + g_p\,\sigma\,\bar{p}\,p$. Throughout this paper we will use $G = 300$ MeV (see Ref. [3]) and $g_p = 7$ (see, e.g., [9]) as benchmark values. It is important to bear in mind that both these parameters and λ_3 and λ_4 are all uncertain at the factor of 2 level.

The contribution of critical fluctuations to the 2-, 3- and 4- particle correlators due to σ-exchanges is given by (see Refs. [10], [3] and [8])

$$\langle \delta n_{\mathbf{k}_1} \delta n_{\mathbf{k}_2} \rangle_\sigma = \frac{d^2}{VT} g^2 \xi^2 \frac{v_{\mathbf{k}_1}^2}{\gamma_{\mathbf{k}_1}} \frac{v_{\mathbf{k}_2}^2}{\gamma_{\mathbf{k}_2}}, \quad \langle \delta n_{\mathbf{k}_1} \delta n_{\mathbf{k}_2} \delta n_{\mathbf{k}_3} \rangle_\sigma = \frac{2d^3 \tilde{\lambda}_3}{V^2 T^{3/2}} g^3 \xi^{9/2} \frac{v_{\mathbf{k}_1}^2}{\gamma_{\mathbf{k}_1}} \frac{v_{\mathbf{k}_2}^2}{\gamma_{\mathbf{k}_2}} \frac{v_{\mathbf{k}_3}^2}{\gamma_{\mathbf{k}_3}}, \tag{1}$$

$$\langle \delta n_{\mathbf{k}_1} \delta n_{\mathbf{k}_2} \delta n_{\mathbf{k}_3} \delta n_{\mathbf{k}_4} \rangle_\sigma = \frac{6d^4}{V^3 T^2} \left(2\tilde{\lambda}_3^2 - \tilde{\lambda}_4\right) g^4 \xi^7 \frac{v_{\mathbf{k}_1}^2}{\gamma_{\mathbf{k}_1}} \frac{v_{\mathbf{k}_2}^2}{\gamma_{\mathbf{k}_2}} \frac{v_{\mathbf{k}_3}^2}{\gamma_{\mathbf{k}_3}} \frac{v_{\mathbf{k}_4}^2}{\gamma_{\mathbf{k}_4}}. \tag{2}$$

Equations (1) - (2) apply to both protons (with $g = g_p$) and pions (with $g = g_\pi = G/m_\pi$). The degeneracy factor d is 2 for both protons and pions since the coupling to the σ-field is both spin and charge "blind". The variance of the fluctuating occupation number distribution is denoted by $v_{\mathbf{k}}^2$ and $\gamma_{\mathbf{k}}$ is the relativistic gamma-factor of the particle with a given momentum $\mathbf{k}$.

Mixed pion-proton correlators can also be obtained from (1) and (2), with small modifications. For example, the 2 pion - 2 proton correlator is given by

$$\langle \delta n_{\mathbf{k}_1}^\pi \delta n_{\mathbf{k}_2}^\pi \delta n_{\mathbf{k}_3}^p \delta n_{\mathbf{k}_4}^p \rangle_\sigma = \frac{6d_\pi^2 d_p^2}{V^3 T^2} \left(2\tilde{\lambda}_3^2 - \tilde{\lambda}_4\right) g_\pi^2 g_p^2 \xi^7 \frac{v_{\mathbf{k}_1}^{\pi\,2}}{\gamma_{\mathbf{k}_1}^\pi} \frac{v_{\mathbf{k}_2}^{\pi\,2}}{\gamma_{\mathbf{k}_2}^\pi} \frac{v_{\mathbf{k}_3}^{p\,2}}{\gamma_{\mathbf{k}_3}^p} \frac{v_{\mathbf{k}_4}^{p\,2}}{\gamma_{\mathbf{k}_4}^p}. \tag{3}$$

In order to obtain the critical contribution to the cumulants of the particle multiplicity distributions using the correlators given above, each momentum index needs to be integrated over using $V \int d^3\mathbf{k}/(2\pi)^3$. A cumulant with i protons and j pions will be denoted by $\kappa_{ipj\pi}$. In order to remove the V dependence of the cumulants we choose to normalize them as

$$\omega_{ipj\pi} \equiv \frac{\kappa_{ipj\pi}}{\langle N_p \rangle^{i/i+j} \langle N_\pi \rangle^{j/i+j}}. \tag{4}$$

We can see from expressions (1) - (2) that higher cumulants are proportional to higher powers of ξ and thus increase by a larger factor near the critical point where ξ becomes large.

In the case of free particles in the classical Boltzmann regime, with no critical fluctuations, the fluctuations of any particle number obey Poisson statistics. The Poisson contribution to ω_{ip} and $\omega_{i\pi}$ is 1, and zero for the mixed cumulants $\omega_{ipj\pi}$. In reality, this 1 of Poisson statistics gets few percent contributions from Bose-Einstein statistics, from system size fluctuations, from initial state correlations that are incompletely washed out, and from interactions other than those with the fluctuations that are enhanced near the critical point. We are ignoring all of these non-critical corrections to the Poissonian 1. Existing data on the ratio of the fourth to second net proton cumulants, $\kappa_{4(p-\bar{p})}/\kappa_{2(p-\bar{p})}$, at $\sqrt{s} = 19.6$, 62.4 and 200 GeV [11] confirm that the non-critical corrections to the Poissonian 1 are indeed small.

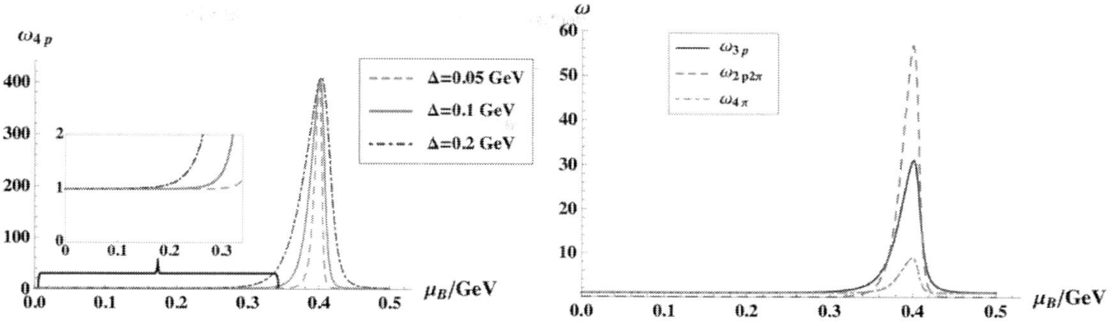

Figure 1. The μ_B-dependence of selected normalized cumulants. We only include the Poisson and critical contributions to the cumulants. In the left panel we show ω_{4p} and vary Δ. In the right panel we show other examples of ω's with $\Delta = 100$ MeV. See text for details.

2.2. Possible experimental outcomes

We now illustrate *possible* experimental outcomes of measurements of the cumulants, assuming that the matter produced at the freezeout point of the fireball evolution for some collision energy $\sqrt{s}$ is near the critical point. To start, let us assume that the critical point occurs at $\mu_B^c = 400$ MeV and that $\xi_{\max} = 2$ fm. The simplest ansatz for $\xi(\mu_B)$ that we have found that incorporates the above physics and has roughly the correct scaling behavior from universality of critical phenomena is

$$\xi(\mu_B) = \frac{\xi_{\max}}{\left[1 + \frac{(\mu_B - \mu_B^c)^2}{W(\mu_B)^2}\right]^{1/3}}, \tag{5}$$

with $W(\mu_B)$ chosen in order to reflect the asymmetry in $\xi(\mu_B)$ when approaching the critical point from the left (smaller μ_B) or from the right. We shall define the width Δ as the width in μ_B within which $\xi > 1$ fm. We also use our benchmark values of the four nonuniversal parameters that determine the $\omega_{ipj\pi}$ for a given ξ, namely $g_\pi = G/m_\pi = 2.1$, $g_p = 7$, $\tilde{\lambda}_3 = 4$ and $\tilde{\lambda}_4 = 12$ and we have allowed for the fact that the chemical freeze-out temperature T decreases somewhat with increasing μ_B by using an empirical parametrization of heavy ion collision data from Ref. [12].

Numerical results for some examples of normalized cumulants are shown in Fig. 1. We can clearly see the peak in all the normalized cumulants near the critical point. The results indicate that the more protons are involved in the observation measure, the easier it is to identify the critical contribution. It is readily apparent that the measurement of these observables in heavy ion collisions at a series of collision energies is very well suited to ruling out (or discovering) the presence of the QCD critical point.

2.3. Ratios of cumulants

In order to locate the critical point, experimental results on multiplicity cumulants will need to be compared to the theoretical predictions of the critical contribution to these cumulants. But, recall that we had to choose benchmark values for four parameters: g_p, g_π, $\tilde{\lambda}_3$, $\tilde{\lambda}_4$ and also use an ansatz for $\xi(\mu_B)$. These parameters are not known reliably or accurately enough to permit a quantitative prediction for the effect of the critical point on any one of the cumulants we have described. In this Section, we suppose that at some $\sqrt{s}$ there is experimental data showing several of the cumulants significantly exceeding their Poisson values. We ask how ratios of cumulants can be used to extract information on ξ and the values of the four parameters. And,

Table 1. Parameter dependence of the contribution of critical fluctuations to various particle multiplicity cumulant ratios. We have subtracted the Poisson contribution from each cumulant before taking the ratio. The table shows the power at which the parameters enter in each case.

ratio	g_p	g_π	$\tilde{\lambda}_3$	$\tilde{\lambda}'_4$	ξ
$\kappa_{2p2\pi} N_\pi / \kappa_{4\pi} \kappa_{2p}$	-	-2	-	-	-2
$\kappa_{2p2\pi} N_p^2 / \kappa_{4p} N_\pi^2$	-2	2	-	-	-
$\kappa_{3p} N_p^{3/2} / \kappa_{2p}^{9/4} N_\pi^{1/4}$	$-3/2$	-	1	-	-
$\kappa_{2p} \kappa_{4p} / \kappa_{3p}^2$	-	-	-2	1	-
$\kappa_{3p} \kappa_{2\pi}^{3/2} / \kappa_{3\pi} \kappa_{2p}^{3/2}$	-	-	-	-	-
$\kappa_{4p} \kappa_{2\pi}^2 / \kappa_{4\pi} \kappa_{2p}^2$	-	-	-	-	-
$\kappa_{4p}^3 \kappa_{3\pi}^4 / \kappa_{4\pi}^3 \kappa_{3p}^4$	-	-	-	-	-
$\kappa_{2p2\pi}^2 / \kappa_{4\pi} \kappa_{4p}$	-	-	-	-	-
$\kappa_{2p1\pi}^3 / \kappa_{3p}^2 \kappa_{3\pi}$	-	-	-	-	-

we construct ratios of cumulants that are independent of ξ and all the parameters, allowing us to make robust predictions for the contribution of critical fluctuations to these ratios. In Table 1 we present the parameter dependence of various cumulant ratios (with Poisson subtracted before taking the ratios), where we defined $2\tilde{\lambda}_3^2 - \tilde{\lambda}_4 \equiv \tilde{\lambda}'_4$.

The correlation length ξ and the four nonuniversal parameters always appear in certain combinations in the multiplicity cumulants and it turns out that we can only constrain four independent combinations. The first four rows of Table 1 show ratios that can be used to constrain one example of four such combinations. The last five entries in the table show parameter-independent ratios. They have no ξ-dependence, no dependence on the four poorly known parameters, and no μ_B-dependence. We find that these five ratios are all precisely 1.

Now let us see how we can use these five ratios in order to locate the critical point. Suppose that as you change the center of mass energy $\sqrt{s}$ of the collisions, experimental evidence for peaks shown in Fig. 1 begin to emerge. How do you check in a parameter-independent fashion whether the behavior seen in experimental data is consistent with the hypothesis that it is due to critical fluctuations? You first subtract the Poisson contributions, and then construct the last five ratios in Table I. If the fluctuations seen in this hypothetical data are in fact due to the proximity of the critical point, all five of these ratios will be equal to 1, with no theoretical uncertainties arising from uncertainty in the values of the parameters. This would be strong evidence indeed for the discovery of the QCD critical point.

[1] M. A. Stephanov, Prog. Theor. Phys. Suppl. **153**, 139 (2004) [Int. J. Mod. Phys. A **20**, 4387 (2005)] [arXiv:hep-ph/0402115].
[2] M. A. Stephanov, K. Rajagopal and E. V. Shuryak, Phys. Rev. Lett. **81**, 4816 (1998) [arXiv:hep-ph/9806219].
[3] M. A. Stephanov, K. Rajagopal and E. V. Shuryak, Phys. Rev. D **60**, 114028 (1999) [arXiv:hep-ph/9903292].
[4] B. Berdnikov and K. Rajagopal, Phys. Rev. D **61**, 105017 (2000) [arXiv:hep-ph/9912274].
[5] D. T. Son and M. A. Stephanov, Phys. Rev. D **70**, 056001 (2004) [arXiv:hep-ph/0401052].
[6] C. Nonaka and M. Asakawa, Phys. Rev. C **71**, 044904 (2005) [arXiv:nucl-th/0410078].
[7] M. M. Tsypin, Phys. Rev. Lett. **73**, 2015 (1994); M. M. Tsypin, arXiv:hep-lat/9601021.
[8] M. A. Stephanov, Phys. Rev. Lett. **102**, 032301 (2009) [arXiv:0809.3450 [hep-ph]].
[9] J. I. Kapusta and C. Gale, "Finite-Temperature Field Theory Principles and Applications," Cambridge, UK: Cambridge University Press (2006)
[10] M. A. Stephanov, Phys. Rev. D **65**, 096008 (2002) [arXiv:hep-ph/0110077].
[11] M. M. Aggarwal *et al.*, arXiv:1004.4959 [nucl-ex].
[12] J. Cleymans, H. Oeschler, K. Redlich and S. Wheaton, Phys. Rev. C **73**, 034905 (2006) [arXiv:hep-ph/0511094].

Journal of Physics: Conference Series **270** (2011) 012052 doi:10.1088/1742-6596/270/1/012052

CGC and initial state effects in Heavy Ion Collisions

Javier L. Albacete

Institut de Physique Théorique - CEA/Saclay, 91191 Gif-sur-Yvette cedex, France.

E-mail: `javier.lopez-albacete@cea.fr`

Abstract. A brief review of the phenomenological studies in the field of heavy ion collisions based on the Color Glass Condensate theory and, in particular, of those relying in the use of the BK equation including running coupling effects is presented.

1. Introduction

The main goal of the experimental programs on ultra-relativistic heavy ion collisions at RHIC and the LHC is the production and characterization of the Quark Gluon Plasma (QGP). A crucial ingredient for the proper understanding of such collisions is the knowledge of the structure of the wave functions of the colliding nuclei, typically encoded in their parton distributions. At high energies, nuclei do not behave as a mere incoherent superposition of their constituents nucleons. Rather, coherence effects are important and modify not only their partonic content, but also the underlying dynamics of particle production in scattering processes. A careful distinction of such *initial state* effects from those originating from the presence of a QGP, or *final state* effects, is of vital importance for a proper characterization of the matter produced in heavy ion collisions, as they may lead sometimes to qualitatively similar phenomena in observables of interest.

The Color Glass Condensate effective theory provides a consistent framework to study QCD scattering at high energies (for a review see e.g. [1, 2]). The main physical ingredient in the CGC is the inclusion of unitarity effects through the proper consideration of non-linear *recombination* effects, both at the level of particle production and also in the quantum evolution of hadronic wave functions. Such effects are expected to be relevant when nuclei (or hadrons, in full generality) are proven at small enough values of Bjorken-x. In that regime gluon occupation numbers are very large and gluon self-interactions become highly probable, thus taming, or *saturating*, further growth of the gluon densities. While the need for unitarity effects comprised in the CGC is, at a theoretical level, clear, the real challenge from a phenomenological point of view is to assess to what extent they are present in available data. Such is a difficult task, since different physical mechanisms concur in data, and also because the limit of asymptotically high energy in which the CGC formalism is developed may not be realized in current experiments. In that sense, the calculation of higher order corrections to the CGC formalism has supposed important leap forward in sharpening the CGC as an useful phenomenological tool.

2. Some recent developments in the CGC: Running coupling BK equation.

The leading order BK-JIMWLK equations resums soft gluon emission in the leading logarithmic (LL) approximation in $\alpha_s \ln 1/x$ to all orders, besides of including non-linear terms required by unitarity. At such degree of accuracy, the theory is incompatible with data. Indeed, the x-dependence of the saturation scale emerging from the LL equations is $Q_s^2(x) \approx Q_0^2(x_0/x)^{\lambda^{LL}}$, with $\lambda^{LL} \sim 4.8\alpha_s$ [3], while phenomenological studies of electron-proton scattering and heavy ion collisions indicate that an evolution speed $\lambda \sim 0.2 \div 0.3$ is preferred by data. Such insufficiency of the theory has been partially fixed by the calculation of running coupling corrections to the BK-JIMWLK equations through the inclusion of quark loops to all orders [4, 5]. Among other interesting dynamical effects, running coupling effects tame the growth of the saturation scale down to values compatible with experimental data [6]. Due to the complexity of the JIMWLK equations, in phenomenological works it is more feasible to solve the BK equation, more tractable numerically, which corresponds to their large-N_c limit. It reads

$$\frac{\partial \mathcal{N}(r,Y)}{\partial Y} = \int d^2\mathbf{r_1}\, K^{\mathrm{run}}(\mathbf{r},\mathbf{r_1},\mathbf{r_2})\, [\mathcal{N}(r_1,Y) + \mathcal{N}(r_2,Y) - \mathcal{N}(r,Y) - \mathcal{N}(r_1,Y)\mathcal{N}(r_2,Y)] \,, \quad (1)$$

where $\mathcal{N}(r,Y)$ is the dipole scattering amplitude on a dense target, $Y = \ln x_0/x$ the rapidity, r the dipole transverse size and $r_2 = r - r_1$. It turns out that running coupling effects can be incorporated to the evolution equation through just a modification of the evolution kernel, referred to as K^{run} in Eq. (1). As discussed in [6], there are several ways of doing so, since the separation of running coupling and conformal terms is not uniquely defined. The evolution kernel proposed in [5] minimizes the role of higher order corrections, making it better suited for phenomenological applications (see [6] for an extended discussion on the subject). Finally, Eq. (1) needs to be suplemented with initial conditions, which can be choosen to be of the McLerran-Venugopalan type [7]. This introduces two free parameters: The value x_0 where the evolution starts and the initial saturation scale Q_0. Finally, the unintegrated gluon distribution entering the different production processes discussed below is related to the dipole amplitude in Eq. (1) through a Fourier transform (see Eq. (5)). Besides of the works in heavy ions described below, the rcBK equation has been shown to successfully account for the x-dependence of structure functions measured in electron-proton scattering [8].

3. Total multiplicities in heavy ion colliisions

The smaller than expected total multiplicities observed at RHIC suggest strong coherence effects in the mechanism for particle production in heavy ion collisions at high energies. The CGC offers a natural explanation of this observation: The total flux of scattering centers (gluons) entering the collisions is significantly reduced due to saturation effects in the wave function of the colliding nuclei. Such idea is realized in the phenomenological KLN model [9], which relies in the use of k_t-factorization. There the total multiplicities in central collisions rise proportional to the saturation scale of the colliding nuclei, $dN/d\eta \propto Q_{sA}^2$. Even though the use of k_t-factorization in A+A collisions is not justified, the good description of the energy, rapidity and centrality of multiplicity data yielded by the model lends support to the underlying physical picture. An important improvement of the KLN model was presented in [10], where the phenomenological model for the unintegrated gluon distributions was replaced by exact solutions of the running ccoupling BK equation, also getting a very good description of RHIC data, as seen in Fig. 1. Besides of reducing the degree of modeling needed to describe data, the use of well controlled QCD tools as the rcBK equation allows for a better controlled extrapolation to LHC energies. Taking the centrality class 0−6% at RHIC as a baseline, the following mid-rapidity multiplicities at the two expected LHC energies in most central Pb+Pb collisions were obtained in [10]:

$$\left.\frac{dN^{ch}_{Pb+Pb}}{d\eta}\right|_{\eta=0} (\sqrt{s_{NN}} = 5.5\,TeV) = 1390 \pm 95\,; \quad \left.\frac{dN^{ch}_{Pb+Pb}}{d\eta}\right|_{\eta=0} (\sqrt{s_{NN}} = 2.76\,TeV) = 1175 \pm 75\,. \quad (2)$$

The error bands in Eq. (2) originate mostly from the uncertainties in the initial conditions for the rcBK equation.

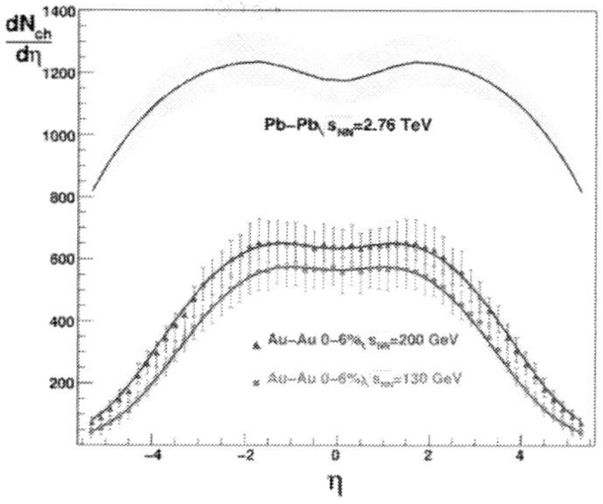

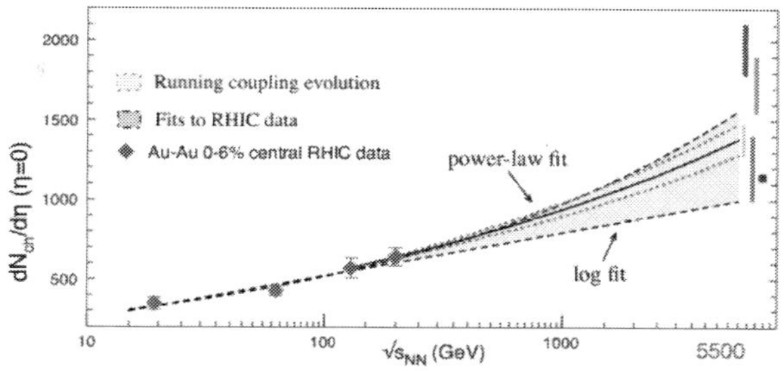

Figure 1. Charged particle multiplicity densities in Au+Au collisions at RHIC (data from [11]) and predictions for Pb+Pb collisions at $\sqrt{s_N N} = 2.76\, TeV$ (left). Mid-rapidity charged multiplicity as a function of the collision energy (left).

4. Single inclusive particle production and nuclear modification factors

Particle spectra in A+A collisions are expected to be strongly modified due to the presence of a QGP. Thus, proton (deuteron) nucleus collisions, where no QGP formation is expected, are better suited for the exploration of initial state effects. For any production process, nuclear effects are typically evaluated in terms of the nuclear modification factors:

$$R_{pA} = \frac{\frac{dN^{pA}}{dyd^2p_t}}{N_{coll}\frac{dN^{pp}}{dyd^2p_t}}\,, \tag{3}$$

where N_{coll} is the number of collisions. If high-energy nuclear reactions were a mere incoherent superposition of nucleon-nucleon collisions, then the observed R_{pA} should be equal to unity. However, RHIC measurements in d+Au collisions (or peripheral Au+Au collisions) [12, 13] feature two distinct regimes: At mid rapidities the nuclear modification factors exhibit an enhancement in particle production at intermediate momenta $p_t \sim 2 \div 4$ GeV. In turn, a suppression at smaller momenta is observed. However, at more forward rapidities such Cronin enhancement disappears, turning into an almost homogeneous suppression for all the measured values of p_t. According to $2 \to 1$ kinematics, the x-values probed in the projectile and target are $x_{1(2)} = (m_t/\sqrt{s})\, e^{\pm y}$. Thus, for RHIC energies, mid rapidity implies moderate $x \sim 0.1 \div 0.01$, probably not small enough to justify the applicability of the CGC. Indeed, mid-rapidity data has been analyzed through different formalisms and techniques, from collinear perturbation theory [14] to independent multiple scatterings [15] or CGC approaches [16], all of them reaching a reasonably good agreement with data. Data collected in the deuteron fragmentation region offer a cleaner opportunity to explore CGC effects, since the nucleus is proben at muchsmaller values of x. There, the CGC formulation of single particle production takes on a relatively simple form [17]:

$$\frac{dN_h}{dy_h\, d^2p_t} = \frac{K}{(2\pi)^2}\sum_q \int_{x_F}^1 \frac{dz}{z^2}\, \Big[x_1 f_{q\,/\,p}(x_1,p_t^2)\, \tilde{N}_F\left(x_2, \frac{p_t}{z}\right)\, D_{h\,/\,q}(z,p_t^2)$$
$$+\, x_1 f_{g\,/\,p}(x_1,p_t^2)\, \tilde{N}_A\left(x_2, \frac{p_t}{z}\right)\, D_{h\,/\,g}(z,p_t^2)\Big]\,, \tag{4}$$

where p_t and y_h are the transverse momentum and rapidity of the produced hadron, and $f_{i/p}$ and $D_{h/i}$ refer to the parton distribution function of the incoming proton and to the final-state

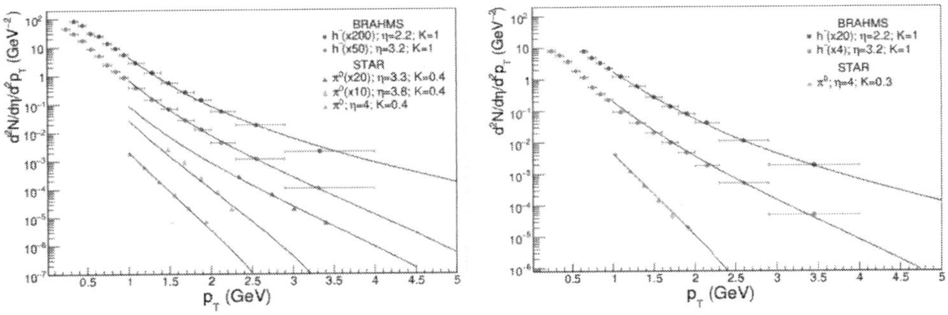

Figure 2. Negative charged hadron and neutral pions at forward rapidities measured by the BRAHMS [12] and STAR [13] Collaborations in p+p (left) and d+Au minimum bias (right) collisions compared to our calculation [18].

hadron fragmentation function respectively. Thus, in the forward region the projectile is in the dilute regime and characterized by its parton distribution functions, while the nucleus is deep in the saturation region and characterized by unintegrated gluon distributions taken from the solutions of the rcBK equation:

$$\tilde{N}_{F(A)}(x,k) = \int d^2\mathbf{r}\, e^{-i\mathbf{k}\cdot\mathbf{r}} \left[1 - \mathcal{N}_{F(A)}(r, Y = \ln(x_0/x))\right], \tag{5}$$

where k refers to transverse momentum. With this set up we reach a very good description of forward neutral pions and negatively charged hadrons yields as measured by the STAR and BRAHMS Collaborations respectively in d+Au minimum bias and in p+p collisions, as shown in Fig 2. All the details of the calculation and fit parameters can be found in [18].

By simply taking the ratios of the corresponding spectra, we get a very good description of the nuclear modification factors at forward rapidities. It should be noted that we use the same normalization as the experimentalist do in their analyses of minimum bias d+Au collisions, i.e. we fix $N_{coll} = 7.2$. Physically, the observed suppression is due to the relative enhancement of non-linear terms in the small-x evolution of the nuclear wave function with respect to that of a proton. However, it has been argued that the observed suppression at forward rapidities is not an effect associated to the small values proven in the nuclear wave function but, rather, to energy-momentum conservation corrections relevant for $x_F \to 1$ Such corrections are not present in the CGC, built upon the eikonal approximation. Thus, the energy degradation of the projectile parton through either elastic scattering or induced gluon brehmstralung would be larger in a nucleus than in proton on account of the stronger color fields of the former, resulting in the relative suppression observed in data. A successful description of forward ratios based on the energy loss calculation was presented in [19].

Thus, at the level of only this observable, one is in an inconclusive situation. Disentangling the underlying dynamics of the suppression requires either data at larger energies or, alternatively, the study of more exclusive observables. The huge leap forward in collision energy reached at the LHC allows for an exploration of small-x effects already at mid-rapidity. There, both the target and projectile x are small, and energy loss effects associated to large-x_F effects are expected to be small. In Fig 4. we present our CGC predictions for the nuclear modification factor for negative charged hadrons in p+Pb collisions at two LHC energies. Our curves correspond to rapidities $y = 2$ and larger. Technical difficulties related to the intrinsic asymmetry of the formalism used for particle production prevent us from calculating the ratios at mid-rapidity. However, the

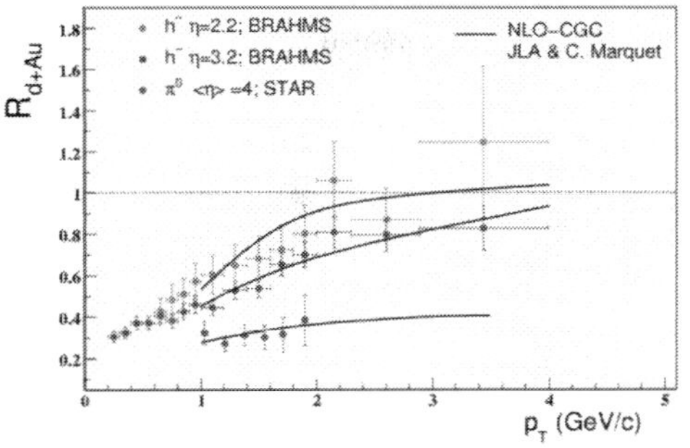

Figure 3. Nuclear modification factors at forward rapidities in minimum bias d+Au collisions in the CGC [18].

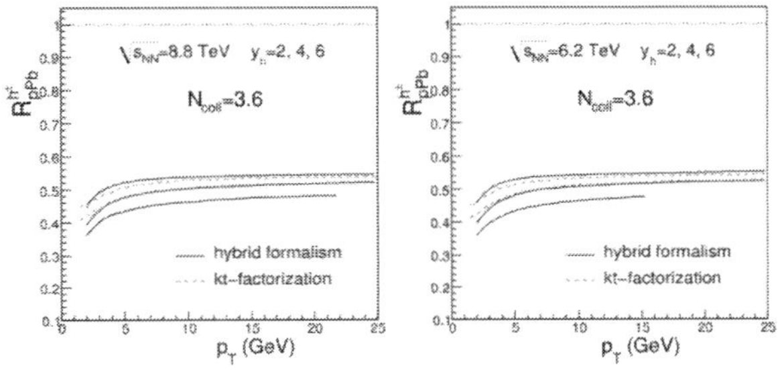

Figure 4. CGC predictions for the nuclear modification factor in p+Pb collisions at two LHC energies and rapidities $y_h = 2, 4$ and 6.

smooth rapidity dependence suggests that a large suppression ~ 0.6 is also expected at mid-rapidity in the LHC. It should be taken into account that the normalization taken to produce the curves in Fig was $N_{coll} = 3.6$.

5. Forward di-hadron correlations

As said before, disentangling CGC from other dynamical effects demands the analysis of more exclusive observables. The opportunity is provided by the recent measurement of forward di-pions correlations in d+Au collisions by the STAR collaboration [20]. The experimental data for the coincidence probability $CP(\Delta\phi)$ feature a clear suppression of the away-side correlations with respect to the ones observed p+p collisions. This suppression is naturally explained in the CGC: The two pions experimentally observed originate from a valence quark-gluon system in perfect back-to-back correlation. The quark-gluon system is put on-shell through the interaction with the nucleus, as a result of which the quark and gluon also acquire a transverse momentum of the order of the saturation scale of the nucleus. When that scale, which marks the onset of non-linear effects, is comparable to the initial transverse momenta of the quark and gluon, their intrinsic angular correlation is washed out. Finally, the outgoing quark and gluon fragment independently into hadrons. A recent calculation CGC [21] provides a very good comparison with available data. There, the required information on the nuclear wave function is built, through the use of the gaussian approximation, from the two-point function Eq. (5) constrained by the analyses of single inclusive spectra, making it a parameter-free calculation. This study provides, arguably, the most compelling evidence for the presence of CGC effects in available

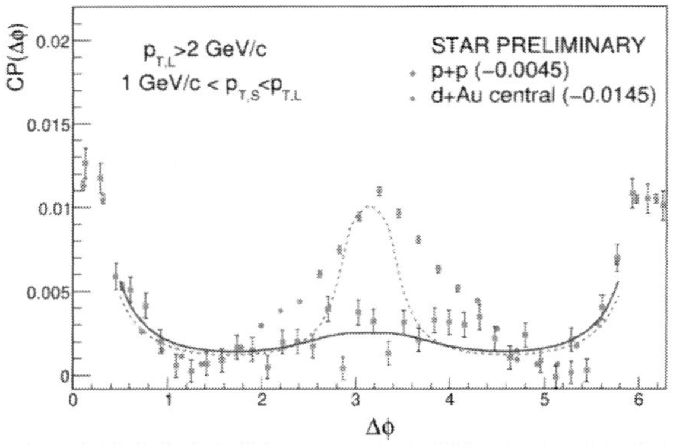

Figure 5. Comparison of the CGC calculation of [21] with data for the coincidence probability at forward rapidities in d+Au (blue) and p+p collisions (red).

data.

Acknowledgments

I would like to thank the organizers of the Hot Quarks 2010 conference for their invitation to such a nice meeting. I am also grateful to Yuri Kovchegov and Cyrille Marquet, with whom I collaborated in obtaining a good part of the results discussed in the manuscript. This work is supported by a Marie Curie Intra-European Fellowship (FP7- PEOPLE-IEF-2008), contract No. 236376.

Bibliography

[1] Gelis F, Iancu E, Jalilian-Marian J and Venugopalan R 2010 (*Preprint* 1002.0333)
[2] Weigert H 2005 *Prog. Part. Nucl. Phys.* **55** 461–565 (*Preprint* hep-ph/0501087)
[3] Albacete J L, Armesto N, Milhano J G, Salgado C A and Wiedemann U A 2005 *Phys. Rev.* **D71** 014003 (*Preprint* hep-ph/0408216)
[4] Kovchegov Y and Weigert H 2007 *Nucl. Phys.* **A 784** 188–226 (*Preprint* hep-ph/0609090)
[5] Balitsky I I 2007 *Phys. Rev. D* **75** 014001 (*Preprint* hep-ph/0609105)
[6] Albacete J L and Kovchegov Y V 2007 *Phys. Rev.* **D75** 125021 (*Preprint* arXiv:0704.0612 [hep-ph])
[7] McLerran L D and Venugopalan R 1998 *Phys. Lett.* **B424** 15–24 (*Preprint* nucl-th/9705055)
[8] Albacete J L, Armesto N, Milhano J G and Salgado C A 2009 *Phys. Rev.* **D80** 034031 (*Preprint* 0902.1112)
[9] Kharzeev D, Levin E and Nardi M 2005 *Nucl. Phys.* **A747** 609–629 (*Preprint* hep-ph/0408050)
[10] Albacete J L 2007 *Phys. Rev. Lett.* **99** 262301 (*Preprint* 0707.2545)
[11] Back B B *et al.* 2005 *Nucl. Phys.* **A757** 28–101 (*Preprint* nucl-ex/0410022)
[12] Arsene I *et al.* (BRAHMS) 2004 *Phys. Rev. Lett.* **93** 242303 (*Preprint* nucl-ex/0403005)
[13] Adams J *et al.* (STAR) 2006 *Phys. Rev. Lett.* **97** 152302 (*Preprint* nucl-ex/0602011)
[14] Eskola K J, Paukkunen H and Salgado C A 2009 *JHEP* **04** 065 (*Preprint* 0902.4154)
[15] Accardi A and Gyulassy M 2004 *Phys. Lett.* **B586** 244–253 (*Preprint* nucl-th/0308029)
[16] Kharzeev D, Kovchegov Y V and Tuchin K 2004 *Phys. Lett.* **B599** 23–31 (*Preprint* hep-ph/0405045)
[17] Dumitru A, Hayashigaki A and Jalilian-Marian J 2006 *Nucl. Phys.* **A765** 464–482 (*Preprint* hep-ph/0506308)
[18] Albacete J L and Marquet C 2010 *Phys. Lett.* **B687** 174–179 (*Preprint* 1001.1378)
[19] Kopeliovich B Z, Nemchik J, Potashnikova I K, Johnson M B and Schmidt I 2005 *Phys. Rev.* **C72** 054606 (*Preprint* hep-ph/0501260)
[20] Braidot E and collaboration f t S 2010 (*Preprint* 1005.2378)
[21] Albacete J L and Marquet C 2010 (*Preprint* 1005.4065)

Journal of Physics: Conference Series **270** (2011) 012053 doi:10.1088/1742-6596/270/1/012053

Testing collinear factorization and nuclear parton distributions with pA collisions at the LHC

Paloma Quiroga-Arias[1], **José Guilherme Milhano**[2,3] **and Urs Achim Wiedemann**[3]

[1] Departamento de Física de Partículas and IGFAE, Universidade de Santiago de Compostela 15706 Santiago de Compostela, Spain

[2] CENTRA, Departamento de Física, Instituto Superior Técnico (IST), Av. Rovisco Pais 1, P-1049-001 Lisboa, Portugal

[3] Physics Department, Theory Unit, CERN, CH-1211 Genève 23, Switzerland

E-mail: `pquiroga@fpaxp1.usc.es`

Abstract. Global perturbative QCD analyses, based on large data sets from electron-proton and hadron collider experiments, provide tight constraints on the parton distribution function (PDF) in the proton. The extension of these analyses to nuclear parton distributions (nPDF) has attracted much interest in recent years. nPDFs are needed as benchmarks for the characterization of hot QCD matter in nucleus-nucleus collisions, and attract further interest since they may show novel signatures of non- linear density-dependent QCD evolution. However, it is not known from first principles whether the factorization of long-range phenomena into process-independent parton distribution, which underlies global PDF extractions for the proton, extends to nuclear effects. As a consequence, assessing the reliability of nPDFs for benchmark calculations goes beyond testing the numerical accuracy of their extraction and requires phenomenological tests of the factorization assumption. Here we argue that a proton-nucleus collision program at the LHC would provide a set of measurements allowing for unprecedented tests of the factorization assumption underlying global nPDF fits.

1. Introduction

Parton distribution functions (PDFs), which define the flux of quarks and gluons ($i = q, g$) in hadrons $f_{i/h}(x, Q^2)$, play a central role in the study of high energy collisions involving hadronic projectiles. For protons, sets of collinearly factorized universal PDFs are obtained in global pQCD analyses based on data from DIS and DY production, as well as W/Z and jet production at hadron colliders. These data provide tight constraints on PDFs over logarithmically wide ranges in Q^2 and x, and have allowed precision testing of linear pQCD evolution. However, our understanding of parton distribution functions in nuclei of nucleon number A, $f_{i/A}(x, Q^2)$, is much less mature. Knowledge of nuclear parton distribution functions (nPDFs) is important in heavy ion collisions at RHIC and at the LHC for a quantitative control of hard processes, which are employed as probes of dense QCD matter. Characterizing nPDFs is also of great interest in its own right, since the nuclear environment is expected to enhance parton density-dependent effects, which can reveal qualitatively novel, non-linear features of QCD evolution.

Paralleling the determination of proton PDFs, several global QCD analyses of nPDFs have been made within the last decade [1–5] based, up until recently, solely on fixed-target nuclear

DIS and DY data which, compared to the data constraining proton PDFs, are of lower precision and cover a much more limited range of Q^2 and x. Such limitation manifests in a particularly poor gluon distribution function, since it can only be obtained from logarithmic Q^2-evolution, for which a wide Q^2-range is mandatory (see [6] for future prospects in DIS on nuclei).

Moreover, in contrast to the theoretical basis for global analyses of proton PDFs, the separability of nuclear effects into process-independent nPDFs and process-dependent but A-independent hard processes is not established within the framework of collinear factorized QCD.

In view of the importance of nPDFs for characterizing benchmark processes in heavy ion collisions, it is thus desirable to look for both stringent phenomenological tests of the working assumption of global nPDF fits and experimental observables that can be used to constrain nPDFs. To this end, we argue that a program of hadron-nucleus collisions at the LHC would provide for such tests with unprecedented quality [7].

2. Collinearly factorized single inclusive approach to particle production

To improve on the insufficiency of nuclear DIS and DY data, recent global nPDF analysis [1, 2] have included for the first time data from inclusive high-p_T hadron production in hadron-nucleus scattering measured at RHIC [8–10]. This is the observable we focus on in order to test collinear factorization and to determine nPDFs.

In the factorized QCD ansatz to hadron-nucleus collisions by construction the entire nuclear dependence of the cross section for hadro-production resides in the nPDF (see eq.(1) in [7]). It is customary to characterize nuclear effects in the parton distribution functions by the ratios

$$R_i^A(x, Q^2) \equiv f_{i/A}(x, Q^2)/f_{i/p}(x, Q^2)\,, \tag{1}$$

which show characteristic deviations from unity in global nPDF analysis for all scales of Q^2 tested so far. These effects are typically referred to as nuclear shadowing ($x < 0.01$), anti-shadowing ($0.01 < x < 0.2$), EMC effect ($0.2 < x < 0.7$) and Fermi motion ($x > 0.7$).

Nuclear effects on single inclusive hadron production are typically characterized by the nuclear modification factor R^h_{pA}

$$R^h_{pA}(p_T, y) = \frac{d\sigma^{pA\to h+X}}{dp_T^2\, dy} \Bigg/ N^{pA}_{\text{coll}} \frac{d\sigma^{pp\to h+X}}{dp_T^2\, dy}\,. \tag{2}$$

We calculate the single inclusive π^0 production at both RHIC and the LHC within the collinear factorization approach. All our calculations use LO PDFs from CTEQ6L [11] with nuclear modifications EPS09LO [1] and the KKP fragmentation functions [12]. When the collinear factorization ansatz is adopted, the p_T-dependence of R^h_{pA} traces the x-dependence of nPDFs. Although the precise kinematics is complicated by the convolution of the distributions, the qualitative dependence of the Bjorken-x of partons inside the nucleus with p_T of the final hadron and rapidity reads $x \sim p_T e^{-y}/\sqrt{s}$.

2.1. From RHIC to the LHC: testing collinear factorization

Fig. 1 shows the nuclear modification factor $R^{\pi^0}_{dAu}$ for the production of neutral pions in $\sqrt{s_{\text{NN}}} = 200$ GeV deuteron-gold collisions at RHIC, calculated within the factorized ansatz at leading order (LO). Results shown in Fig. 1 are also consistent with the NLO-calculation of $R^{\pi^0}_{dAu}$ in [1]. The RHIC data [8] in Fig. 1 have been used in constraining the nPDF analysis EPS09 [1] but they were not employed in a closely related nPDF fit [3], which provides an equally satisfactory description of these RHIC data. Therefore, the agreement of data and calculation in Fig. 1 is in support of collinear factorization: the enhancement of $R^{\pi^0}_{dAu}$ in the region around $p_T \simeq 4$ GeV at mid-rapidity tests momentum fractions in the anti-shadowing region.

Journal of Physics: Conference Series **270** (2011) 012053 doi:10.1088/1742-6596/270/1/012053

However, qualitatively different explanations of the $R^{\pi^0}_{d\,Au}$ measured at RHIC are conceivable [13]. While the above calculation accounted for $R^{\pi^0}_{d\,Au}(\eta = 0)$ in terms of a nuclear modification of the *longitudinal* parton momentum distribution only, such nuclear effects can be accommodated within a variety of assumptions, namely initial state parton multiple scattering (k_T broadening) [14] and non-linear low-x QCD evolution [15].

Yet, RHIC data being inconclusive about the dynamical explanation of the nuclear modification of the spectrum, we argue that repeating the same measurement at the LHC at a much higher energy and in the much wider available p_T range, will help to disentangle which of the theories is correct, or in which kinematical range each one is valid.

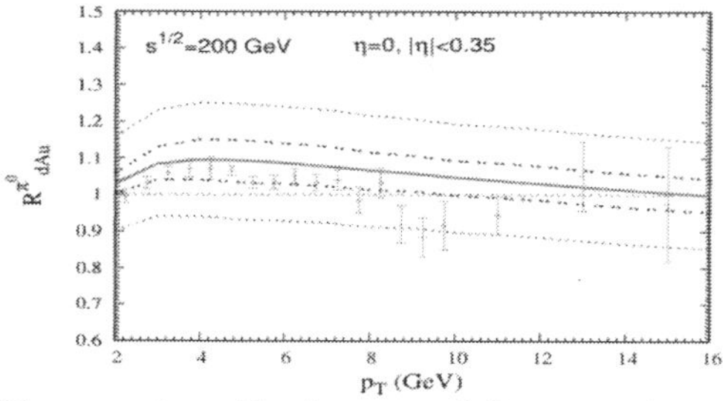

Figure 1. Nuclear modification factor (2) at RHIC mid-rapidity. Data points are PHENIX [8]. Solid line represents the EPS09 LO calculation. The dashed lines are the EPS09 uncertainty [1] (inner) plus data normalization uncertainty (outer).

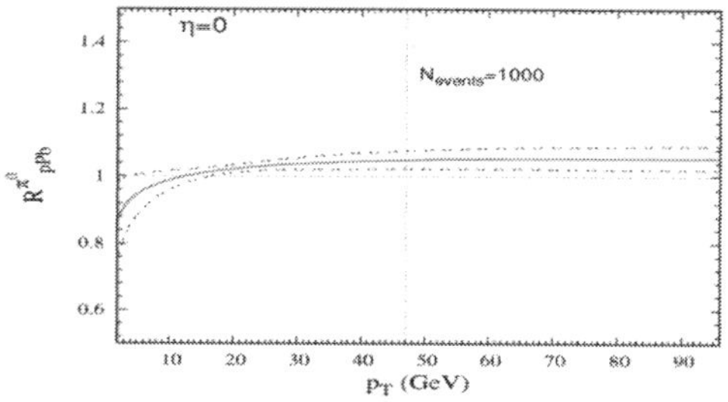

Figure 2. EPS09 LO prediction for the nuclear modification factor of neutral pion single inclusive production in proton-lead collisions at LHC mid-rapidity. All LHC rapidities are in the center-of-mass (shift to lab $\Delta y \sim 0.47$).

As one can see in Fig. 2, the shape of the nuclear modification factor that would be measured at LHC mid-rapidity, if the entire nuclear effect in pPb collisions can be factorized into nPDFs, is qualitatively different from that observed at RHIC: at more than 40 times higher center of mass energy, final state hadrons at the same transverse momentum test $\mathcal{O}(40)$ times smaller momentum fractions x_i. Even tough a shift of the maximum of $R^{\pi^0}_{pPb}(\eta = 0)$ to such high values of p_T is a natural consequence in collinear factorization reflecting nuclear modifications of PDFs, it cannot be accounted for in terms of k_T-broadening, since such models predict a mild dependence of the spectrum with the center-of-mass energy [14].

When the single-inclusive spectrum is calculated for different values of rapidity the results shown in figure Fig. 3 are obtained, to learn that the rapidity dependence of R^h_{pPb} allows one to scan the main qualitatively different ranges of standard nPDFs in an unprecedented way.

The behavior of the nuclear modification factor when saturation is taken into account [15] is rather different: a strong suppression factor is found which is incompatible with any of the existing nuclear parton distribution sets. Thus, a measurement at the LHC which shows the rapidity dependent behavior in Fig. 2 and 3 would be in strong support of collinear factorization as opposed to initial state multiple scattering and CGC models.

2.2. Constraining nPDFs with the LHC

Despite the perspectives for qualitative tests of collinear factorization, current global analyses of nuclear parton distribution functions come with significant uncertainties as we can see in Fig. 4. We have compared in Fig. 5 the nuclear modification factor for two nPDF sets, which show marked differences: in contrast to EPS09 the gluon distribution of HKN07 [5] does not show an anti-shadowing peak but turns for $x > 0.2$ from suppression to strong enhancement at initial scale $Q^2 = 1\,\mathrm{GeV}^2$. Fig. 5 thus illustrates that within the validity of a collinearly factorized approach, LHC data can resolve the qualitative differences between existing nPDF analyses

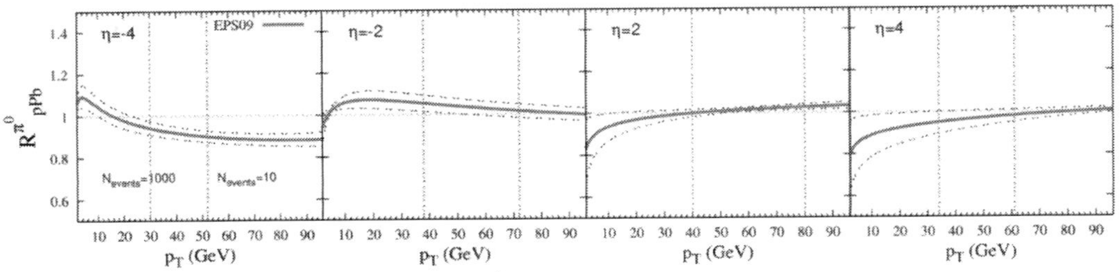

Figure 3. Rapidity dependence of $R^{\pi^0}_{pPb}$ (2) for $\sqrt{s_{\rm NN}} = 8.8$ TeV pPb (LHC). The different plots scan the dependence from $y = -4$ (close to Pb projectile rapidity) up to $y = 4$ (close to proton projectile rapidity). Labels indicate whether the nuclear modification originates mainly from the shadowing (S), anti-shadowing (AS) or EMC regime. Vertical lines illustrate the rapidity-dependent p_T range, which can be accessed experimentally with more than $N_{\rm events} = 1000$ (= 10) per GeV-bin within one month of running at nominal luminosity.

and can improve significantly and within a nominally perturbative regime on our knowledge of nuclear gluon distribution functions.

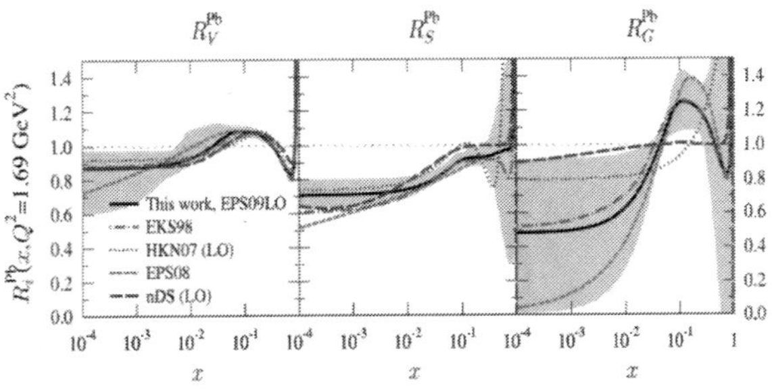

Figure 4. The nuclear modification factor (1) in lead R_i^{Pb} for three different flavours as given by the available nPDF parametrization at LO [1].

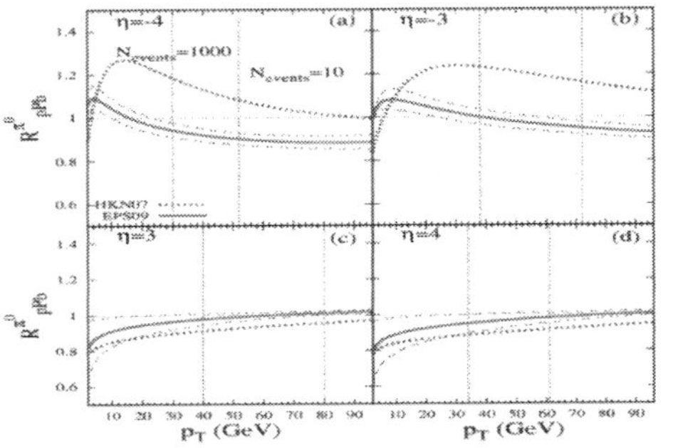

Figure 5. $R^{\pi^0}_{pPb}$ (2) for $\sqrt{s_{\rm NN}} = 8.8$ TeV pPb (LHC), from two different sets of nPDFs.

Acknowledgments

We acknowledge support from MICINN (Spain) under project FPA2008-01177 and FPU grant; Xunta de Galicia (Consellería de Educación) and through grant PGIDIT07PXIB206126PR, the Spanish Consolider- Ingenio 2010 Programme CPAN (CSD2007-00042) and Marie Curie MEST-CT-2005-020238-EUROTHEPHY (PQA); and Fundação para a Ciência e a Tecnologia (Portugal) under project CERN/FP/109356/2009 (JGM).

References

[1] Eskola K J, Paukkunen H and Salgado C A 2009 *JHEP* **04** 065 (*Preprint* 0902.4154)
[2] Eskola K J, Paukkunen H and Salgado C A 2008 *JHEP* **07** 102 (*Preprint* 0802.0139)
[3] Eskola K J, Kolhinen V J and Salgado C A 1999 *Eur. Phys. J.* **C9** 61–68 (*Preprint* hep-ph/9807297)
[4] de Florian D and Sassot R 2004 *Phys. Rev.* **D69** 074028 (*Preprint* hep-ph/0311227)
[5] Hirai M, Kumano S and Nagai T H 2007 *Phys. Rev.* **C76** 065207 (*Preprint* 0709.3038)
[6] Lamont M 2010 *these proceedings*
[7] Quiroga-Arias P, Milhano J G and Wiedemann U A 2010 (*Preprint* 1002.2537)
[8] Adler S S *et al.* (PHENIX) 2007 *Phys. Rev. Lett.* **98** 172302 (*Preprint* nucl-ex/0610036)
[9] Adams J *et al.* (STAR) 2006 *Phys. Lett.* **B637** 161–169 (*Preprint* nucl-ex/0601033)
[10] Arsene I *et al.* (BRAHMS) 2004 *Phys. Rev. Lett.* **93** 242303 (*Preprint* nucl-ex/0403005)
[11] Pumplin J, Belyaev A, Huston J, Stump D and Tung W K 2006 *JHEP* **02** 032 (*Preprint* hep-ph/0512167)
[12] Kniehl B A, Kramer G and Potter B 2000 *Nucl. Phys.* **B582** 514–536 (*Preprint* hep-ph/0010289)
[13] Albacete J L 2010 *these proceedings*
[14] Zhang Y, Fai G I, Papp G, Barnafoldi G G and Levai P 2002 *Phys. Rev.* **C65** 034903 (*Preprint* hep-ph/0109233)
[15] Albacete J L and Marquet C 2010 *Phys. Lett.* **B687** 174–179 (*Preprint* 1001.1378)

Journal of Physics: Conference Series **270** (2011) 012054 doi:10.1088/1742-6596/270/1/012054

Inclusive and diffractive DIS at low x from HERA to the EIC

Konrad Tywoniuk

Departamento de Física de Partículas,
Universidade de Santiago de Compostela,
15706 Santiago de Compostela, Spain

E-mail: konrad.tywoniuk@gmail.com

Abstract. We describe a model for low x DIS off nucleons and nuclei which provides initial conditions for QCD evolution that satisfy unitarity at low-x. We discuss it's extension to the perturbative regime and it's predictions for several DIS observables. We emphasize the well-known connection between rise of the diffractive cross section at high energies and nuclear shadowing.

1. A unitarized model for low x DIS

The Regge limit of QCD ($s \to \infty$ at constant Q^2) is of great interest both from the perspective of high-energy collider experiments and inasmuch as it poses a challenge for the theoretical understanding of the hadronic wave-function. The rapid growth of cross sections observed at high Q^2 and fairly low x is expected to slow down due to unitarity as $x \to 0$. Compared to the nucleon case, in $\gamma^* A$ collisions these effects are enhanced by the nuclear thickness factor $\sim A^{1/3}$ causing a depletion of the nuclear structure function observed at $x < 0.1$, compared to the incoherent superposition of A $\gamma^* p$ cross sections, called nuclear shadowing

So far, all data from DIS experiments off both nucleons and nuclei can be well described by perturbative QCD (pQCD) with universal parton distribution functions (PDFs) incorporating scaling violations. But in spite of its success, this scheme has no predictive power on the energy dependence of PDFs at a given initial Q_0^2 rendering them uncertain, especially for the gluon sector, in the kinematical regime away from present day experiments, particularly at low x. Models that aim at predicting the low-x behaviour of the structure functions have to include unitarity conserving mechanisms that tame the growth of cross sections at high energies. Recently, attempts to derive these corrections within pQCD have made significant progress.

In a frame where the target nucleus is at rest, these effects arise due to multiple scattering of the projectile. The large probability of rescattering and, thus, a large probability of diffraction dissociation, is known to arise from large, and thus principally non-perturbative, partonic fluctuations. In this sense, conservation of unitarity and growth of the probability of diffraction is intimately linked.

A suitable framework, although not rigorously established within QCD, to treat these configurations is provided by the reggeon calculus [1], where rescattering of the projectile wave function is accounted for by including multi-reggeon exchanges. In the absence of a unified QCD approach to the entirety of $\gamma^* N$ and $\gamma^* A$ processes, this framework can serve as a useful guidance for investigating the connection between non-perturbative and perturbative aspects of DIS valid for extrapolations to extremely small momentum fractions.

2. DIS off protons

In a particular realization within this framework one distinguishes explicitly between a large (L) and a small (S) component in the γ^* wave-function [2, 3]. The former interacts strongly even at high-Q^2 but is quite rare, while the latter, cast in the form of a dipole model in [3], interacts according to $r \sim 1/Q$. Thus both components have a leading $1/Q^2$ dependence of the total cross section while the L-component gives the leading contribution to diffraction ($1/Q^2$ vs. $1/Q^4$). Large-mass diffraction is included through triple-reggeon interactions. The model gave a simultaneous description of inclusive F_2 and diffraction in the region of $0 < Q^2 \leq 5 - 10$ GeV^2, and was used to predict the structure functions at very low x. With the advent of high-energy colliders the need for low-x structure functions for nucleons and nuclei at high-Q^2 have arisen, which motivates an extension of the model [3] to the perturbative regime by the inclusion of QCD scaling violations.

In [4] we describe a prescription for extracting the initial conditions at leading order for the DGLAP equations from the non-perturbative model both for inclusive F_2 and diffraction. In the former case, this procedure does not involve new parameters. The situation for the inclusive diffractive cross section is more complex, because it involves both more complicated reggeon exchanges and additional variables in the problem. For the proper description of data in the whole β and $x_{I\!P}$ region we identify explicitly pomeron and reggeon contributions to diffraction. One can then invoke a supplementary factorization of variables, the so-called Regge factorization, which allows for a comprehensible QCD analysis. In the reggeon case, important for small-mass diffraction, the diagrams not taken into account in the original formulation are included by a pion PDF.

Thus, equipped with properly unitarized initial conditions for the DGLAP evolution equations we obtain leading-order structure functions and PDFs for the proton down to $x \sim 10^{-8}$ at high-

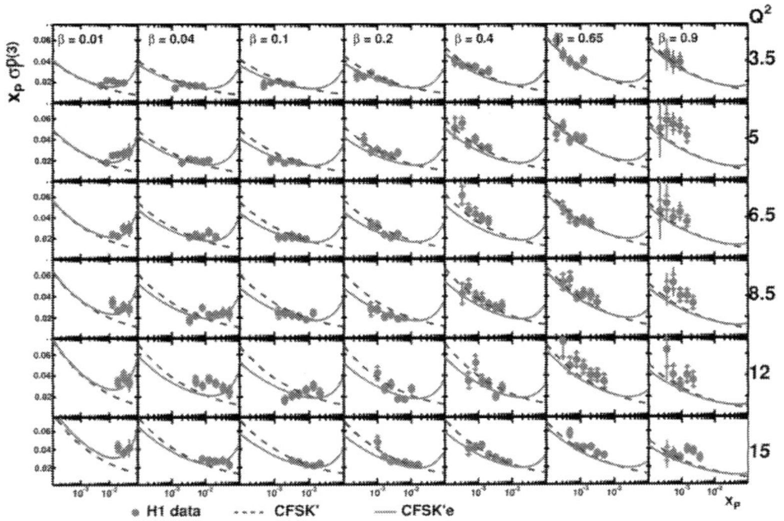

Figure 1. The proton diffractive structure function obtained from the CFSK model with QCD evolution at LO compared to a set of data from the H1 experiment (see [4] for details).

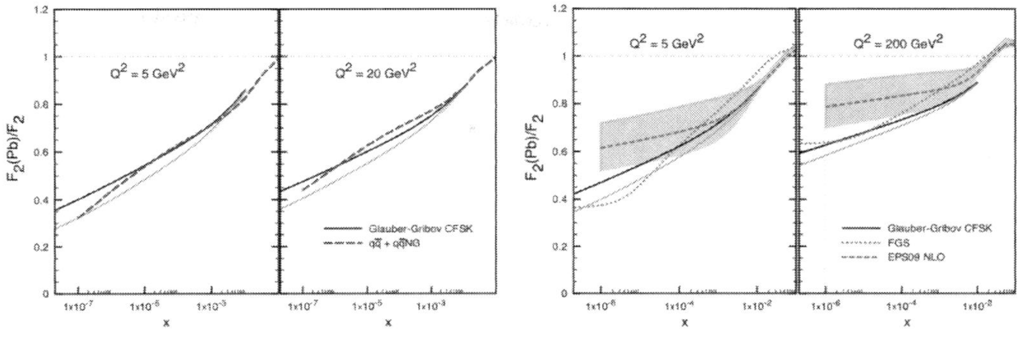

Figure 2. Results for $F_2(\mathrm{Pb})/F_2$ compared to the dipole calculation in [8] (left), the EPS09 NLO parameterization [9] and the FGS model [10] (right). Solid upper (black) curves are calculated in the Schwimmer model, lower (orange) ones are calculated in the eikonal model.

Q^2. The resulting F_2 and $x_{I\!P}F_{2D}^{(3)}$ are in good agreement with the most recent experimental data, with $\chi^2 \gtrsim 1.5$ in both cases. In Fig. 1 we show a comparison with some of the diffractive data. We have also computed the longitudinal structure function within the dipole model using the perturbative gluon PDF thus obtained. Comparisons made with the recently computed solution of the running-coupling BK equation [5] show a large discrepancy of the predicted F_L behaviour at low Q^2 [4].

3. DIS off nuclei

In the context of DIS off nuclei, there emerges a critical length scale related to a change of the underlying space-time picture of the collision. The coherence length (or life-time) of a given fluctuation of the incoming projectile is given by

$$l_C = \frac{1}{Q}\frac{E_{LAB}}{Q} \simeq \frac{1}{2m_N x}\,, \tag{1}$$

in the limit $2m_N E_{LAB} = W^2 \gg Q^2$, where x is the Bjorken variable. At low energies, where l_C is of the order of the internucleon distance, the projectile undergoes incoherent multiple scattering off the target. Remarkably, all higher-order rescatterings cancel and the total γ^*A cross section is simply given as a superposition of γ^*p collisions. The critical value is reached when the coherence length becomes of the order of the nuclear radius. For $l_C > R_A$, i.e. at $x < 1/2m_N R_A$, the projectile scatters coherently off all constituents of the nucleus at some given impact parameter. Despite the non-local nature of the interactions, the total cross section can be written in the form of a multiple scattering series, now including corrections from higher-order rescattering diagrams which lead to an overall depletion of the total cross section, called Gribov inelastic shadowing. This formalism relates the inclusive and diffractive γ^*p cross sections to the corresponding γ^*A ones by means of the AGK cutting rules [6], thus providing a neat connection between diffraction and shadowing. The multiple scattering has been truncated within fan and eikonal diagram re-summations. Thus our calculations for nuclei, based on the model presented in the previous subsection [4] can be extrapolated down to very low x and thanks to the inclusion of scaling violations coming from QCD evolution it can also be used at high Q^2 [7]. Extension of the model to higher x (and low-mass diffraction) is under way.

We show the results of our calculations for F_2^{Pb}/F_2 in Fig. 2, where also a comparison to a recent dipole model calculation [8] (right) and a NLO pQCD fit [9] together with a

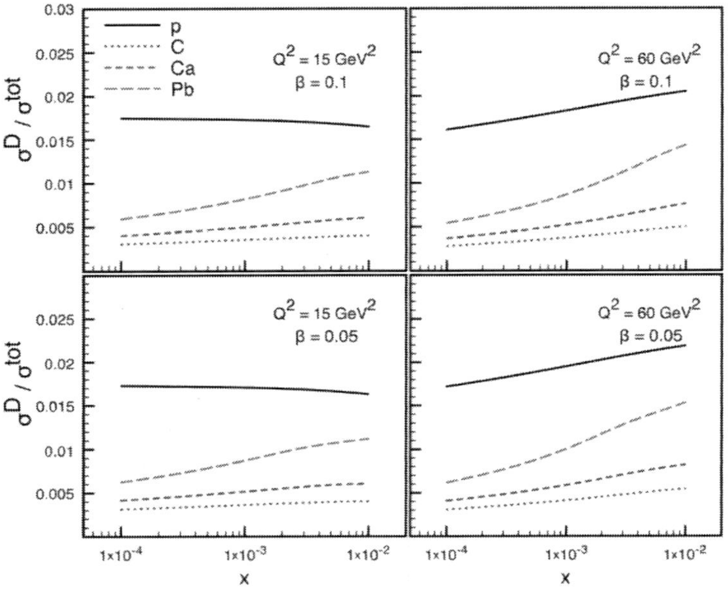

Figure 3. Nuclear diffractive-to-inclusive ratio in the Schwimmer model with CFSK PDFs.

similar calculation to ours [10] (left) is presented. At low x, a large deviation from the pQCD parameterization, which can be traced back to the choice of initial condition, is apparent. Needless to say, the lox-x region in nuclei is of extreme importance for calculating e.g. multiplicity distributions in heavy-ion collisions

Additionally, in [7] we calculated high-mass diffraction off nuclei, noting that Regge factorization is broken in this case due to the large rescattering. We compare the ratio of total and diffractive cross section for protons and nuclei (calculated for fan diagram re-summation) in Fig. 3.

Acknowledgments

The work presented here was done in collaboration with N. Armesto, C. Salgado and A. B. Kaidalov, who I would like to thank for immense support and illuminating discussions. I would like to thank the organizers of the conference for financial support. The work is dedicated to A. B. Kaidalov, who sadly passed away in July.

References

[1] Gribov V N 1968 *Sov. Phys. JETP* **26** 414
[2] Capella A, Ferreiro E G, Salgado C A and Kaidalov A B 2001 *Nucl. Phys.* B **593** 336
[3] Capella A, Ferreiro E G, Salgado C A and Kaidalov A B 2001 *Phys. Rev.* D **63** 054010
[4] Armesto N, Kaidalov A B, Salgado C A and Tywoniuk K 2010 *Phys. Rev.* D **81** 074002
[5] Albacete J L, Armesto N, Milhano J G and Salgado C A 2009 *Phys. Rev.* D **80** 034031
[6] Abramovsky V A, Gribov V N and Kancheli O V 1973 *Yad. Fiz.* **18** 595
[7] Armesto N, Kaidalov A B, Salgado C A and Tywoniuk K 2010 *Eur. Phys. J.* C **68** 447
[8] Kopeliovich B Z, Nemchik J, Potashnikova I K and Schmidt I 2008 *J. Phys.* G **35** 115010
[9] Eskola K J, Paukkunen H and Salgado C A 2009 *J. High Energy Phys.* JHEP0904(2009)065
[10] Frankfurt L, Guzey V and Strikman M 2005 *Phys. Rev.* D **71** 054001

Hot Quarks 2010 IOP Publishing
Journal of Physics: Conference Series **270** (2011) 012055 doi:10.1088/1742-6596/270/1/012055

Gluon correlations in the glasma

T. Lappi

Physics Department, P.O. Box 35, 40014, University of Jyväskylä, Finland and
Helsinki Institute of Physics, P.O. Box 64, 00014 University of Helsinki, Finland

E-mail: `tuomas.lappi@jyu.fi`

Abstract. The physics of the initial conditions of heavy ion collisions is dominated by the nonlinear gluonic interactions of QCD. These lead to the concepts of parton saturation and the Color Glass Condensate (CGC). We discuss recent progress in calculating multi-gluon correlations in this framework, prompted by the observation that these correlations are in fact easier to compute in a dense system (nucleus-nucleus) than a dilute one (proton-proton).

1. Introduction

Bulk particle production in relativistic collisions around midrapidity originates from small x degrees of freedom, predominantly gluons, in the wavefunctions of the colliding hadrons or nuclei. At large energies these gluons form a dense system characterized by a *saturation scale* Q_s. The degrees of freedom with $p_T \lesssim Q_s$ are fully nonlinear Yang-Mills fields with large field strength $A_\mu \sim 1/g$ and occupation numbers $\sim 1/\alpha_s$; they can therefore be understood as classical fields radiated from the large x partons. Because of their large longitudinal momentum, the large x degrees of freedom are effectively "frozen" during the interaction. They can be described as random color charges drawn from a classical probability distribution $W_y[\rho]$ that depends on the rapidity cutoff $y = \ln 1/x$ separating the large and small x degrees of freedom. The dependence of $W_y[\rho]$ on y is described by a Wilsonian renormalization group equation known by the acronym JIMWLK. Note that while this description is inherently nonperturbative, it is still based on a weak coupling argument, because the classical approximation requires $\alpha_s(Q_s)$ to be small and therefore $Q_s \gg \Lambda_{QCD}$. The Color Glass Condensate (CGC) is a systematic effective theory (effective because the large x part of the wavefunction is integrated out) description of the classical small x degrees of freedom.

The term glasma [1] refers to the coherent, classical field configuration resulting from the collision of two such objects CGC. The glasma fields are initially longitudinal, whence the "glasma flux tube" [2, 3] picture. More importantly for computing multigluon correlations, they are boost invariant (to leading order in the QCD coupling) and depend on the transverse coordinate with a characteristic correlation length $1/Q_s$. There are several signals in the RHIC data [4] that point to strong correlations originating from the initial stage of the collision. The glasma fields provide a natural framework for understanding these effects, although much work is still left to do in understanding the interplay with purely geometrical effects from the fluctuating positions of the nucleons in the colliding nuclei [5].

We shall first describe some general observations on computing multigluon correlations in the glasma, arguing in Sec. 2 that they are in some sense simpler to compute in a collision of two dense, saturated nuclear wavefunctions than in the dilute limit (see Ref. [6] for a more formal discussion). We shall then, in Sec. 3 discuss one application of these ideas to computing the multiplicity distribution of gluons in the collision before moving to the leading $\ln 1/x$ rapidity dependence of the correlation in Sec. 4.

2. Multigluon correlations in the glasma

The gluon fields in the glasma are nonperturbatively strong, $A_\mu \sim 1/g$. This means that the gluon multiplicity is $N \sim 1/\alpha_s$. For a fixed configuration of the classical color sources it is well known that the multiplicity distribution of produced gluons is Poissonian, i.e. $\langle N^2 \rangle - \langle N \rangle^2 = \langle N \rangle$. In this case the correlations and fluctuations in the gluon multiplicity are all quantum effects that appear only starting from the one-loop level, i.e. suppressed by a power of the coupling constant α_s. The computation in the CGC framework does not end here, however. To calculate the moments of the gluon multiplicity distribution one must first calculate the gluon spectra for fixed configuration of the color charges ρ and then average over the probability distribution $W_y[\rho(\boldsymbol{x}_\perp)]$. For the nth moment of the multiplicity distribution, i.e. an n-gluon correlation, the leading order result is

$$\left\langle \frac{\mathrm{d}N}{\mathrm{d}^3\boldsymbol{p}_1} \cdots \frac{\mathrm{d}N}{\mathrm{d}^3\boldsymbol{p}_n} \right\rangle = \left[\int_{[\rho]} W[\rho_1(y)] W[\rho_2(y)] \left. \frac{\mathrm{d}N}{\mathrm{d}^3\boldsymbol{p}_1} \right|_{\mathrm{LO}} \cdots \left. \frac{\mathrm{d}N}{\mathrm{d}^3\boldsymbol{p}_n} \right|_{\mathrm{LO}} \right] \tag{1}$$

This averaging, even after the subsequent subtraction of the appropriate disconnected contributions, introduces a correlation already at the leading order in α_s, i.e. enhanced by an additional $1/\alpha_s$ compared to the quantum correlations. A natural example is the negative binomial distribution that we shall discuss below, with variance $\langle N^2 \rangle - \langle N \rangle^2 = \langle N \rangle^2/k + \langle N \rangle$. One must emphasize here that although these contributions arise as formally classical correlations in the effective theory that is the CGC, they are physically also quantum effects, where the weak coupling is compensated by a large logarithm of the energy that has been resummed into the probability distribution $W_y[\rho(\boldsymbol{x}_\perp)]$. In this sense the leading correlations are present already in the wavefunctions of the colliding objects.

3. Multiplicity distribution

We can then apply this formalism to the calculation of the probability distribution of the number of gluons in the glasma [7]. We shall assume the "AA" power counting of sources that are parametrically strong in g, but nevertheless work to the lowest nontrivial order in the color sources. Formally this would correspond to a power counting $\rho \sim g^{\varepsilon - 1}$ with a small $\epsilon > 0$. In this limit, as we have discussed, the dominant contributions to multiparticle correlations come from diagrams that are disconnected for fixed sources and become connected only after averaging over the color charge configurations. The corresponding two gluon correlation function was computed in Ref. [2] and generalized to a three gluons in Ref. [8]. We shall here sketch the derivation [7] of the general n-gluon correlation in this simplified limit.

Working with the MV model Gaussian probability distribution

$$W[\rho] = \exp\left[- \int \mathrm{d}^2\boldsymbol{x}_\perp \frac{\rho^a(\boldsymbol{x}_\perp)\rho^a(\boldsymbol{x}_\perp)}{g^4\mu^2} \right] \tag{2}$$

computing the correlations and the multiplicity distribution in the linearized approximation is a simple combinatorial problem. It can be expressed in terms of two parameters, the mean multiplicity $\bar{n}$, and a parameter k describing the width of the distribution. The result of the combinatorial exercise is that the qth factorial moment m_q (defined as $\langle N^q \rangle$ minus the

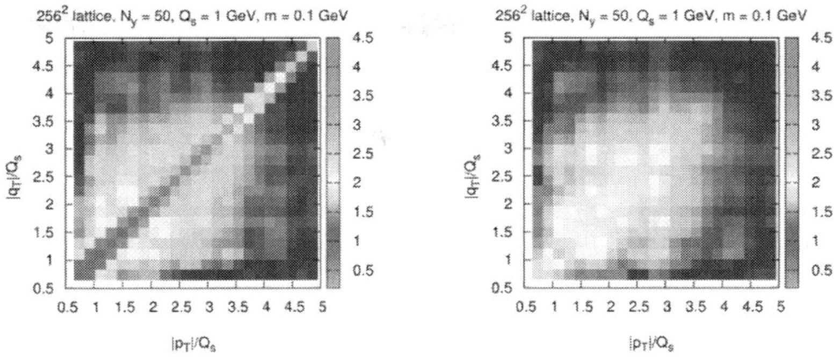

Figure 1. Numerical evaluation of the two-gluon correlation strength in the MV model as a function of the two transverse momenta $p_\perp$ and $q_\perp$. Left: the correlation strength on the "near side" $|\varphi_p - \varphi_q| < \pi/2$, right: the "away side" $|\varphi_p - \varphi_q| > \pi/2$.

corresponding disconnected contributions) is, with $S_\perp = \pi R_A^2$:

$$m_q = (q-1)!\, k\left(\frac{\bar{n}}{k}\right)^q \quad \text{with} \quad k \approx \frac{({N_c}^2 - 1)Q_s^2 S_\perp}{2\pi} \quad \text{and} \quad \bar{n} = f_N \frac{1}{\alpha_s} Q_s^2 S_\perp. \tag{3}$$

These moments define a *negative binomial* distribution with parameters k and $\bar{n}$, known as a phenomenological observation in high energy hadron and nuclear collisions already for a long time. In terms of the glasma flux tube picture this result has a natural interpretation. The transverse area of a typical flux tube is $1/Q_s^2$, and thus there are $Q_s^2 S_\perp = N_{\mathrm{FT}}$ independent ones. Each of these radiates particles independently into ${N_c}^2 - 1$ color states in a Bose-Einstein distribution (see e.g. [9]). A sum of $k \approx N_{\mathrm{FT}}({N_c}^2 - 1)$ independent Bose-Einstein-distributions is precisely equivalent to a negative binomial distribution with parameter k. A numerical evaluation [10] of the second moment of the distribution, parametrized in terms of

$$\kappa_2(\boldsymbol{p}_\perp, \boldsymbol{q}_\perp) = Q_s^2 S_\perp \left(\frac{\mathrm{d}^2 N}{\mathrm{d}^2\boldsymbol{p}_\perp \mathrm{d}^2\boldsymbol{q}_\perp} - \frac{\mathrm{d}N}{\mathrm{d}^2\boldsymbol{p}_\perp}\frac{\mathrm{d}N}{\mathrm{d}^2\boldsymbol{q}_\perp}\right) \Big/ \frac{\mathrm{d}N}{\mathrm{d}^2\boldsymbol{p}_\perp}\frac{\mathrm{d}N}{\mathrm{d}^2\boldsymbol{q}_\perp} \tag{4}$$

is shown in Fig. 1. It confirms the expectations of [7] that this ratio is of order one and depends only weakly on the momenta $\boldsymbol{p}_\perp, \boldsymbol{q}_\perp$.

4. Rapidity dependence

The general discussion of Sec. 2 on the different nature of multigluon correlations in the "AA" case applies also to the rapidity dependence. Until now we have only been discussing gluon production in a rapidity interval smaller than $1/\alpha_s$. For this we needed only the correlations between the color charges $\rho(\boldsymbol{x}_\perp)$ measured at this same rapidity. To understand the rapidity dependence of the correlations one needs also the correlation between color charges at different rapidities, $\langle \rho_y(\boldsymbol{x}_\perp)\rho_{y'}(\boldsymbol{y}_\perp)\rangle$. Also this information is contained in the JIMWLK renormalization group evolution, at least to leading $\ln 1/x$ accuracy [12]. An intuitive description of the resulting correlations is provided by the formulation of JIMWLK as a Langevin equation in the space of Wilson lines formed from the color charges. In this picture the evolution proceeds in individual trajectories along an increasing rapidity. A first attempt of a realistic estimate of the rapidity dependence of two-gluon correlations is performed in Ref. [11]. Evaluating the two gluon correlation in a dilute limit in a $k_\perp$-factorized approximation, but keeping the general structure

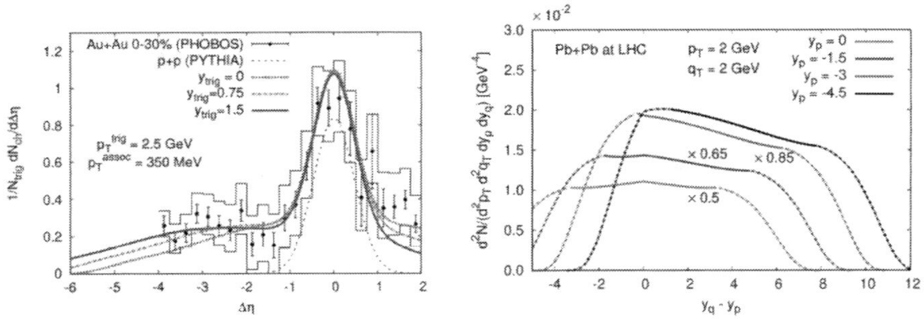

Figure 2. Left: Comparison of a two-particle correlation with PHOBOS data. The curves are eq. (5) supplemented with a short-range correlation contribution from PYTHIA. Right: Rapidity correlation at LHC energies $k_\perp$-factorization approximation. Plots from[11].

resulting from the JIMWLK evolution leads to the following expression:

$$C(\boldsymbol{p}, \boldsymbol{q}) \sim \int_{\boldsymbol{k}_\perp} \left\{ \Phi^2_{A_1,y_p}(\boldsymbol{k}_\perp)\Phi_{A_2,y_p}(\boldsymbol{p}_\perp - \boldsymbol{k}_\perp)\Phi_{A_2,y_q}(\boldsymbol{q}_\perp + \boldsymbol{k}_\perp) + [\boldsymbol{k}_\perp \leftrightarrow -\boldsymbol{k}_\perp] + [A_1 \leftrightarrow A_2] \right\}. \quad (5)$$

Note the very different structure of this correlation compared to one where the gluons would be produced from the same diagram for fixed sources. The two gluon correlation function is proportional to the product of *four* unintegrated gluon distributions, with three of them evaluated at the rapidity of one of the produced gluons and only one at the other. This structure is a direct consequence of the nature of JIMWLK evolution. The resulting correlation is compared to PHOBOS data in fig. 2.

Acknowledgements The author is supported by the Academy of Finland, contract 126604.

References

[1] T. Lappi and L. McLerran, *Nucl. Phys.* **A772** (2006) 200 [arXiv:hep-ph/0602189].
[2] A. Dumitru, F. Gelis, L. McLerran and R. Venugopalan, *Nucl. Phys.* **A810** (2008) 91 [arXiv:0804.3858 [hep-ph]].
[3] S. Gavin, L. McLerran and G. Moschelli, *Phys. Rev.* **C79** (2009) 051902 [arXiv:0806.4718 [nucl-th]].
[4] J. Putschke, *J. Phys.* **G34** (2007) S679 [arXiv:nucl-ex/0701074]; **STAR** collaboration, M. Daugherity, *J. Phys.* **G35** (2008) 104090 [arXiv:0806.2121 [nucl-ex]]; **PHOBOS** collaboration, B. Alver *et. al.*, *J. Phys.* **G35** (2008) 104080 [arXiv:0804.3038 [nucl-ex]]; **STAR** collaboration, B. I. Abelev *et. al.*, *Phys. Rev. Lett* **103** (2009) 172301 [arXiv:0905.0237 [nucl-ex]].
[5] V. P. Konchakovski, M. Hauer, G. Torrieri, M. I. Gorenstein and E. L. Bratkovskaya, *Phys. Rev.* **C79** (2009) 034910 [arXiv:0812.3967 [nucl-th]]; A. Bzdak, *Phys. Rev.* **C80** (2009) 024906 [arXiv:0902.2639 [hep-ph]]; T. Lappi and L. McLerran, *Nucl. Phys.* **A832** (2010) 330 [arXiv:0909.0428 [hep-ph]]; B. Alver and G. Roland, *Phys. Rev.* **C81** (2010) 054905 [arXiv:1003.0194 [nucl-th]].
[6] F. Gelis, T. Lappi and R. Venugopalan, *Phys. Rev.* **D78** (2008) 054019 [arXiv:0804.2630 [hep-ph]]; F. Gelis, T. Lappi and R. Venugopalan, *Phys. Rev.* **D78** (2008) 054020 [arXiv:0807.1306 [hep-ph]].
[7] F. Gelis, T. Lappi and L. McLerran, *Nucl. Phys.* **A828** (2009) 149 [arXiv:0905.3234 [hep-ph]].
[8] K. Dusling, D. Fernandez-Fraile and R. Venugopalan, *Nucl. Phys.* **A828** (2009) 161 [arXiv:0902.4435 [nucl-th]].
[9] K. Fukushima, F. Gelis and T. Lappi, *Nucl. Phys.* **A831** (2009) 184 [arXiv:0907.4793 [hep-ph]].
[10] T. Lappi, S. Srednyak and R. Venugopalan, *JHEP* **01** (2010) 066 [arXiv:0911.2068 [hep-ph]].
[11] K. Dusling, F. Gelis, T. Lappi and R. Venugopalan, *Nucl. Phys.* **A836** (2010) 159 [arXiv:0911.2720 [hep-ph]].
[12] F. Gelis, T. Lappi and R. Venugopalan, *Phys. Rev.* **D79** (2008) 094017 [arXiv:0810.4829 [hep-ph]]; T. Lappi, *Acta Phys. Polon.* **B40** (2009) 1997 [arXiv:0904.1670 [hep-ph]].

Hot Quarks 2010 IOP Publishing
Journal of Physics: Conference Series **270** (2011) 012056 doi:10.1088/1742-6596/270/1/012056

A predictive phenomenological tool at small Bjorken-x

José Guilherme Milhano[1,2], Javier L. Albacete[3], Néstor Armesto[4], Paloma Quiroga-Arias[4], and Carlos A. Salgado[4]

[1] CENTRA, Departamento de Física, Instituto Superior Técnico (IST), Av. Rovisco Pais 1, P-1049-001 Lisboa, Portugal
[2] Physics Department, Theory Unit, CERN, CH-1211 Genève 23, Switzerland
[3] Institut de Physique Théorique, CEA/Saclay, 91191 Gif-sur-Yvette cedex, France. URA 2306, unité de recherche associée au CNRS.
[4] Departamento de Física de Partículas and IGFAE, Universidade de Santiago de Compostela 15706 Santiago de Compostela, Spain.

E-mail: guilherme.milhano@ist.utl.pt

Abstract. We present the results from global fits of inclusive DIS experimental data using the Balitsky-Kovchegov equation with running coupling.

On rather general physical grounds, one expects that the large gluon density that develops in the proton (nucleus) wave function with decreasing Bjorken-x will lead to deviations from standard collinear factorization. The necessary unitarity of the theory requires the rate of growth of the gluon density to be limited. Such a taming arises naturally once non-linearities, due to gluon overlap, are accounted for in the evolution.

The Colour Glass Condensate (CGC) [1–3], along with its B-JIMWLK evolution equations [4,5], provides a solid theoretical framework in which to address the non-linear dynamics driving the small-x evolution of the proton (nucleus) wave function. The mathematical complexity of the CGC evolution equations has hindered its *direct* phenomenological application. In face of such difficulties, the phenomenological use of the CGC has proceeded via dipole-models in which the general physical properties of the CGC have been, in one form or another, implemented in an effective manner. The behaviour at small-x of the parton distribution functions plays an important role in the computation of many observables in proton-proton, nucleus-nucleus, and proton-nucleus at the LHC. Given that evidence for small, yet systematic, deviations from NLO-DGLAP evolution have been identified in HERA DIS data [6], it is mandatory that the best available and numerically tractable non-linear small-x evolution tools replace *CGC-motivated* models.

The simplest theoretically sound realization of the CGC, the Balitsky-Kovchegov (BK) equation [5, 7], although tractable from the numerical point of view, fails to reproduce experimental data unless rather unphysical assumptions are made. Early heuristic modifications of the BK equation [8] indicated that NLO corrections would drive the evolution towards compatibility with experimental data. The explicit computation of the full NLO-BK [9–11] paved the way to the rigourous and phenomenologically relevant use the CGC. Although these corrections present a rather complicated structure and are not presently amenable to an efficient numerical implementation, it was shown [12] that considering only running coupling corrections to the BK equation accounts for most of the NLO effects. The ability of the rcBK equation to correctly describe experimental data was shown in [13]. There, a global fit of the inclusive DIS structure function led to the public release of a parametrization of the dipole scattering cross section at small values of x. Thus, running coupling BK (rcBK), the most reliable and phenomenologically usable non-linear small-x evolution tool, has become the standard choice for CGC computation of experimental observables (e.g., [14–16]).

The H1 and ZEUS combined data set for the DIS reduced cross section [17] poses a series of new challenges and opportunities. On the one hand, the high-accuracy of the data puts the rcBK equation under much more stringent test conditions and will eventually lead to a better constrained parametrization. On the other hand, the data being given directly in terms of the reduced cross section, the experimentally measured quantity, eliminates the theoretical bias associated with the extraction of F_2 and F_L from data. Here, we present the first results obtained both from the global fit of these data within the setup of [13] where only light quarks were taken as contributing to the DIS cross section, and from a global fit in which available F_2^c data is used to constrain the heavy quark contribution to the DIS cross section.

In the kinematical domain ($Q^2 < 50$ GeV, $x < 0.01$) considered in this work, the reduced DIS cross-section σ_r can be written in terms of the virtual photon-proton cross-section $\sigma_{T,L}$, with T (transverse) and L (longitudinal) the polarization of the virtual photon, as

$$\sigma_r(x,y,Q^2) = \frac{Q^2}{4\,\pi^2\alpha_{em}}\left(\sigma_T + \frac{2(1-y)}{1+(1-y)^2}\sigma_L\right). \qquad (1)$$

Here, $y = Q^2/sx$ ($\sqrt{s}$ is the center of mass collision energy) is the inelasticity variable. In the dipole formulation of QCD, valid for small-x, one has

$$\sigma_{T,L}(x,Q^2) = \sum_f \sigma_{0,f}\int_0^1 dz\, d\mathbf{r}\, |\Psi^f_{T,L}(e_f, m_f, z, Q^2, \mathbf{r})|^2 \mathcal{N}(\mathbf{r},x)\,, \qquad (2)$$

where $\Psi^f_{T,L}$ is the light-cone wave function for a virtual photon to fluctuate into a quark-antiquark dipole of quark flavor f (with mass m_f and electric charge e_f). $\mathcal{N}(\mathbf{r},x)$ is the imaginary part of the dipole-target scattering amplitude averaged over impact parameter, with $\mathbf{r}$ the transverse dipole size. $\sigma_{0,f}$ is (half) the transverse area over which quarks of a given flavour are distributed. Light quarks are taken to be identically distributed ($\sigma_{0,f=u,d,s} = \sigma_{0,light}$). When accounting for heavy flavour contributions in (1) we allow $\sigma_{0,f=c,b} = \sigma_{0,heavy}$ to be different from $\sigma_{0,light}$. These are free fit parameters.

The evolution of $\mathcal{N}(\mathbf{r},x)$ is given by the rcBK equation:

$$\frac{\partial\mathcal{N}(r,x)}{\partial\ln(x_0/x)} = \frac{N_c\,\alpha_s(r^2)}{2\pi^2}\int d\mathbf{r_1}\left[\frac{r^2}{r_1^2\,r_2^2} + \frac{1}{r_1^2}\left(\frac{\alpha_s(r_1^2)}{\alpha_s(r_2^2)} - 1\right) + \frac{1}{r_2^2}\left(\frac{\alpha_s(r_2^2)}{\alpha_s(r_1^2)} - 1\right)\right]$$
$$\cdot\left[\mathcal{N}(r_1,x) + \mathcal{N}(r_2,x) - \mathcal{N}(r,x) - \mathcal{N}(r_1,x)\mathcal{N}(r_2,x)\right]. \qquad (3)$$

where $\mathbf{r_2} = \mathbf{r} - \mathbf{r_1}$, and x_0 (in our case $x_0 = 0.01$) is the value of x where the evolution starts (the highest value of x in the data included in the fits). The kinematic cost of the virtual photon to fluctuate into a dipole of a given mass is accounted for by $x = x_{exp}\,(1 + 4\,m_f^2/Q^2)$. The running coupling in (3) is evaluated at 1-loop accuracy in coordinate space

$$\alpha_{s,n_f}(r^2) = \frac{4\pi}{\beta_{0,n_f}\ln\left(\frac{4C^2}{r^2\Lambda^2_{n_f}}\right)}\,, \qquad \beta_{0,n_f} = 11 - \frac{2}{3}n_f\,, \qquad (4)$$

where the constant C (a fit parameter) accounts for the uncertainty in the Fourier transform from momentum to coordinate space. When only light flavours ($f = u, d, s$) are taken as contributing to (1), the coupling (4) is evaluated with $n_f = 3$ (*light only*). Once the contribution of heavy flavours ($f = c, b$) is accounted for, the coupling (4) is evaluated in a variable flavour scheme matched on the quark mass thresholds (*light+heavy*). Since in the rcBK equation (3) all dipoles sizes are explored, an infrared regulation is called for: $\alpha_s(r^2 > r_{fr}^2) = \alpha_{fr} = 0.7$ where r_{fr}^2 is the dipole size at which the coupling reaches α_{fr}. Although we have explored [18] a number of different initial conditions, the results show a negligible dependence on specific choice. For brevity, only results obtained with the GBW initial condition

$$\mathcal{N}^{GBW}(r, x=x_0) = 1 - \exp\left[-\frac{(r^2\, Q_{s\,0}^2)^{\gamma}}{4}\right], \qquad (5)$$

are shown here. The fit parameters $Q_{s\,0}^2$ and γ are, respectively, the saturation scale and the characteristic fall-off of the dipole scattering amplitude with decreasing r at the initial x_0.

Fig. 1 shows a comparison of data for the reduced cross section (selected Q^2 data bins) with the results of a *light only* fit using all combined H1/ZEUS data (196 points) within cuts $Q^2 < 50$ GeV^2 and $x < 0.01$. The mass of the light flavours was kept fixed, $m_{light} = 0.14$ GeV. The obtained fit parameters — $\sigma_{0,light} = 33.1$ mb, $Q_{s\,0,light}^2 = 0.26$ GeV^2, $\gamma_{light} = 0.97$, and $C = 2.3$ — give χ^2/d.o.f. $= 206/192 = 1.07$.

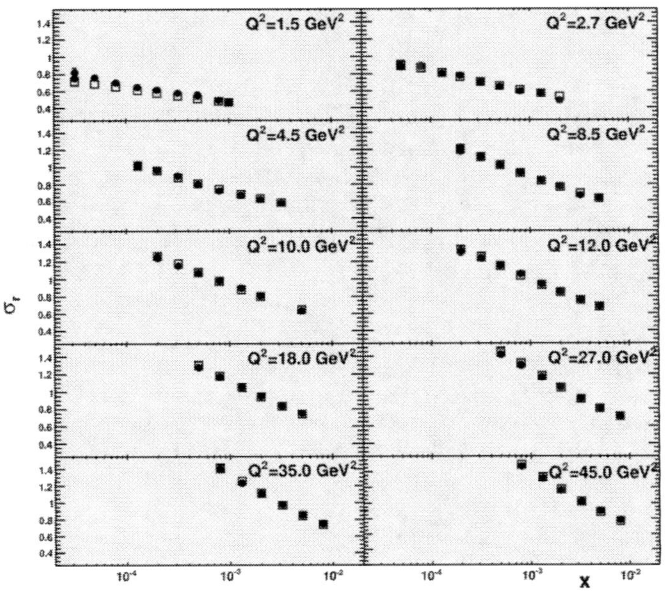

Figure 1. Comparison of reduced cross section experimental data (full circles) with our (*light only*) fit results (open squares). Notice that the experimental error is often smaller that size of the symbols used for the data points.

Fig. 2 shows a comparison of data on F_2^c (the charm contribution to F_2) with the results of a *light+heavy* fit. In this case, all existing data on F_2^c (within the same cuts as above) is added to the H1/ZEUS sample (a total of 265 points). We allow for independent light and heavy initial conditions and for a free light flavour mass. The heavy quark masses are set to $m_c = 1.4$ GeV and $m_b = 4.75$ GeV. The χ^2/d.o.f. $= 312/257 = 1.2$ is obtained for the fit parameters $\sigma_{0,light} = 32.7$ mb, $Q_{s\,0,light}^2 = 0.26$ GeV^2, $\gamma_{light} = 1.24$, $\sigma_{0,heavy} = 28.4$ mb, $Q_{s\,0,heavy}^2 = 0.18$ GeV^2, $\gamma_{heavy} = 0.95$, $C = 3.3$, and $m_{light} = 0.104$ GeV.

The ability of the rcBK equation to accurately describe the high precision H1/ZEUS combined sample and the charm component of the structure function strengthens its role as a viable

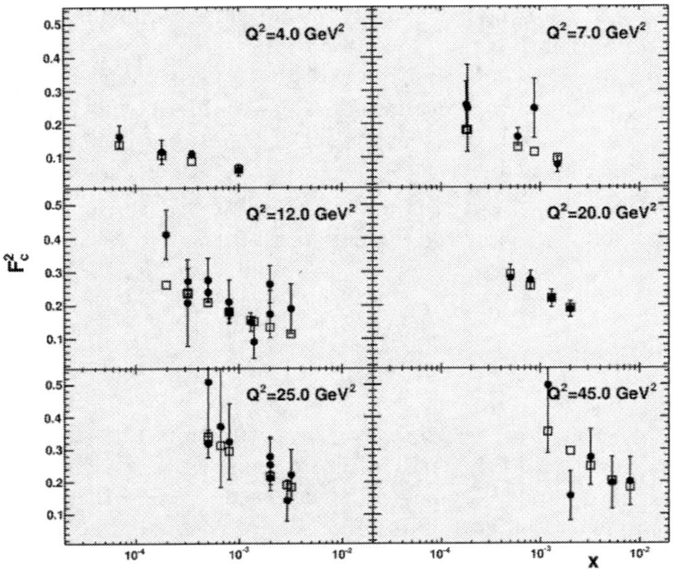

Figure 2. Comparison of F_2^c data (full circles) with our (*light+heavy*) fit results (open squares). *light+heavy* fits give description of reduced cross section similar to that shown in Fig. 1

phenomenological tool to address the small-x behaviour of the proton structure. Further details, along with a parametrization of the dipole-proton cross section for use in the computation of physical observables, will be given in [18].

Acknowledgments
We acknowledge support from MICINN (Spain) under project FPA2008-01177 and FPU grant; Xunta de Galicia (Conselleria de Educacion) and through grant PGIDIT07PXIB206126PR, the Spanish Consolider- Ingenio 2010 Programme CPAN (CSD2007-00042) and Marie Curie MEST-CT-2005-020238-EUROTHEPHY (PQA); and Fundação para a Ciência e a Tecnologia (Portugal) under project CERN/FP/109356/2009 (JGM).

References

[1] Kovner A, Milhano J G and Weigert H 2000 *Phys. Rev.* **D62** 114005 (*Preprint* hep-ph/0004014)
[2] Iancu E, Leonidov A and McLerran L D 2001 *Nucl. Phys.* **A692** 583–645 (*Preprint* hep-ph/0011241)
[3] Ferreiro E, Iancu E, Leonidov A and McLerran L 2002 *Nucl. Phys.* **A703** 489–538 (*Preprint* hep-ph/0109115)
[4] Jalilian-Marian J, Kovner A, Leonidov A and Weigert H 1999 *Phys. Rev.* **D59** 014014 (*Preprint* hep-ph/9706377)
[5] Balitsky I 1996 *Nucl. Phys.* **B463** 99–160 (*Preprint* hep-ph/9509348)
[6] Caola F, Forte S and Rojo J 2010 *Phys. Lett.* **B686** 127–135 (*Preprint* 0910.3143)
[7] Kovchegov Y V 1999 *Phys. Rev.* **D60** 034008 (*Preprint* hep-ph/9901281)
[8] Albacete J L, Armesto N, Milhano J G, Salgado C A and Wiedemann U A 2005 *Phys. Rev.* **D71** 014003 (*Preprint* hep-ph/0408216)
[9] Balitsky I 2007 *Phys. Rev.* **D75** 014001 (*Preprint* hep-ph/0609105)
[10] Kovchegov Y V and Weigert H 2007 *Nucl. Phys.* **A784** 188–226 (*Preprint* hep-ph/0609090)
[11] Balitsky I and Chirilli G A 2008 *Phys. Rev.* **D77** 014019 (*Preprint* 0710.4330)
[12] Albacete J L and Kovchegov Y V 2007 *Phys. Rev.* **D75** 125021 (*Preprint* 0704.0612)
[13] Albacete J L, Armesto N, Milhano J G and Salgado C A 2009 *Phys. Rev.* **D80** 034031 (*Preprint* 0902.1112)
[14] Albacete J L 2007 *Phys. Rev. Lett.* **99** 262301 (*Preprint* 0707.2545)
[15] Albacete J L and Marquet C 2010 *Phys. Lett.* **B687** 174–179 (*Preprint* 1001.1378)
[16] Albacete J L and Marquet C 2010 (*Preprint* 1005.4065)
[17] Aaron F D *et al.* (H1) 2010 *JHEP* **01** 109 (*Preprint* 0911.0884)
[18] Albacete J L, Armesto N, Milhano J G, Quiroga-Arias P and Salgado C A *in preparation*

Journal of Physics: Conference Series **270** (2011) 012057 doi:10.1088/1742-6596/270/1/012057

Quantifying the glue - understanding the initial conditions at RHIC and the LHC

Matthew A. C. Lamont for the BNL EIC Science Task Force
Brookhaven National Lab, Upton, NY 11777, USA
E-mail: macl@bnl.gov

Abstract. Recent results on the suppression of hadrons at forward rapidities from the STAR collaboration indicate that high gluon densities play an important role in the initial conditions of heavy-ion collisions. The study of these initial conditions is therefore crucial in the next stage of understanding and quantifying the data from RHIC and the LHC. The best way to perform these measurements is through Deep Inelastic Scattering measurements on nuclei. Currently, there are no accelerator facilities which are able to perform this measurement. In this paper, I will outline the latest developments of the eRHIC proposal at BNL as well as preliminary designs of a new detector as well as an assessment of the current detectors (PHENIX and STAR) abilities to run in an eRHIC era.

1. Introduction

One of the principal results emanating from the heavy-ion programme at RHIC was of the energy loss of high momentum particles at mid-rapidity ($|y| < 1$), referred to as *jet quenching*. This energy loss, manifested as the absence of a back-to-back peak in azimuthal correlations of high momentum particles separated by 180° in azimuth, was only present in the most central (head-on) heavy-ion collisions. In peripheral (large impact parameter) Au+Au collisions and in d+Au collisions, this correlation looked like that observed in p+p collisions. Therefore, the interpretation of this effect was that it was a final-state effect, with the suppression occurring due to final-state interactions of the fast-moving partons in the de-confined medium (Quark-Gluon Plasma) produced in central heavy-ion collisions.

However, recent results from the STAR experiment have taken this one step further. Instead of performing the measurement at mid-rapidity as in the past, the same azimuthal correlation analysis was performed with both particles being at forward rapidity, with the measurement taking place in the forward meson spectrometer (FMS) [1]. The acceptance of the FMS detector is $2.3 < \eta < 4.0$, which means that for this measurement, the particles have $< \eta >\sim 3.1$. This time, the correlations showed a suppression in central d+Au collisions, whereas previously no suppression had been observed in d+Au collisions at mid-rapidity. As a cross-check, this analysis at forward rapidities was also performed in p+p collisions and no suppression was observed. The difference between these two measurements is then simply the value of x (the fractional value of the hadron momentum carried by the interacting parton), given by the relation:

$$x = \frac{2p_T}{\sqrt{s}} e^{-\eta}$$

Therefore, at mid-rapidity, the collision is happening at a much higher value of x than those at forward rapidities.

2. Gluons at small-x

In order to fully understand the momentum distributions of partons inside hadrons as a function of x, it is necessary to perform Deep-Inelastic Scattering (DIS) measurements of electrons on hadrons. These offer a clean probe of the hadron. Whilst hadron-hadron interactions provide a complementary probe, additional colour interactions between the partons make interpreting the phenomena difficult whilst at the same time, they offer no access to the collision kinematics.

DIS measurements on protons were performed at the HERA collider by the H1 and ZEUS collaborations. The invariant cross-section in DIS can be written as:

$$\frac{d^2\sigma^{ep\to eX}}{dxdQ^2} = \frac{4\pi\alpha_{e.m.}^2}{xQ^4}\left[\left(1 - y + \frac{y^2}{2}\right)F_2(x,Q^2) - \frac{y^2}{2}F_L(x,Q^2)\right]$$

where y is the fraction of the energy lost by the lepton in the rest frame of the nuclei. F_2 represents the quark and anti-quark structure function and F_L represents that of the gluons. At HERA, the gluon distribution in nucleons was measured indirectly via the scaling violation of the F_2 structure function. By fitting these distributions with a pQCD fit based on the linear DGLAP model, it was found to be dominated by gluons for $x < 10^{-1}$ and increases rapidly with decreasing x [2]. This is believed to occur due to gluon Bremsstrahlung, where large-x gluons radiate smaller-x gluons and was observed to grow more rapidly with increasing Q^2. However, the Froissart Unitarity Bound relation places a limit on the size of the overall cross-section and if this is to be adhered to, then the Bremsstrahlung cannot continue unabated to smaller and smaller x. Therefore, a competing mechanism must tame the growth of the gluon cross-section. One such explanation is that of recombination of gluons, whereby small x gluons can recombine to form higher x gluons depending on the density. As such, at small enough x, the competing processes cancel each other out and there will be a saturation of the gluon densities. These processes are described by the Colour Glass Condensate (CGC) effective field theory and the relevant non-linear JIMWLK equations [3]. A CGC-inspired model is so far the only one able to reproduce the quenching of high p_T azimuthal correlations at forward rapidities in d+Au collisions, giving hints that the onset of saturation in nuclei may be observed at RHIC [4].

3. Gluon distributions in nuclei and the nuclear enhancement factor

As outlined in other presentations at this conference [5], whilst the distribution of sea and valence quarks in nuclei, relative to those in nucleons, are well constrained, very little is known about the gluon distribution for large ranges of x. Therefore, it is desirable to perform DIS experiments on nuclei. Secondly, it is also desirable to perform these experiments in order to search for saturation.

In DIS collisions on nucleons at HERA, the x range was not large enough to explore the saturation regime, which has a characteristic scale associated with it referred to as Q_S. However, simple geometric considerations give Q_S^A as the following:

$$(Q_S^A)^2 \approx c\ Q_0^2(\frac{A}{x})^{1/3}$$

meaning that Q_S scales with the nuclear size ($A^{1/3}$). This is confirmed by rigorous calculations which increase the dependence on A further for large nuclei. This has a significant impact, meaning that for large nuclei, the effective value of x which is being probed is an order of magnitude and more smaller than in e+p collisions at the same energies.

4. The eRHIC proposal at Brookhaven National Lab

Although DIS experiments on nuclei are desirable for many reasons, there is currently no facility in the World which provides for these collisions. However, there is an ever growing section of the community interested in this and one such proposal is to build an electron-ion collider (EIC), namely eRHIC at Brookhaven National Lab in New York, USA. This proposal will build upon the current complex which houses an ion-ion collider and will build an electron ring to provide both polarised $e+p$ and unpolarised $e+A$ collisions.

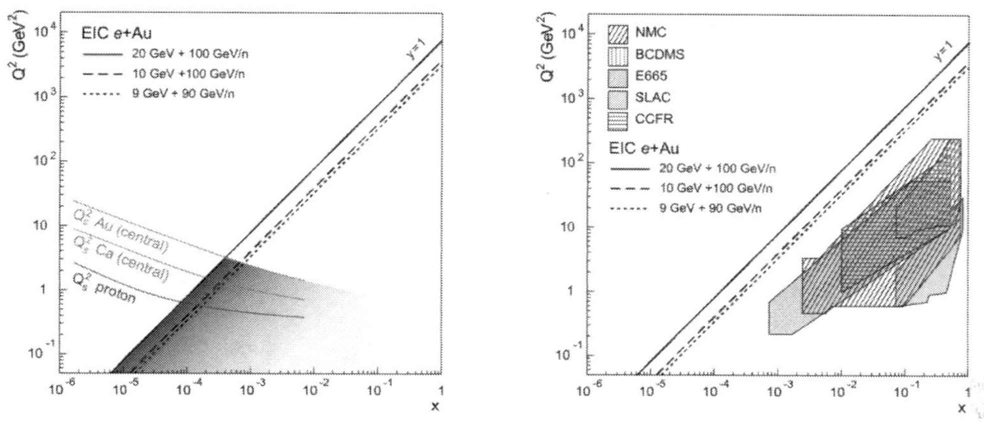

Figure 1. Left: the coverage of the EIC and the projected values of Q_s for different nuclei. Right: the coverage of previous l+A experiments together with that projected for various combinations of EIC energies.

The electron ring will provide energies in the range 5 - 30 GeV and will collide with p (50-325 GeV) and A (5-130 GeV) in the same interaction points as are currently available at RHIC. The x and Q^2 phase-space for various energies are shown in Figure 1, together with the ranges for existing l+A experiments.

A schematic of the eRHIC design is given in Figure 4, which shows that the electron acceleration will occur in two energy recovery linacs and collisions will be possible in either PHENIX, STAR or a new collision point. Whilst eRHIC will be able to run $p+p$ and e+h collisions, it will not be able to do so simultaneously and hence it is a worthwhile exercise to determine whether or not STAR and PHENIX, which were designed specifically for heavy-ion collisions, will be able to run in an eRHIC mode.

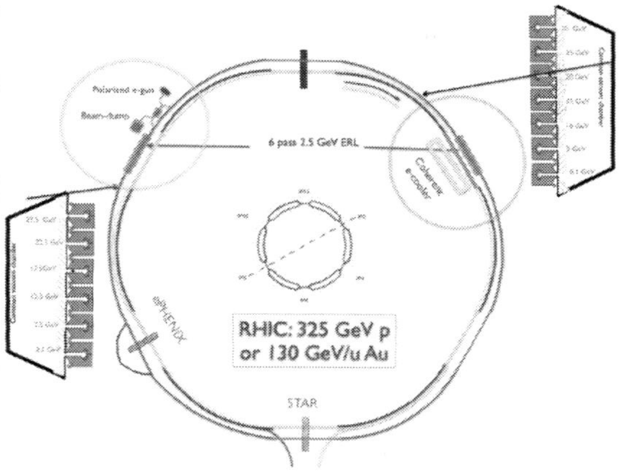

Figure 2. The eRHIC complex, comprising two energy recovery linacs for accelerating the electrons and capabilities of providing $e+p$/A collisions at PHENIX, STAR or a new IP.

The PHENIX experiment has been specifically designed to be a high-rate and low-acceptance experiment. However, a high-acceptance experiment is optimal in DIS and hence the PHENIX experiment, as it currently stands, would not be optimal for this. The STAR experiment, on the other hand, is a lower-rate but high-acceptance spectrometer. Additionally, not only is the mid-rapidity region well instrumented, but one of the forward regions has large coverage with electro-magnetic calorimetry. Figure 3 shows the capability of STAR to measure both the outgoing electron and jet produced in the collision respectively, depending on whether or not the hadron beam will be in the anti-clockwise or clockwise direction.

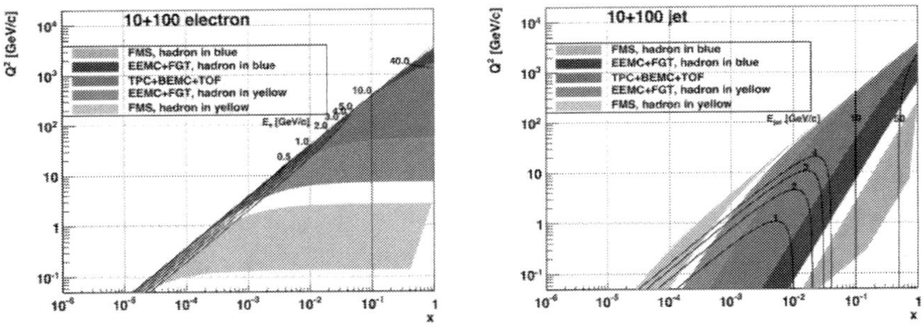

Figure 3. The coverage in x and Q^2 of the different detectors in STAR for 10+100 GeV $e + h$ collisions for measuring the scattered e^- (left) and jet (right) respectively. In each case, the two different options for the hadron beam being in the blue beam pipe (clockwise) or yellow beam pipe (anti-clockwise) are given.

5. A dedicated detector for eRHIC

Despite the large acceptance coverage of STAR, it is desirable to be able to design and build a detector for eRHIC from scratch. For example, the measurement of exclusive vector meson production (which is very sensitive to the gluon distribution) is significantly easier if the struck nucleon/nucleus is measured.

In order to perform this measurement, it is therefore desirable to have as hermetic as possible, detector coverage in the forward region (both tracking and calorimetry) coupled with Roman-Pot type detectors to measure nuclear break-up. A sketch of a proposed detector is shown in Fig 4 which outlines both the detector technology and the magnet design currently under investigation. This is ongoing work and there are many open questions to be resolved (such as is a DIRC necessary or is a barrel time-of-flight detector desirable?).

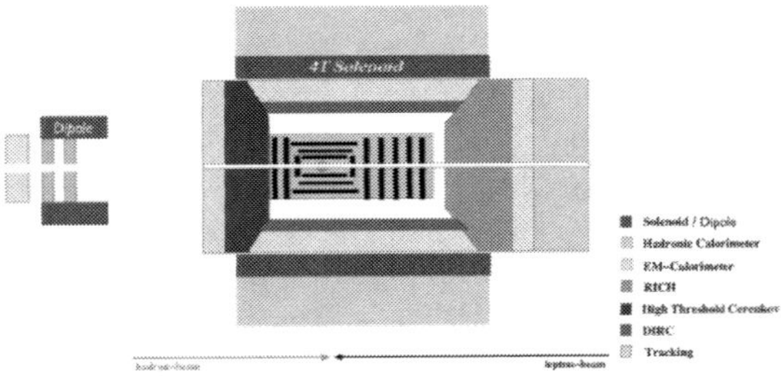

Figure 4. A sketch of a proposed detector for eRHIC.

References

[1] E. Braidot for the STAR collaboration, arXiv:1005.2378.

[2] F. D. Aaron *et al.* (H1 and ZEUS Collaborations), JHEP 1001 (2010) 109.

[3] E. Iancu and R. Venugopalan, review for QGP3, arXiv:hep-ph/0303204.

[4] J. Albacete and C. Marquet, arXiv:1005.4065v1.

[5] P. Quiroga Arias, these proceedings.

Journal of Physics: Conference Series **270** (2011) 012058 doi:10.1088/1742-6596/270/1/012058

Azimuth Quadrupole Systematics in Au-Au Collisions

David Kettler (STAR Collaboration)
University of Washington, Seattle, USA
E-mail: dkettler@u.washington.edu

Abstract. We have measured p_t-dependent two-particle number correlations on azimuth and pseudorapidity for eleven centralities of $\sqrt{s_{NN}} = 62$ and 200 GeV Au-Au collisions at STAR. 2D fits to these angular correlations isolate the azimuth quadrupole amplitude, denoted $2v_2^2\{2D\}(p_t)$, from localized same-side correlations. Event-plane $v_2(p_t)$ measurements within the STAR TPC acceptance can be expressed as a sum of the azimuth quadrupole and the quadrupole component of the same-side peak. $v_2\{2D\}(p_t)$ can be transformed to reveal quadrupole p_t spectra which are approximately described by a fixed transverse boost and universal Lévy form nearly independent of centrality. A parametrization of $v_2\{2D\}(p_t)$ can be factored into centrality and p_t-dependent pieces with a simple p_t dependence above 0.75 GeV/c. Results from STAR are compared to published data and model predictions.

1. Introduction

The collective behavior of particles produced in heavy-ion collisions is one of the major topics of study at RHIC, and a large quadrupole component in the distribution on azimuthal angle ϕ has been observed [1]. The large quadrupole azimuthal distribution in non-central collisions is usually described by elliptic flow, in which the different pressure gradients in and out of plane turn the initial position-space anisotropy of the colliding nuclei into a momentum-space anisotropy [2].

The azimuthal anisotropy is typically measured with the quantity v_2, the second Fourier component of the distribution of particle emission on azimuth with respect to the angle of the reaction plane [3]. In real events this reaction plane angle is not known so many methods exist to estimate v_2. The present analysis uses a method based on fitting two-dimensional angular correlations similar to [4] but expanded to include the dependence on transverse momentum as well as centrality. A simultaneous analysis of centrality and transverse momentum dependence of v_2 in the absence of "nonflow" from the same-side 2D peak (jet-like correlations) is presented.

One method of measuring v_2 is the event-plane method, $v_2\{\text{EP}\}$, in which particles in the region of interest are used to estimate the reaction plane [3]. A related approach, $v_2\{2\}$, calculates v_2 directly from two-particle correlations [5]. These two methods are essentially equivalent ($v_2\{\text{EP}\} \sim v_2\{2\}$ to 5%) [6]. This analysis, denoted by $v_2\{\text{2D}\}$, uses an approach similar to $v_2\{2\}$ but includes the dependence of the two-particle distribution on pseudorapidity as well as azimuth. Significant jet-like correlations are observed at all centralities [7, 8] and their 2D angular dependence can be used to distinguish them from the azimuth quadrupole component, which within the STAR TPC acceptance has no significant η dependence [9]. Jet-like correlations are separated from $v_2\{\text{2D}\}$ by 2D model fits described in Sec. 3.

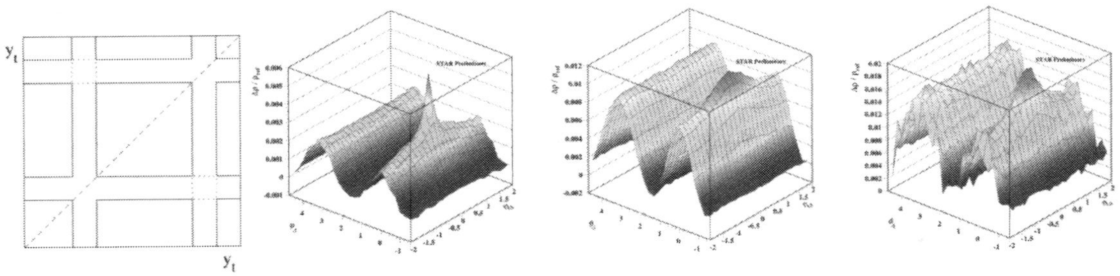

Figure 1. Left Panel: Two examples of marginal distribution cuts in $y_t \times y_t$ space. Remaining panels: An example of the y_t evolution of correlation structures for 40-50% central collisions. The plots correspond to y_t bins of $1.4 < y_t < 1.8$, $3.0 < y_t < 3.4$, and $3.8 < y_t < 4.2$.

2. Correlations

Angular correlations without a trigger particle are constructed by considering all pairs of particles in an event. For each particle the azimuth angle ϕ, pseudorapidity η and transverse momentum p_t are measured, defining a six dimensional two-particle space. Integrating over p_t and projecting the angular variables onto their difference axes $\eta_\Delta \equiv \eta_1 - \eta_2$ and $\phi_\Delta \equiv \phi_1 - \phi_2$ yields 2D angular autocorrelations [4, 7, 10].

The correlation measure is constructed from pair densities. Sibling pairs from the same event contain the primary signal. Mixed pairs from different events are useful as a reference for removing detector artifacts. The basic correlation measure we use is $\Delta\rho/\rho_{ref}$, the ratio of the sibling minus the mixed pairs to the mixed pairs, a per-pair measure. It is also sometimes desirable to use a two-particle measure like $\Delta\rho/\sqrt{\rho_{ref}}$ [11]. However, this analysis uses $\Delta\rho/\rho_{ref}$ which is easily converted to v_2: $2v_2^2\{2D\} \equiv \Delta\rho[2]/\rho_{ref}$, where the [2] indicates the quadrupole term of the fit model in Eq. 1 below.

When we consider momentum dependence it is desirable to use transverse rapidity instead of transverse momentum: $y_t = \ln\{(p_t + m_t)/m_0\}$. Transverse rapidity is the analogue to longitudinal rapidity but in the transverse direction. For unidentified particles we make the approximation $m_0 = m_\pi$ to define a regularized $\ln(p_t)$ measure.

We construct marginal distributions by restricting the y_t range of one of the particles in the two-particle correlation to a certain interval. Due to the inherent diagonal symmetry in $y_t \times y_t$ space this corresponds to making cross-shaped cuts as seen in Fig. 1 (left panel). These cuts are analogous to the procedure in a standard event-plane $v_2(p_t)$ analysis [3] in which particles from the full momentum range are used to estimate the event plane and particles in the momentum interval are correlated to it. The marginal distributions allow us to study correlations at large p_t values and are easily converted into the usual $v_2(p_t)$ measure for comparison with published $v_2(p_t)$ data.

In this analysis there is a minimum p_t cut for all particles of 150 MeV/c which corresponds to a pion y_t of 0.93. There are nine y_t bins. The first y_t bin includes all particles with y_t less than 1.4 and the seven subsequent bins are 0.4 units of y_t wide. The final bin includes all particles with y_t greater than 4.2. This corresponds to particles above ~ 5 GeV/c.

3. Fit Parameters

This analysis is based on 14.5 million Au-Au collisions at $\sqrt{s_{NN}} = 200$ GeV and 6.7 million Au-Au collisions at $\sqrt{s_{NN}} = 62$ GeV observed with the STAR TPC. The acceptance was defined by transverse momentum $p_t > 0.15$ GeV/c, $|\eta| < 1$ and 2π azimuth. Minimum-bias event samples were divided into 11 centrality bins: nine $\sim 10\%$ bins from near 100% to 10%,

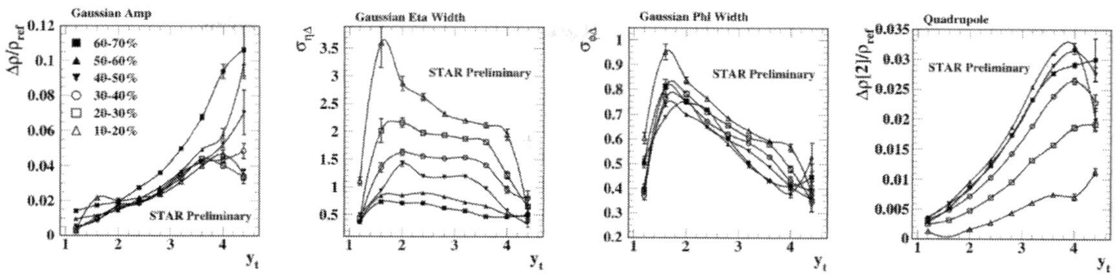

Figure 2. The amplitude, width in η_Δ, and width in ϕ_Δ of the 2D Gaussian and the amplitude of the quadrupole component of the fit for Au-Au 200 GeV 60-70%, 50-60%, 40-50%, 30-40%, 20-30%, and 10-20% collisions. Error bars are for fit errors only.

the last 10% divided into two $\sim 5\%$ bins.

Two-particle correlation histograms are constructed for each of the 11 centrality and 9 y_t bins described above for a total of 99 bins at each $\sqrt{s_{NN}}$. Example histograms are shown in Fig. 1. Fig. 2 shows the evolution of the most relevant fit parameters at $\sqrt{s_{NN}} = 200$ GeV with y_t for six centrality bins. Systematics errors are still under study. 62 GeV results have similar trends as 200 GeV.

In order to distinguish the quadrupole signal from η-localized (jet-like) correlations we will fit the major structures in the 2D histograms. The fitting model used is similar to that used in [4], but the sharp peak from electron pairs is excluded rather than fit. The histograms are constructed with 24 bins on ϕ_Δ with centers ranging from $-\pi/2$ to $17\pi/12$ in steps of $\pi/12$ and 25 bins on η_Δ with centers ranging from -1.92 to 1.92. A total of seven bins are excluded from the fit, where $\eta_\Delta = 0$ and $\phi_\Delta = 0, \pm\pi/12$ and where $\phi_\Delta = 0$ and $\eta_\Delta = \pm 0.08, \pm 0.16$.

The complete fitting function is then

$$F = A_{\phi_\Delta}\cos(\phi_\Delta) + A_{2\phi_\Delta}\cos(2\phi_\Delta) + A_0 e^{-\frac{1}{2}\left(\frac{\eta_\Delta}{\sigma_0}\right)^2} + A_1 e^{-\frac{1}{2}\left\{\left(\frac{\phi_\Delta}{\sigma_{\phi_\Delta}}\right)^2 + \left(\frac{\eta_\Delta}{\sigma_{\eta_\Delta}}\right)^2\right\}} + A_2. \quad (1)$$

There are 5 terms and 8 parameters. Two of the terms are sinusoids on ϕ_Δ, one is a constant offset, and one models only structure on η_Δ. Thus, the remaining term—the two-dimensional Gaussian—is solely responsible for describing "nonflow" [6].

Fig. 3 (first three panels) shows measured quadrupole results from this analysis vs. p_t compared to previously published $v_2\{\mathrm{EP}\}$ results for 30-40%, 5-10% and 0-5% central collisions. In the case of 0-5% the quadrupole component was found to be too small to measure reliably so the cross-hatched region represents an upper bound. We calculate the contribution of the 2D Gaussian peak to the second Fourier component using the measured amplitudes and widths in the curve labeled 'jets'. We find $v_2\{EP\}$ seems to be dominated by the same-side peak structure in 5-10% collisions and entirely due to it in 0-5% collisions. This has important implications for the interpretation of published v_2 data.

4. Quadrupole Descriptions

Consider a model in which the single-particle density on y_t and ϕ and corresponding transverse boosts can be decomposed into azimuth-independent ρ_0 and quadrupole ρ_2 pieces [12]. Define $V_2(y_t, b) \equiv \rho_0(y_t, b)v_2(y_t, b)$ and invoke a blastwave model to get the relation $V_2(y_t, b) \approx p_t\Delta y_{t2}(b)\rho_2(y_t, b)/(2T_2)$ where Δy_{t2} is the quadrupole component of the radial boost distribution, ρ_2 is the quadrupole spectrum, and T_2 is a blastwave parameter. Now construct the unit-integral quantity $Q(y_t, b) \equiv [V_2(y_t, b)/p_t]/[V_2(b)\langle 1/p_t\rangle] \approx [\rho_2(y_t, b)]/[\rho_2(b)]$. Most of the parameters drop out and this variable directly relates measurable quantities to the quadrupole spectrum model.

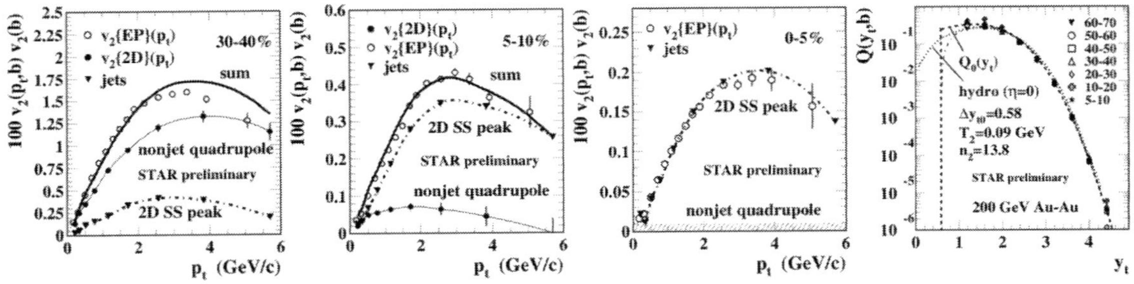

Figure 3. First three panels: Comparison of published $v_2\{\mathrm{EP}\}$ for Au-Au 200 GeV 30-40%, 5-10% and 0-5% central collisions [9] to quadrupole and same-side (SS) peak results from this analysis. Errors are fit errors only. Right Panel: The quadrupole spectrum data $Q(y_t, b)$ and fit $Q_0(y_t)$. The dotted curve is a zero-viscosity limit of a viscous hydro theory defined in [13].

There is little centrality dependence to the quantity $Q(y_t, b)$. A single Lévy distribution $Q_0(y_t)$ can describe the data within 10% with a few simple parameters: a y_t boost $\Delta y_{t0} = 0.58$ and parameters $T_2 = 0.09$ GeV and $n_2 = 13.8$. There is deviation from this curve for very central collisions at low-p_t. A similar analysis was made in [12] for minimum-bias identified particles without centrality dependence and the observed boost values agree. It is notable that the boost is a single sharp value and not a distribution as would be expected from transverse Hubble expansion.

A related and more accurate parametrization has been made. At higher p_t it is observed that $v_2\{2\mathrm{D}\}(p_t, b)$ is proportional to $p_t \exp(-p_t/4)$ so we derive the form:

$$v_2\{2\mathrm{D}\}(p_t, b) \approx \langle 1/p_t \rangle v_2\{2\mathrm{D}\}(b) p_t \exp(-p_t/4) \times f(p_t, b), \quad (2)$$

where $\langle 1/p_t \rangle \approx 2.1\,(\mathrm{GeV}/c)^{-1}$ and $f(p_t, b)$ is a dimensionless factor necessary for the full p_t and centrality dependence. It is fit to data to get $f(p_t, b) = 1 + C(b)[\mathrm{erf}(y_t - 1.2) - \mathrm{erf}(1.8 - 1.2)]$ and $C(b) = 0.12 - (\nu - 3.4)/5 - [(\nu - 3.4)/2]^5$.

Using 2D correlations we observe $v_2(p_t)$ for a wide range of centralities without significant "nonflow" contamination and conclude that published $v_2(p_t, b)\{\mathrm{EP}\}$ results [9] contain a significant contribution from the same-side peak correlations.

We have measured the quadrupole on y_t as a function of b and deduced the quadrupole spectrum variation with centrality. This is the first complete factorization of the scaling properties of the quadrupole. A detailed study of the systematics is in progress. When the quadrupole spectrum is isolated we also observe a single b-independent y_t boost value that is consistent with the minimum-bias results for different particle species [12]. We developed an accurate parametrization of $v_2\{2\mathrm{D}\}$ over a large kinetic domain and surprisingly simple trends are observed.

[1] J. Adams *et al.* (STAR Collaboration), Nucl. Phys. A **757**, 102 (2005).
[2] D. Teaney, J. Lauret, and E.V. Shuryak, Phys. Rev. Lett. **86**, 4783 (2001).
[3] A.M. Poskanzer, S.A. Voloshin, Phys. Rev. C **58**, 1671 (1998).
[4] D.T. Kettler (STAR Collaboration), Eur. Phys. J. C **62**, 175 (2008).
[5] N. Borghini, P.M. Dinh and J.Y. Ollitraut, Phys. Rev. C **64**, 054901 (2001).
[6] T.A. Trainor and D.T. Kettler, Int. J. Mod. Phys. E **17**, 1219 (2008).
[7] M. Daugherity (STAR Collaboration), J. Phys. G **35**, 104090 (2008).
[8] T.A. Trainor, Phys. Rev. C **80**, 044901 (2009).
[9] J. Adams *et al.* (STAR Collaboration), Phys. Rev. C **72**, 014904 (2005).
[10] J. Adams *et al.* (STAR Collaboration), Phys. Rev. C **73**, 64907 (2006).
[11] T.A. Trainor, R.J. Porter, D.J. Prindle, J. Phys. G **31**, 809 (2005).
[12] T.A. Trainor, Phys. Rev. C **78**, 064908 (2008).
[13] P. Romatschke and U. Romatschke, Phys. Rev. Lett. **99**, 172301 (2007).

Journal of Physics: Conference Series **270** (2011) 012059 doi:10.1088/1742-6596/270/1/012059

Non-equilibrium fluctuations at the QCD phase transition

M. Nahrgang, M. Bleicher

Institut für Theoretische Physik & Frankfurt Institute for Advanced Studies (FIAS), Goethe-Universität, Max-von-Laue-Str. 1, D-60438 Frankfurt am Main, Germany

E-mail: nahrgang@th.physik.uni-frankfurt.de

Abstract. We study the chiral phase transition by the non-equilibrium propagation of the sigma field. A quark fluid acts as a heat bath in local thermal equilibrium and evolves fluid dynamically. We allow for dissipative processes and fluctuations since the sigma field is propagated according to a Langevin-type equation of motion. Non-equilibrium fluctuations at the first order phase transition lead to an increase in the intensity of sigma excitations.

1. Introduction

The QCD phase transition is subject to intensive theoretical and experimental research. Besides theoretical indications of a QCD critical point [1, 2] experimental signatures are proposed to discover the suggested critical point in experiment [3, 4]. The fast dynamics during the expansion of the system created in a collision makes it important to consider non-equilibrium phenomena. At a critical point relaxation times become infinite. A system that cools rapidly through a second order phase transition is necessarily driven out of equilibrium. Any signal for a critical point is crucially weakened [5]. A non-equilibrium situation, however, leads to interesting phenomena at a first order phase transition, such as nucleation [6] and spinodal decomposition [7]. We embed a field theoretical model with a phase transition into a realistic expansion of the hot and dense matter and extend similar models [8, 9] by a complete treatment of fluctuations and dissipation of the chiral order parameter coupled to the fluid dynamic expansion. While the chiral field is damped energy dissipates into the quark fluid. Random kicks by the noise term also contribute to the energy and momentum exchange between the field and fluid. We introduce a stochastic source term into the fluid dynamic expansion to conserve energy of the whole system.

2. Chiral fluid dynamics

2.1. Semiclassical equation of motion for the sigma field including dissipation and noise terms

We use the linear sigma model with consituent quarks as a low energy effective model of QCD. It exhibits a chiral phase transition at different strengths, a crossover, a critical point and a first order phase transition [10]

$$\mathcal{L} = \overline{q}\left[\mathrm{i}\gamma\partial_\mu - \mathrm{g}\left(\sigma + \mathrm{i}\gamma_5\tau\vec{\pi}\right)\right]q + \frac{1}{2}(\partial_\mu\sigma\partial^\mu\sigma) + \frac{1}{2}(\partial_\mu\vec{\pi}\partial^\mu\vec{\pi}) - U(\sigma,\vec{\pi})\,, \tag{1}$$

with the constituent quark field $q = (u, d)$, the coupling g between the quarks and the chiral fields. The potential is given by

$$U(\sigma, \vec{\pi}) = \frac{\lambda^2}{4}\left(\sigma^2 + \vec{\pi}^2 - \nu^2\right)^2 - h_q\sigma - U_0\,, \tag{2}$$

where $\langle\sigma\rangle = f_\pi = 93$ MeV and $\langle\vec{\pi}\rangle = 0$. The explicit symmetry breaking term is $h_q\sigma = f_\pi m_\pi^2$ with $m_\pi = 138$ MeV. Thus $\nu^2 = f_\pi^2 - m\pi^2/\lambda^2$. $\lambda^2 = 20$ yields a sigma mass $m_\sigma^2 = 2\lambda^2 f_\pi^2 + m_\pi^2 \approx$ 604 MeV. For zero potential energy in the ground state $U_0 = m_\pi^4/(4\lambda^2) - f_\pi^2 m_\pi^2$.

In the rest of this paper we will concentrate on the order parameter of the chiral phase transition, the sigma field and set $\pi = \langle\pi\rangle = 0$ for all times. The equation of motion for the sigma field including dissipation and noise is given by

$$\partial_\mu\partial^\mu\sigma + \frac{\delta U}{\delta\sigma} + g\langle\bar{q}q\rangle_\sigma + \eta\partial_t\sigma = \xi\,. \tag{3}$$

Here, the chiral condensate to one-loop level is

$$\langle\bar{q}q\rangle_\sigma = 2gd_q\sigma\int\frac{d^3p}{(2\pi)^3}\frac{1}{E}n_F(E) \tag{4}$$

$n_F(p)$ is the Fermi-Dirac distribution, $d_q = 12$ the degeneracy factor and $E = \sqrt{p^2 + g^2\sigma^2}$ the energy of the quarks.

Our choice of the damping coefficient η and the noise term ξ is motivated by the results of two-loop calculations in a simplified $\mathcal{O}(4)$ chiral model [11, 12]. In a first approximation we assume that η is independent of temperature and mass changes and that the noise is white. Inspired by [13] we use $\eta = 2.2$/fm. The dissipation-fluctuation theorem then gives

$$\langle\xi(t)\rangle = 0 \quad\text{and}\quad \langle\xi(t)\xi(t')\rangle = \frac{2T}{V}\eta\delta(t - t')\,. \tag{5}$$

2.2. The effective potential and the equation of state
The pressure of the quark fluid is given by the thermodynamic potential. To one-loop level it is

$$V_{\text{eff}}(\sigma, \vec{\pi}, T) = -\frac{T}{V}\log\mathcal{Z} = -2d_qT\int\frac{d^3p}{(2\pi)^3}\log(1 + e^{-\frac{E}{T}}) + U(\sigma, \vec{\pi})\,. \tag{6}$$

At $\mu_B = 0$ the strength of the phase transition can be tuned by varying the strength of the coupling g. For a small coupling the transition is a crossover, for $g = 3.63$ the potential becomes flat as for a continuous transition. And for $g = 5.5$ the phase transition is discontinuous.

The equation of state for the fluid dynamic expansion is obtained from

$$p(\sigma, \vec{\pi}, T) = -V_{\text{eff}}(\sigma, \vec{\pi}, T) + U(\sigma, \vec{\pi}) \tag{7}$$

$$e(\sigma, \vec{\pi}, T) = T\frac{\partial p(\sigma, \vec{\pi}, T)}{\partial T} - p(\sigma, \vec{\pi}, T)\,. \tag{8}$$

2.3. The stochastic source term
The equations of relativistic fluid dynamics are

$$\partial_\mu(T^{\mu\nu}_{\text{fluid}} + T^{\mu\nu}_{\text{field}}) = 0 \tag{9}$$

By using the explicit form of the equation of motion (3) we derive for the source term

$$S^\nu = -\partial_\mu T^{\mu\nu}_{\text{field}} = -(\partial_\mu\partial^\mu\sigma + \frac{\delta U}{\delta\sigma})\partial^\nu\sigma = -(-g\langle\bar{q}q\rangle_\sigma - \eta\partial_t\sigma + \xi)\partial^\nu\sigma \tag{10}$$

The damping and the noise term enter the source term which is therefore also stochastic.

3. Numerical results

We solve the fluid dynamic equations (9) by a SHASTA code and the field equations (3) by a staggered leap-frog algorithm. The numerical equilibrium solution of equations like (3) depends on the lattice spacing [14]. It turns out, however, that the results presented here depend only slightly on the lattice spacing, for which we use $\Delta = 0.2$ fm. The energy density is initiated as the equilibrium energy density at $T = 160$MeV, ellipsoidal in x-y-plane and homogenous in z-direction, smoothed by a Woods Saxon type distribution. The sigma field is initiated in equilibrium with the quark fluid. All observations shown below are purely due to non-equilibrium effects at the phase transition.

The intensity of the sigma fluctuations is given by

$$\frac{\mathrm{d}N_\sigma}{\mathrm{d}^3k} = \frac{a_k^\dagger a_k}{(2\pi)^3 2\omega_k} = \frac{1}{(2\pi)^3 2\omega_k}(\omega_k^2|\sigma_k|^2 + |\dot{\sigma}_k|^2) \tag{11}$$

The time evolution of (11) is shown in figure (1) for a first order phase transition and in figure (3) for a critical point.

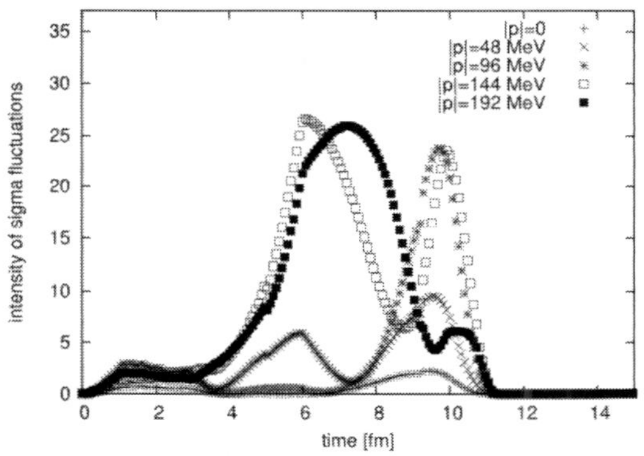

Figure 1. Time evolution of the intensity of sigma fluctuations at a first order phase transition.

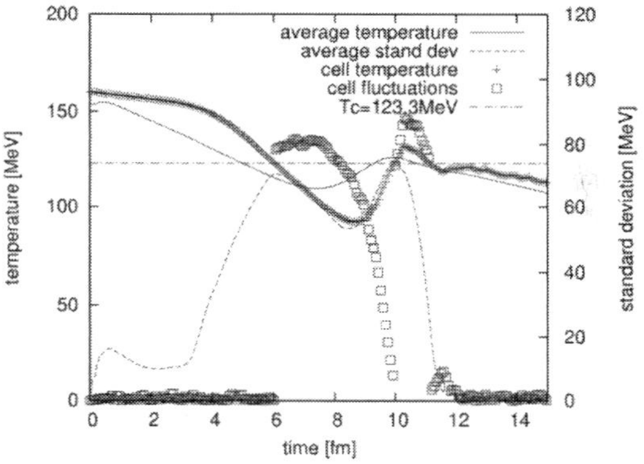

Figure 2. Temperature of the quark fluid and standard deviation of the sigma field at a first order phase transition.

In a first order phase transition the intensity of sigma fluctuations between 4 and 11 fm is much larger than at a critical point. This can be better understood by looking at the evolution of the temperature and the size of the fluctuations. We show these values in a single cell in the middle of the grid and averaged over a volume in the hot region of the fireball ($V = 0.24$ fm^3). The standard deviation of the fluctuations of the sigma field in V is $\Delta\sigma^2 = 1/V \sum_V \Delta^3(\sigma_{\mathrm{eq}} - \sigma(x))^2$. For a first order phase transition the intensity of sigma fluctuations increases around $t = 5$ fm when the average temperature drops below the critical temperature, in figure (2). Here the standard deviation increases to almost 70 MeV. In one particular cell in the middle of the hot region one sees that the size of the fluctuation corresponds to the difference between the two minima at $\langle\sigma\rangle_{T>T_C} \simeq 10$ MeV and $\langle\sigma\rangle_{\mathrm{vac}} = f_\pi = 93$ MeV. Obviously the fluctuations in a critical point scenario, figure (4), are smaller. For a first order phase transition we find a second rise in the intensity of sigma fluctuations at around $t = 10$ fm. While the sigma field relaxes from the unstable minimum towards its vacuum value energy is transfered to the quark fluid where re-heating occurs. This effect is so strong that the average temperature again crosses the critical temperature leading to an increase in the fluctuations. Thereafter the sigma field relaxes slowly towards $\langle\sigma\rangle_{\mathrm{vac}}$.

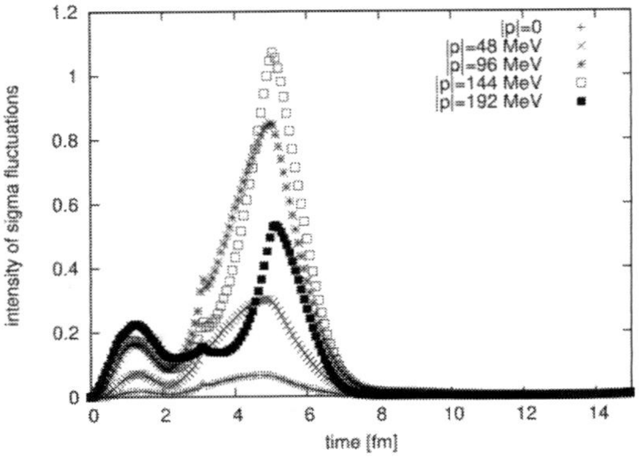

Figure 3. Time evolution of the intensity of sigma fluctuations at a critical point.

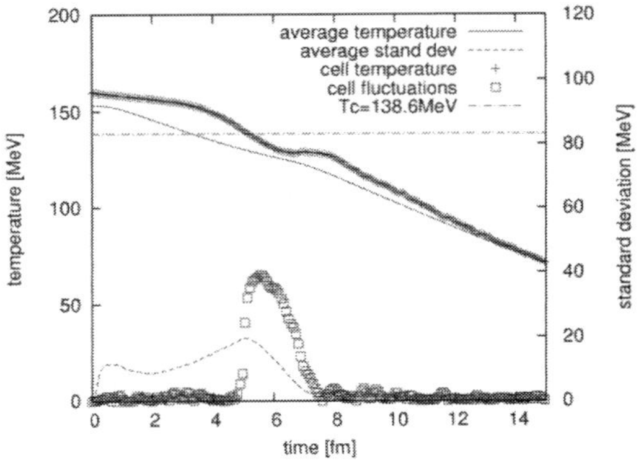

Figure 4. Temperature of the quark fluid and standard deviation of the sigma field at a critical point.

4. Summary
We presented a dynamic model of phase transitions in heavy ion collisions. We included damping and noise originating from the interaction of the chiral fields with the quark fluid. To ensure energy and momentum conservation of the coupled system we suggested to introduce a stochastic source term into the equations of fluid dynamic. The number of coherently produced soft sigmas is much larger at a first order phase transition than compared to a critical point transition. Non-equilibrium effects like supercooling and reheating are observed.

4.1. Acknowledgments
The authors thank Carsten Greiner, Stephan Leupold, Igor Mishustin and Christoph Herold for fruitful discussions and Dirk Rischke for providing the hydro code. This work was supported by the Hessian LOEWE initiative Helmholtz International Center for FAIR. M. Nahrgang acknowledges finanical support from the Stiftung Polytechnische Gesellschaft Frankfurt.

References
[1] Z. Fodor and S. D. Katz, JHEP **0203** (2002) 014
[2] C. Schmidt [for RBC-Bielefeld Collaboration], J. Phys. G **35** (2008) 104093
[3] M. A. Stephanov, K. Rajagopal and E. V. Shuryak, Phys. Rev. Lett. **81** (1998) 4816
[4] M. A. Stephanov, K. Rajagopal and E. V. Shuryak, Phys. Rev. D **60** (1999) 114028
[5] B. Berdnikov and K. Rajagopal, Phys. Rev. D **61**, 105017 (2000)
[6] Ph.Chomaz, M.Colonna, J.Randrup Phys. Rev. **389** (2004) 263
[7] I. N. Mishustin, Phys. Rev. Lett. **82** (1999) 4779
[8] I. N. Mishustin and O. Scavenius, Phys. Rev. Lett. **83** (1999) 3134
[9] K. Paech, H. Stoecker and A. Dumitru, Phys. Rev. C **68** (2003) 044907
[10] O. Scavenius, A. Mocsy, I. N. Mishustin and D. H. Rischke, Phys. Rev. C **64** (2001) 045202
[11] Z. Xu and C. Greiner, Phys. Rev. D **62** (2000) 036012
[12] C. Greiner and B. Muller, Phys. Rev. D **55** (1997) 1026
[13] T. S. Biro and C. Greiner, Phys. Rev. Lett. **79** (1997) 3138
[14] N. C. Cassol-Seewald, R. L. S. Farias, E. S. Fraga, G. Krein, R. O. Ramos arXiv:0711.1866 [hep-ph]

Hot Quarks 2010 IOP Publishing
Journal of Physics: Conference Series **270** (2011) 012060 doi:10.1088/1742-6596/270/1/012060

Modeling the jet quenching, thermal resonance production and hydrodynamical flow in relativistic heavy ion collisions

I P Lokhtin[1,*], A V Belyaev[1], L V Malinina[1], S V Petrushanko[1], A M Snigirev[1], I Arsene[2] and E E Zabrodin[3,1]

[1] D.V. Skobeltsyn Institute of Nuclear Physics, M.V. Lomonosov Moscow State University, Moscow, Russia

[2] Extreme Matter Institute EMMI, GSI Helmholtzzentrum fur Schwerionenforchung GmbH, Darmstadt, Germany

[3] The Department of Physics, University of Oslo, Norway

E-mail: * `Igor.Lokhtin@cern.ch`

Abstract.
Physical observables in relativistic heavy ion collisions are determined by various multi-particle production mechanisms. The simultaneous model treatment of different collective nuclear effects at high energies (such as a hard multi-parton fragmentation in hot QCD-matter, thermal resonance production, hydrodynamical flows, etc.) is actually a rather complicated task. We discuss the simulation of the above effects by means of the Monte-Carlo model HYDJET++.

1. Introduction

Ongoing and future experimental studies of relativistic heavy ion collisions in a wide range of beam energies require the development of new Monte-Carlo (MC) event generators and improvement of existing ones. A realistic MC event generator should include a maximum possible number of observable physical effects which are important to determine the event topology: from the bulk properties of soft hadroproduction (domain of low transverse momenta $p_T < 1 \div 2$ GeV/c) such as thermal resonance production and collective flows, to hard multi-parton fragmentation in hot and dense QCD-matter, which reveals itself in the spectra of high-p_T particles and hadronic jets. The HYDJET++ event generator [1, 2] includes detailed treatment of soft hadroproduction as well as hard multi-parton production, and takes into account medium-induced parton rescattering and energy loss. The heavy ion event in HYDJET++ is the superposition of two independent components: the soft, hydro-type state and the hard state resulting from multi-parton fragmentation. Note that a conceptually similar approximation has been developed in [3, 4]. The HYDJET++ model is the development and continuation of the HYDJET event generator [5], and it contains important additional features for the soft component: resonance decays and more detailed treatment of thermal and chemical freeze-out hypersurfaces [6, 7]. The details on physics model and simulation procedure can be found in the HYDJET++ manual [1], the main features of the model being listed only very briefly below.

2. HYDJET++ model

The model for the hard multi-parton part of a HYDJET++ event is the same as for the HYDJET event generator, and it is based on the PYQUEN partonic energy loss model [5]. The approach to the description of multiple scattering of hard partons in the dense QCD-matter (such as quark-gluon plasma) is based on the accumulated energy loss via the gluon radiation being associated with each parton scattering in the expanding quark-gluon fluid and includes the interference effect (for the emission of gluons with a finite formation time) using the modified radiation spectrum dE/dl as a function of decreasing temperature T. The model takes into account radiative and collisional energy loss of hard partons in longitudinally expanding quark-gluon fluid, as well as realistic nuclear geometry. The event generator for single hard nucleon-nucleon sub-collisions PYQUEN was constructed as a modification of the jet event obtained with the generator of hadron-hadron interactions PYTHIA_6.4 [8]. The event-by-event simulation procedure in PYQUEN includes *1)* generation of initial parton spectra with PYTHIA and production vertices at given impact parameter; *2)* rescattering-by-rescattering simulation of the parton path in a dense zone and its radiative and collisional energy loss; *3)* final hadronization according to the Lund string model for hard partons and in-medium emitted gluons. Then the PYQUEN multi-jets generated according to the binomial distribution are included in the hard part of the event. The mean number of jets produced in an AA event is the product of the number of binary NN sub-collisions at a given impact parameter and the integral cross section of the hard process in NN collisions with the minimum transverse momentum transfer $p_T^{\rm min}$. In order to take into account the effect of nuclear shadowing on parton distribution functions, the impact parameter dependent parameterization obtained in the framework of Glauber-Gribov theory [9] is used.

The soft part of a HYDJET++ event is the "thermal" hadronic state generated on the chemical and thermal freeze-out hypersurfaces obtained from the parametrization of relativistic hydrodynamics with preset freeze-out conditions (the adapted C++ code FAST MC [6, 7]). Hadron multiplicities are calculated using the effective thermal volume approximation and Poisson multiplicity distribution around its mean value, which is supposed to be proportional to the number of participating nucleons for a given impact parameter of a AA collision. The fast soft hadron simulation procedure includes *1)* generation of the 4-momentum of a hadron in the rest frame of a liquid element in accordance with the equilibrium distribution function; *2)* generation of the spatial position of a liquid element and its local 4-velocity in accordance with phase space and the character of motion of the fluid; *3)* the standard von Neumann rejection/acceptance procedure to account for the difference between the true and generated probabilities; *4)* boost of the hadron 4-momentum in the center of mass frame of the event; *5)* the two- and three-body decays of resonances with branching ratios taken from the SHARE particle decay table [10]. The high generation speed in HYDJET++ is achieved due to almost 100% generation efficiency of the "soft" part because of the nearly uniform residual invariant weights which appear in the freeze-out momentum and coordinate simulation.

Note that although HYDJET++ is optimized for very high energies of RHIC and LHC colliders (c.m.s. energies of heavy ion beams $\sqrt{s} = 200$ and $2760 \div 5500$ GeV per nucleon pair respectively), it can also be used for studying the particle production in a wider energy range down to $\sqrt{s} \sim 10$ GeV per nucleon pair at future facilities FAIR and NICA. As one moves from very high to moderately high energies, the contribution of the hard part of the event becomes smaller, while the soft part turns into just a multi-parameter fit to the data.

3. Some applications of HYDJET++ at RHIC and LHC

It was demonstrated in [1] that the HYDJET++ model can describe the bulk properties of the hadronic state created in Au+Au collisions at RHIC at $\sqrt{s} = 200A$ GeV (hadron spectra and ratios, radial and elliptic flow, femtoscopic momentum correlations), as well as the high-

p_T hadron spectra. A number of input parameters of the model can be fixed from fitting the RHIC data to various physical observables. For example, the thermodynamical potentials and the chemical freeze-out temperature $T^{\rm ch} = 0.165$ GeV have been fixed in HYDJET++ from fitting the RHIC data to hadron ratios near mid-rapidity in central Au+Au collisions. The slopes of transverse mass m_T hadron spectra (π^+, K^+ and p with $m_T < 0.7$ GeV/c^2) near mid-rapidity at different centralities spectra allow the thermal freeze-out temperature $T^{\rm th} = 0.1$ GeV and the maximal radial flow rapidity in central collisions $\rho_{\rm max}(b=0) = 1.1$ to be fixed. The space-time parameters of the thermal freeze-out region can be fixed by means of fitting the three-dimensional correlation functions measured for $\pi^+\pi^+$ pairs and extracting the correlation radii $R_{\rm side}$, $R_{\rm out}$ and $R_{\rm long}$. Rapidity spectra of charged hadrons at different centralities allow us to fix the particle densities in the mid-rapidity region and the maximum longitudinal flow rapidity $\eta_{\rm max} = 3.3$. Since mean "soft" and "hard" hadron multiplicities depend on the centrality in different ways, the relative contribution of soft and hard parts to the total event multiplicity can be fixed through the centrality dependence of $dN/d\eta$. High transverse momentum hadron spectra ($p_T > 2 \div 4$ GeV/c) are sensitive to parton production and jet quenching effect, therefore fitting the measured high-p_T tail allows the extraction of PYQUEN energy loss model parameters. The momentum and azimuthal anisotropy parameters are estimated for different centrality sets by fitting the measured transverse momentum dependence of the elliptic flow coefficient, $v_2(p_T)$.

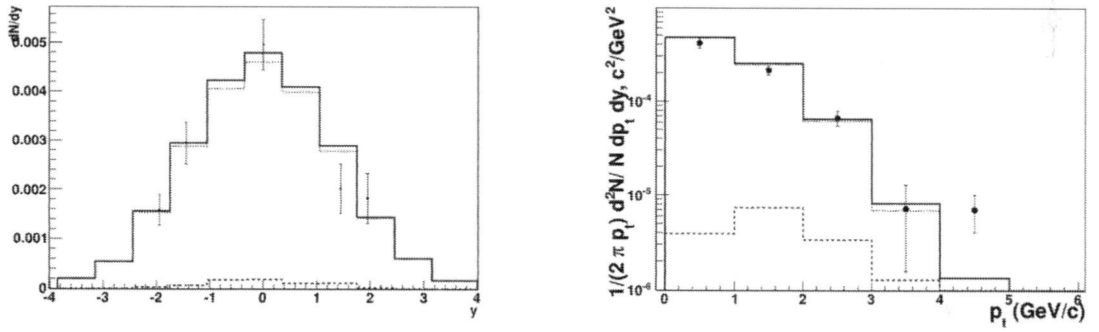

Figure 1. The pseudorapidity (left) and the transverse momentum (right) spectra of $J/\psi-$mesons in Au+Au collisions at $\sqrt{s} = 200A$ GeV for $0 \div 20\%$ centrality. The points are PHENIX data [13], histograms are the HYDJET++ calculations for $\gamma_c = 7$ and $T^{\rm th}(J/\psi) = T^{\rm ch} = 0.165$ GeV (solid – total, dotted – soft component, dashed – hard component).

Thermal charm hadron production was implemented in HYDJET++ recently. D, J/ψ and Λ_c hadrons are generated within the statistical hadronization model [11, 12]. Momentum spectra of charm hadrons are computed according to the thermal distribution, and the multiplicities N_c ($C = D, J/\psi, \Lambda_c$) are calculated through the corresponding thermal numbers $N_c^{\rm th}$ as

$$N_c = \gamma_c^{\rm nc} N_c^{\rm th} \,, \tag{1}$$

where γ_c is the charm enhancement factor (or charm fugacity), and N_c is the number of charm quarks in a hadron C. The fugacity γ_c can be treated as a free parameter of the model or calculated through the number of charm quark pairs obtained from perturbative QCD. HYDJET++ fits measured by PHENIX J/ψ yields in central Au+Au collisions [13] with $\gamma_c = 7$. If we assume that thermal freeze-out for J/ψ-mesons happens at the same temperature as for light hadrons, $T^{\rm th} = 0.1$ GeV ($\eta_{\rm max} = 3.3$, $\rho_{\rm max} = 1.1$), then simulated $y-$ and p_T-spectra are much wider than the data. However if we assume that thermal freeze-out for J/ψ-mesons occurs at the same temperature as chemical freeze-out, $T^{\rm th}(J/\psi) = T^{\rm ch} = 0.165$ GeV, then simulated spectra are capable of fitting the PHENIX data for the values of maximal longitudinal and radial flow rapidities $\eta^{\rm ch}_{\rm max} = 1.1$ and $\rho^{\rm ch}_{\rm max} = 0.5$ respectively (see Fig. 1).

The heavy ion collision energy at LHC will be a factor of $\sim 15 \div 30$ larger then that in RHIC, thereby allows one to probe new frontiers of super-high temperature and (almost) net-baryon free QCD. It is expected that at such ultra-high energies the role of hard and semi-hard particle production may be significant even for the bulk properties of created matter. In particular, the spectacular predictions of HYDJET++ are possible reducing the femtoscopic correlation radii [14] and elliptic flow [15] in heavy ion collisions as one moves from RHIC to LHC energies due to the significant contribution of (semi)hard component to the space-time structure of the hadron emission source.

4. Summary

Among other heavy ion event generators, the Monte-Carlo model HYDJET++ focuses on the detailed simulation of jet quenching effect based on the partonic energy loss model PYQUEN, and also reproduces the main features of nuclear collective dynamics by the parametrization of relativistic hydrodynamics with preset freeze-out conditions (including resonance decays, and separate treatment of thermal and chemical freeze-out hypersurfaces). Thus the final hadronic state in HYDJET++ represents the superposition of two independent components: medium-modified hard multi-parton fragmentation and soft hydro-type part. HYDJET++ is capable of reproducing the bulk properties of heavy ion collisions at RHIC (hadron spectra and ratios, radial and elliptic flow, femtoscopic momentum correlations), as well as hard probes (high-p_T hadron and J/ψ spectra). The simulations for LHC (and also at lower energies of future facilities FAIR and NICA) are in progress.

Acknowledgments

We would like to thank L.V. Bravina, D. d'Enterria, G.Kh. Eyyubova, A.M. Gribushin, V.L. Korotkikh, R. Lednicky, C. Loizides, L.I. Sarycheva, Yu.M. Sinyukov and K. Tywoniuk for joint work and fruitful discussions. I.L. and S.P. wish to express the gratitude to the organizers of the Workshop "Hot Quarks 2010" for the warm welcome and the hospitality. This work was supported by Russian Foundation for Basic Research (grants Nos 08-02-91001, 08-02-92496 and 10-02-93118), Russian Ministry for Education and Science (contract 02.740.11.0244), Grant of President of Russian Federation (No 4142.2010.2) and Dynasty Foundation.

References

[1] Lokhtin I P, Malinina L V, Petrushanko S V, Snigirev A M, Arsene I and Tywoniuk K 2009 *Comput. Phys. Commun.* **180** 779
[2] *http://cern.ch/lokhtin/hydjet++*
[3] Hirano T and Nara T 2004 *Nucl. Phys.* **743** 305
[4] Armesto N, Cacciari M, Hirano T, Nagle J L and Salgado C A 2010 *J. Phys.* G **37** 025104
[5] Lokhtin I P and Snigirev A M 2006 *Eur. Phys. J.* C **45** 211
[6] Amelin N S et al. 2006 *Phys. Rev.* C **74** 064901
[7] Amelin N S et al. 2008 *Phys. Rev.* C **77** 014903
[8] Sjostrand T, Mrenna S and Skands P 2006 *JHEP* **0605** 026
[9] Tywoniuk K, Arsene I C, Bravina L, Kaidalov A B and Zabrodin E 2007 *Phys. Lett.* B **657** 170
[10] Torrieri G et al. 2005 *Comput. Phys. Commun.* **167** 229
[11] Andronic A, Braun-Munzinger P, Redlich K and Stachel J 2003 *Phys. Lett.* B **571** 36
[12] Andronic A, Braun-Munzinger P, Redlich K and Stachel J 2007 *Nucl. Phys.* A **789** 334
[13] Adare A et al. (PHENIX Collaboration) 2007 *Phys. Rev. Let.* **98** 232301
[14] Lokhtin I P, Malinina L V, Petrushanko S V, Snigirev A M, Arsene I and Tywoniuk K 2009 *PoS* **High-pTphysics09** 023 (*Preprint* arXiv:0903.0525)
[15] Eyyubova G et al. 2009 *Phys. Rev.* C **80** 064907

Collective Flows in a Transport Approach

S. Plumari[1,2], V. Baran[3], M. Di Toro[1,2] and V. Greco[1,2]

Department of Physics and Astronomy, University of Catania, Via S. Sofia 64, I-95125 Catania
Laboratorio Nazionale del Sud, INFN-LNS, Via S. Sofia 63, I-95125 Catania
Physics Faculty, University of Bucharest and NIPNE-HH, Bucharest, Romania

E-mail: greco@lns.infn.it

Abstract. We introduce a transport approach at fixed shear viscosity to entropy ratio η/s to study the generation of collective flows in ultra-relativistic heavy-ion collisions. Transport theory supplies a covariant approach valid also at large η/s and at intermediate transverse momentum p_T, where deviations from equilibrium is no longer negligible. Such an approach shows that at RHIC energies a temperature dependent η/s enhances significantly the v_4/v_2^2 respect to the case of constant η/s. Furthermore if NJL chiral dynamics is self-consistently implemented we show that it does not modify the relation between v_2 and η/s.

1. Introduction

The RHIC program at BNL has shown that the azimuthal asymmetry in momentum space, namely the elliptic flow v_2, is the largest ever seen in HIC suggesting that an almost perfect fluid with a very small shear viscosity to entropy density ratio, η/s, has been created [1, 2]. Nonetheless it has been found that dissipative effects cannot be neglected and even a small shear viscosity to entropy ratio η/s produce sizeable effect increasing with the transverse momentum p_T of the particles [3]. This has triggered a lot of activity in developing a relativistic theory of viscous hydrodynamics. Viscous corrections to ideal hydrodynamics are indeed large and a simple relativistic extension of first order Navier Stokes equations is affected by causality and stability pathologies [5]. It is therefore necessary to go to second order gradient expansion, and in particular the Israel-Stewart theory has been implemented to simulate the RHIC collisions providing an upper bound for $\eta/s \leq 0.4$ [4]. Such an approach, apart from the limitation to 2+1D simulations, has the more fundamental problem of a limited range of validity in η/s and in the transverse momentum p_T.

On the contrary a relativistic transport approach has the advantage to be a 3+1D approach not based on a gradient expansion in viscosity that is valid also for large η/s and for out of equilibrium momentum distribution allowing a reliable description also of the intermediate p_T range where the important property of quark number scaling (QNS) of $v_2(p_T)$ has been observed [10]. In this p_T region viscous hydrodynamics breaks its validity because the relative deviation of the equilibrium distribution function $\delta f/f_{eq}$ increases with p_T^2 becoming large already at $p_T \geq 3T \sim 1GeV$ [4]. Furthermore Boltzmann-Vlasov transport theory distinguishes between the short range interaction associated to collisions and long range interaction associated to the field interaction, responsible for the change of the Equation of State (EoS) respect to that of a free gas. This last feature allows to unify two main ingredients that are relevant for the formation of collective flow. In ideal hydrodynamics the $v_2(p_T)$ depends strongly on the EoS namely on the sound velocity $c_s^2 = dP/d\epsilon$ [8], while the mean free path λ is assumed to be vanishing. In the parton cascade approach the EoS is fixed to be the one of a free gas $c_s^2 = 1/3 = P/\epsilon$, on the other hand the mean free path $\lambda = 1/\rho\sigma$ is finite. In the first stage of RHIC the two different approaches were able to account for the large $v_2(p_T)$ observed; in particular the parton cascade with large scattering cross section predicted the saturation of $v_2(p_T)$ vs p_T [9]. Anyway at finite viscosity both λ and the EoS are important for the generation of the momentum anisotropies.

The basic equation of transport for the (anti-) quark phase-space distribution function $f^{\pm}$ under a scalar field interaction can be written as:

$$p^{\mu}\partial_{\mu} f^{\pm}(x,p) + M(x)\partial_{\mu}M(x)\partial_p^{\mu} f^{\pm}(x,p) = \mathcal{C}(x,p) \tag{1}$$

where the first term is related to the free streaming, the second term represents the effect of a scalar field modifying the $\epsilon = 3P$ relation (giving a finite interaction measure, see Fig.2 (left)) and the last term describes the collision dynamics associated to a finite λ, hence a finite η/s.

In the following we will show two results obtained with such an approach: one related with the large v_4/v_2^2 ratio observed at RHIC and the other discussing the role of the NJL chiral dynamics on the relation between $v_2(p_T)$ and η/s. We will show calculations for $Au + Au$ at 200 AGeV; the density distribution in coordinate space is given by the standard Glauber model. The maximum initial temperature is $T = 340$ MeV and the initial time is $\tau_0 = 0.6$ fm/c as usually done also in hydrodynamical calculations.

2. The large v_4/v_2^2 ratio

We first consider the reduction of the transport approach to a cascade, i.e. neglect the field interaction, included in the next section. In such a case the mass $M(x) = 0$ in Eq.(1). It has been shown that a fluid at finite η/s can be studied within a transport approach renormalizing locally the scattering cross section according to $\eta/s \cdot \sigma_{tr} =< p > /15\rho$, for details see Ref.s [6, 7, 5]. The system of course expands and cools and when the energy density is below $\epsilon \sim 2GeV/fm^3$ it starts the cross-over region and then the hadronic stage. During this phase we increase the η/s of the medium according to Fig. 1 (left) in order to have finally in the hadronic phase an $\eta/s = 8/4\pi \sim 0.65$ in agreement with several calculation of η/s.

At RHIC it has been observed $v_4(p_T)/v_2^2(p_T)$ ratio of about 0.8-0.9 [11] quite larger than first ideal hydrodynamical predictions of about 0.5. More realistic hydrodynamical calculations show that the ratio can be even lower than 0.5 and viscosity does not strongly affect such a ratio but can bring it up to about 0.6. Furthermore in Ref.[12] it has been shown that such a ratio is almost independent on the initial eccentricity, on the viscosity, on the particle species. Therefore even if v_4 and v_2 can depend on several details such a ratio is a quite stable one and its large value should be considered as an oppurtunity to have some new information on the system not yet spot. In Ref.[7] we have shown that indeed the formation time of v_4 is larger than the v_2 one and therefore it is more affected by dynamics of the later stage of the reaction. Thanks to the difference in the formation time we argue that such a ratio is significantly affected by the temperature dependence of the η/s. Therefore we have performed two calculations one with a constant $4\pi\eta/s = 1$ in the QGP phase (dashed line) and the other, shown by solid line in Fig. 1 (left), with a T or ϵ energy density dependence $4\pi\eta/s = T/T_0 = (\epsilon/\epsilon_0)^{1/4}$ where $\epsilon_0 = 1.7\,GeV/fm^3$ gives the beginning of the cross over region. The plot in Fig. 1 (left) is just a simple sketch of a possible realistic behavior but does not corresponds to any attempted calculations. However we remind that an increase of η/s is expected moving away from the phase transition region on general considerations [13].

In Fig.1 (right) we show the results of the two calculations for $Au + Au$ at 200 AGeV. For a constant η/s we find a value of about 0.6 while in the case of a temperature dependent η/s, it increases getting very close to experimental data, $v_4/v_2^2 \sim 0.8$. The reason for the sensitivity of this ratio on the T dependence is that the elliptic flow v_2 mainly develops earlier respect to v_4

Journal of Physics: Conference Series **270** (2011) 012061 doi:10.1088/1742-6596/270/1/012061

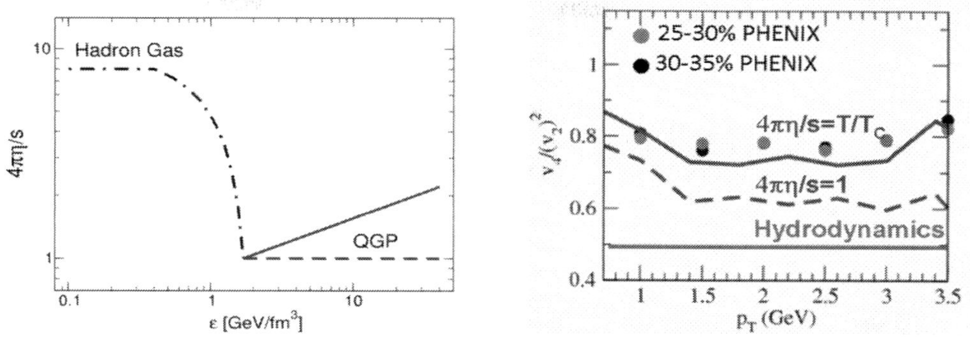

Figure 1. Left: Dependence of η/s on the energy density ϵ implemented in the transport code. Right: v_4/v_2^2 for $Au + Au$ collisions in the mid-rapidity region $|y| < 1$ at b= 8 fm; dashed line is the result for a constant $4\pi\eta/s = 1$, the solid line is the result for a temperature dependence of the η/s like shown by solid line on the left figure.

and therefore on the average at larger η/s (for the solid line case), while v_4 develops more when the η/s is smaller hence is less suppressed by dissipative effects.

3. Impact of NJL Dynamics

To start the investigation of the role of the field interaction we have implemented the NJL dynamics that introduces the generation of a finite mass associated to chiral symmetry breaking. Therefore particles do not have a vanishing mass but a finite one according to the mass gap equation of the NJL model:

$$\frac{M(x) - m}{4\,g\,N_c} = M(x) \int \frac{d^3p}{(2\pi)^3} \frac{1 - f^-(x,p) - f^+(x,p)}{E_p(x)} \tag{2}$$

that determines the local mass $M(x)$ at the space-time point x in terms of the distribution functions $f^{\pm}(x,p)$. For a gas at finite mass $\epsilon - 3P$ is no longer vanishing as shown in Fig.2 (left) for several NJL parametrization.

Eqs.(1) and (2) form a closed system of equations constituting the Boltzmann-Vlasov equation associated to the NJL Lagrangian. The space-time dependence of the mass $M_\alpha(r,t) = m - 2g\langle\bar{\psi}\psi\rangle$ influences the momenta of the particles because the finite gradient of the condensate generates a force which changes the momentum of a particle proportionally to $\vec{\nabla}_r\langle\bar{\psi}\psi\rangle$. The last is negative because the phase transition occurs earlier in the surface of the expanding QGP fireball. Therefore the phase transition which takes place locally results in a negative contribution to the particle momenta that makes the system more sticky respect to a free massless gas. To study a fluid at finite η/s and NJL field interaction one has to extend the simple relation $\sigma_{tr} \cdot \eta/s =< p > /15\rho$ valid for a massless gas to the general case of massive relativistic particles and this has been discussed in Ref.[14].

In Fig.2, we show the elliptic flow as a function of p_T. We find that the presence of an NJL-field that drives the chiral phase transition suppress the $v_2(p_T)$ by about 20% at $p_T > 1$ GeV. This would imply the need of a parton scattering cross section σ_{tr} even larger than that estimated with the cascade model which was already quite larger than the pQCD estimates [9] in order to describe the data. On the other hand the mean field modifies both the local entropy density, reduced by the mass generation, and the shear viscosity that increases respect to the massless case. Considering that one of the main goal is to determine the η/s of the QGP, we have investigated what is the action of the mean field once the η/s of the system is fixed to be the same with and without the NJL field. This of course implies different values of the cross section in the two cases.

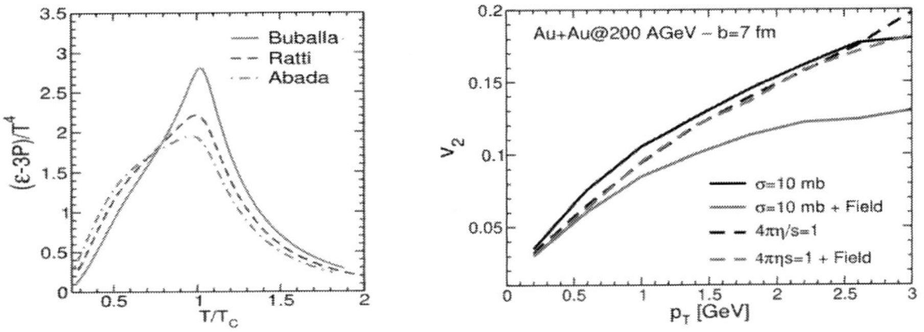

Figure 2. Left: Interaction measure shown as a function of temperature T for three different NJL parameter sets. Right: Average elliptic flow as a function of time for $Au + Au$ collisions in the mid-rapidity region $|y| < 1$ at b= 7 fm; see text.

The results for $4\pi\eta/s = 1$ are shown by dashed lines in Fig. 2 (right), the (black) dashed line is the case with only the collision term (cascade) while the (green) dashed line is the case with the field. We can see that once the η/s is fixed there is essentially no difference in the calculations with and without a field dynamics included. This is a key result that shows that even in a microscopic approach that distinguishes between the mean field and the collisional dynamics the $v_2(p_T)$ is mainly driven by the η/s of the fluid. In other words we have found that in a microscopic approach the η/s is the pertinent parameter and the language of viscous hydrodynamics is appropriate. Of course this does not mean that $v_2(p_T)$ in the transport theory is the same as in viscous hydrodynamics (especially at $p_T > 1.5$ GeV), but that the direct relation between $v_2(p_T)$ and η/s is not modified by the NJL field dynamics. This confirms the validity of the studies pursued till now even if they miss an explicit mean field or chiral dynamics.

Acknoweldgements

This work for V. Baran is supported in part by the Romanian Ministry for Education and Research under the CNCSIS contract PNII ID-946/2007.

References

[1] STAR, J. Adams *et al.*, Nucl. Phys. **A757**, 102 (2005);
[2] PHENIX, K. Adcox *et al.*, Nucl. Phys. **A757**, 184 (2005).
[3] P. Romatschke and U. Romatschke, Phys. Rev. Lett. **99**, 172301 (2007).
[4] H. Song and U. W. Heinz, (2009), 0909.1549.
[5] P. Huovinen and D. Molnar, Phys. Rev. **C79**, 014906 (2009).
[6] G. Ferini, M. Colonna, M. Di Toro, and V. Greco, Phys. Lett. **B670**, 325 (2009).
[7] V. Greco, M. Colonna, M. Di Toro, and G. Ferini, Prog. Part. Nucl. Phys. **62**, 562 (2009).
[8] R. S. Bhalerao, J.-P. Blaizot, N. Borghini, and J.-Y. Ollitrault, Phys. Lett. **B627**, 49 (2005).
[9] D. Molnar and M. Gyulassy, Nucl. Phys. **A697**, 495 (2002).
[10] R. J. Fries, V. Greco, and P. Sorensen, Ann. Rev. Nucl. Part. Sci. **58**, 177 (2008).
[11] A. Adare et al. [The PHENIX Collaboration], arXiv:1003.5586 [nucl-ex]; Yuting Bai, PhD thesis, the University of Utrecht (2007).
[12] M. Luzum, C. Gombeaud and J. Y. Ollitrault, Phys. Rev. C **81** (2010) 054910
[13] L. P. Csernai, J. I. Kapusta and L. D. McLerran, Phys. Rev. Lett. **97** (2006) 152303
[14] S. Plumari, V. Baran, M. Di Toro, G. Ferini and V. Greco, Phys. Lett. B **689** (2010) 18

Viscosities of the quasigluon plasma

M Bluhm[1], B Kämpfer[2,3] and K Redlich[4]

[1] SUBATECH, UMR 6457, Université de Nantes, Ecole des Mines de Nantes, IN2P3/CRNS. 4 rue Alfred Kastler, 44307 Nantes cedex 3, France

[2] Institut für Strahlenphysik, Forschungszentrum Dresden-Rossendorf, PF 510119, 01314 Dresden, Germany

[3] Institut für Theoretische Physik, TU Dresden, 01062 Dresden, Germany

[4] Institute of Theoretical Physics, University of Wroclaw, 50204 Wroclaw, Poland

E-mail: bluhm@subatech.in2p3.fr

Abstract. We investigate bulk and shear viscosities of the gluon plasma within relaxation time approximation to an effective Boltzmann-Vlasov type kinetic theory by viewing the plasma as describable in terms of quasigluon excitations with temperature dependent self-energies. The found temperature dependence of the transport coefficients agrees fairly well with available lattice QCD results. The impact of some details in the quasigluon dispersion relation on the specific shear viscosity is discussed.

1. Introduction

Bulk ζ and shear η viscosities are important properties of strongly interacting matter. Their firm knowledge is a necessary prerequisite for dynamical descriptions of relativistic heavy-ion collisions, but also in cosmology and astrophysics. The success of ideal relativistic hydrodynamic simulations, cf. e. g. [1], in quantitatively describing the collective flow behaviour of the matter produced at the Relativistic Heavy Ion Collider led to the conclusion that at most small dissipative effects are present in the created medium [2, 3, 4]. Later investigations by means of dissipative relativistic hydrodynamics, cf. e. g. [5, 6, 7, 8], confirmed this observation.

However, the quantum mechanical uncertainty principle enforces a fundamental lower bound on the shear viscosity [9] for any physical system. Moreover, unitarity arguments led to a lower limit conjecture for the specific shear viscosity [10, 11] $(\eta/s)_{KSS} \geq \hbar/(4\pi)$ (Kovtun-Son-Starinets bound, s is the entropy density). The bulk viscosity, in contrast, is exactly zero in conformal theories. In QCD, however, it is only (approximately) zero at large temperatures T, while for T close to the deconfinement transition temperature T_c the specific bulk viscosity ζ/s is expected to become large [12, 13, 14].

The viscosity coefficients were recently evaluated by means of non-perturbative first-principle numerical lattice QCD calculations [15, 16, 17, 18]. Other analytic approaches are based on the Kubo-formalism [19] or on linearized Boltzmann kinetic theory [19, 20, 21, 22]. In fact, it was shown that in scalar theory [19, 20] and in hot QED [21, 22] both approaches are equivalent. For small QCD running coupling, perturbative QCD (pQCD) kinetic theory results at leading order for shear and bulk viscosities were reported in [23, 24] and [25], respectively.

2. Bulk and shear viscosity coefficients

We calculate bulk and shear viscosities of the gluon plasma by assuming that it can be described by quasigluon excitations with T-dependent self-energy $\Pi(T)$, cf. [26, 27]. In local thermal equilibrium, this picture is comprised within a quasiparticle model (QPM), where the gluon dispersion relation reads $E^0 = \sqrt{\vec{p}^2 + \Pi(T)}$. $\Pi(T) = T^2 G^2(T)/2$ contains the effective coupling $G^2(T) = 16\pi^2 / \left(11 \ln\left[\lambda(T - T_s)/T_c\right]^2\right)$ with parameters λ and T_s. The QPM works impressively well for describing equilibrium lattice QCD thermodynamics, cf. [28, 29, 30].

By means of an effective kinetic theory [19, 20], the model can be extended to systems facing small deviations from local thermal equilibrium. In fact, as T is space-time dependent in general, so is the dispersion relation $E(x)$. The corresponding energy-momentum tensor reads $T^{\mu\nu}(x) = \int \frac{d^3\vec{p}}{(2\pi)^3 E(x)} p^\mu(x) p^\nu(x) b(x,p) + g^{\mu\nu} B(\Pi(x))$, where $b(x,p)$ represents the quasigluon distribution function entering $E(x)$ and $p^\alpha = (E, \vec{p})$. This expression for $T^{\mu\nu}$ is interwoven with a Boltzmann-Vlasov type kinetic equation [19]. The mean field term $B(\Pi)$ is necessary for satisfying locally energy and momentum conservation, $\partial_\mu T^{\mu\nu} = 0$, established when [26, 27] $\partial B/\partial \Pi = -\frac{1}{2} \int \frac{d^3\vec{p}}{(2\pi)^3 E(x)} b(x,p)$. This also guarantees thermodynamic self-consistency of the approach in thermal equilibrium and $E = \delta T^{00}/\delta b$ as a basic principle of statistical mechanics.

Expanding $b(x,p) = b^0(x,p) + \delta b(x,p)$ for small disturbances δb from equilibrium b^0, $T^{\mu\nu}$ is decomposed in an equilibrium part and first-order corrections thereof. Using the relaxation time approximation for determining δb in the latter, one reads off [27] for η and ζ in the local fluid rest frame by comparison with the phenomenological definitions [31]

$$\eta = \frac{1}{15T} \int \frac{d^3p}{(2\pi)^3} n(T)[1 + d^{-1} n(T)] \frac{\tau}{(E^0)^2} \vec{p}^4 , \qquad (1)$$

$$\zeta = \frac{1}{T} \int \frac{d^3p}{(2\pi)^3} n(T)[1 + d^{-1} n(T)] \frac{\tau}{(E^0)^2} \left\{ \left[(E^0)^2 - T^2 \frac{\partial \Pi}{\partial T^2} \right] \frac{\partial P}{\partial \epsilon} - \frac{1}{3} \vec{p}^2 \right\}^2 , \qquad (2)$$

where $n(T) = d(e^{E^0/T} - 1)^{-1}$ with degeneracy factor $d = 16$, $\partial P/\partial \epsilon$ is the squared speed of sound and τ the relaxation time. These results generalize previous work [32, 33], in which excitations with constant mass M were considered. While the expression for η looks formally the same, cf. also [34], ζ in Eq. (2) contains a combination of E^0 and self-energy derivative in contrast to [32, 33, 34]. It is this particular combination which guarantees $\zeta \to 0$ in the conformal limit.

3. Numerical results

For numerically evaluating Eqs. (1) and (2), we employ as ansatz for the relaxation time $\tau = [a_1 T G^4(T) \ln(a_2/G^2(T))]^{-1}$, which is inspired by considerations in [35]. It turns out that with this ansatz for τ, both η and ζ resemble at large T the behaviour with temperature and coupling known from pQCD [23, 24, 25]. The parameters in $G^2(T)$ are adjusted such that thermal equilibrium lattice QCD results for s/T^3 from [36, 37] are described by the QPM. Local thermal equilibrium quantities, such as s, are obtained by inserting $n(T)$ for $b^0(x,p)$ into $T^{\mu\nu}(x)$, cf. [27]. Furthermore, we fit the parameters in τ as $a_1 = 2.587 \times 10^{-2}$ and $a_2 = (\mu_*/T)^2$ (Fit 1) or as $a_1 = 3.85 \times 10^{-2}$ and $a_2 = 2(\mu_*/T)^2$ (Fit 2), where $\mu_*/T = 2.765$ as in pQCD [24]. QPM results for η/s and the corresponding relaxation times are exhibited in Fig. 1, while results for ζ/s are depicted in Fig. 2.

Good agreement with available lattice QCD results is found. In particular, η/s exhibits a minimum in the vicinity of T_c, which is common for a variety of liquids and gases [11, 40]. We find $(\eta/s)^*_{min} = 0.096$ at $T = 1.22\,T_c$ for Fit 2. The exact location of the minimum is driven by a_2 in τ. With increasing T, η/s increases rather mildly, while for $T \to T_c^+$ it dramatically rises in line with the relaxation time and the phenomenon of critical slowing down.

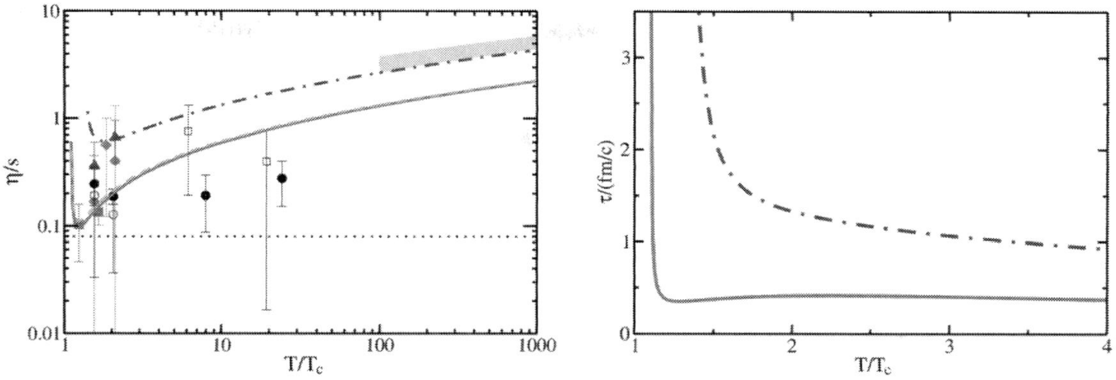

Figure 1. Left: Specific shear viscosity as a function of T/T_c. Dash-dotted and solid curves exhibit QPM results for Fit 1 and Fit 2, respectively, compared with lattice QCD data from [17] (full squares), [15] (diamonds and triangles) and [16] (open squares, open and full circles). The dotted line exhibits the KSS bound. The grey band at large T depicts the pQCD result $\eta_{NLL}/\tilde{s}$, where η_{NLL} is from [24] and $\tilde{s}$ is from [38]. The dashed curve (almost on top of the solid curve) shows η/s for Fit 2 for constant mass $M = 200$ MeV instead of $\Pi(T)$ in E^0. Right: Corresponding relaxation times τ. The behaviour is crucially determined by the parameter a_2.

For comparison, η/s for Fit 2 based on Eq. (1) but for particles with constant T-independent mass $M = 200$ MeV is also exhibited in Fig. 1 (dashed curve). Only the region around the minimum in η/s is affected (the minimal value is about 14 % larger than $(\eta/s)^*_{min}$), while at large T details in E^0 become irrelevant. Increasing M generally leads to a decrease in η/s. For $M = 1$ GeV a minimal value 9 % smaller than $(\eta/s)^*_{min}$ is found.

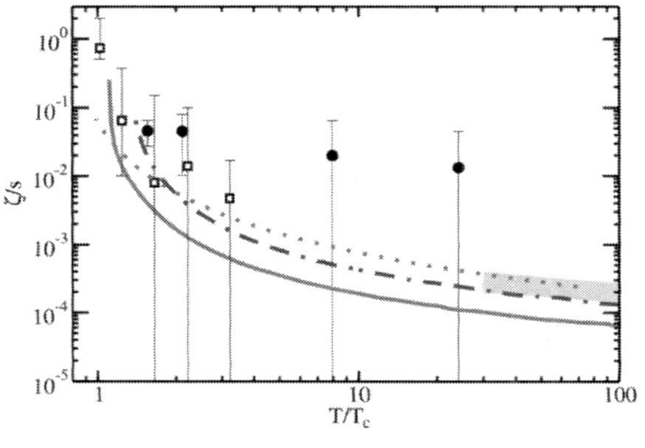

Figure 2. Specific bulk viscosity as a function of T/T_c. Dash-dotted and solid curves exhibit QPM results for Fit 1 and Fit 2, respectively, compared with available lattice QCD data from [18] (squares) and [16] (circles). The dotted curve shows for comparison results from holographic QCD [39]. The grey band at large T depicts the pQCD result $\zeta_{LO}/\tilde{s}$, where ζ_{LO} is from [25] and $\tilde{s}$ is from [38].

4. Conclusion

We derived expressions for bulk and shear viscosities within relaxation time approximation to an effective kinetic theory for the gluon plasma assumed to be composed of quasigluon excitations with T-dependent dispersion relation E^0. Medium effects show up in η only implicitly via E^0, while they are prominently evident in ζ. Specifying the relaxation time and adjusting model parameters to thermal equilibrium lattice QCD results, we find good agreement with available lattice QCD data for η/s and ζ/s for T close to T_c. The influence of moderate constant masses M on η/s is observed to be small and focussed on the region around its minimum. At large T our results resemble the parametric dependence on T and coupling known from pQCD.

Acknowledgments
We acknowledge insightful discussions with S. Jeon, U. Gürsoy, E. Kiritsis, H. B. Meyer, S. Peigne, A. Peshier, C. Sasaki and U. Wiedemann. M.B. and K.R. acknowledge the hospitality of the CERN TH department. This work is supported by 06 DR 9059, GSI-FE, the European Network I3-HP2 Toric and the Polish Ministry of Science.

References

[1] Kolb P F and Heinz U 2004 *Quark Gluon Plasma* vol 3, ed R C Hwa and X-N Wang (Singapoore: World Scientific) p 634
[2] Heinz U and Kolb P F 2002 *Nucl. Phys.* A **702** 269
[3] Lacey R A et al 2007 *Phys. Rev. Lett.* **98** 092301
[4] Drescher H-J, Dumitru A, Gombeaud C and Ollitrault J-Y 2007 *Phys. Rev.* C **76** 024905
[5] Romatschke P and Romatschke U 2007 *Phys. Rev. Lett.* **99** 172301
[6] Dusling K and Teaney D 2008 *Phys. Rev.* C **77** 034905
[7] Song H and Heinz U W 2008 *Phys. Lett.* B **658** 279; 2008 *Phys. Rev.* C **77** 064901; 2008 *Phys. Rev.* C **78** 024902; 2010 *Phys. Rev.* C **81** 024905
[8] Luzum M and Romatschke P 2008 *Phys. Rev.* C **78** 034915; Erratum 2009 *Phys. Rev.* C **79** 039903
[9] Danielewicz P and Gyulassy M 1985 *Phys. Rev.* D **31** 53
[10] Policastro G, Son D T and Starinets A O 2001 *Phys. Rev. Lett.* **87** 081601
[11] Kovtun P, Son D T and Starinets A O 2003 Holography and hydrodynamics: Diffusion on streched horizons *J. High Energy Phys.* JHEP10(2003)064; 2005 *Phys. Rev. Lett.* **94** 111601
[12] Kharzeev D and Tuchin K 2008 Bulk viscosity of QCD matter near the critical temperature *J. High Energy Phys.* JHEP09(2008)093
[13] Karsch F, Kharzeev D and Tuchin K 2008 *Phys. Lett.* B **663** 217
[14] Romatschke P and Son D T 2009 *Phys. Rev.* D **80** 065021
[15] Nakamura A and Sakai S 2005 *Phys. Rev. Lett.* **94** 072305
[16] Sakai S and Nakamura A 2007 Lattice calculation of QGP viscosities: Present results and next project *Proc. Sci.* LAT2007(2007)221
[17] Meyer H B 2007 *Phys. Rev.* D **76** 101701
[18] Meyer H B 2008 *Phys. Rev. Lett.* **100** 162001
[19] Jeon S 1995 *Phys. Rev.* D **52** 3591
[20] Jeon S and Yaffe L G 1996 *Phys. Rev.* D **53** 5799
[21] Gagnon J S and Jeon S 2007 *Phys. Rev.* D **75** 025014; Erratum 2007 *Phys. Rev.* D **76** 089902
[22] Gagnon J S and Jeon S 2007 *Phys. Rev.* D **76** 105019
[23] Arnold P, Moore G D and Yaffe L G 2000 Transport coefficients in high temperature gauge theories: (I) Leading-log results *J. High Energy Phys.* JHEP11(2000)001
[24] Arnold P, Moore G D and Yaffe L G 2003 Transport coefficients in high temperature gauge theories: (II) Beyond leading log *J. High Energy Phys.* JHEP05(2003)051
[25] Arnold P, Dogan C and Moore G D 2006 *Phys. Rev.* D **74** 085021
[26] Bluhm M, Kämpfer B and Redlich K 2009 *Nucl. Phys.* A **830** 737C
[27] Bluhm M, Kämpfer B and Redlich K 2010 Bulk and shear viscosities of the gluon plasma in a quasiparticle description *Preprint* arXiv:1011.5634
[28] Bluhm M, Kämpfer B and Soff G 2005 *Phys. Lett.* B **620** 131
[29] Bluhm M et al 2007 *Phys. Rev.* C **76** 034901
[30] Bluhm M and Kämpfer B 2008 *Phys. Rev.* D **77** 034004; 2008 *Phys. Rev.* D **77** 114016
[31] de Groot S R, van Leeuwen W A and van Weert C G 1980 *Relativistic Kinetic Theory* (Amsterdam: North-Holland)
[32] Hosoya A and Kajantie K 1985 *Nucl. Phys.* B **250** 666
[33] Gavin S 1985 *Nucl. Phys.* A **435** 826
[34] Sasaki C and Redlich K 2009 *Phys. Rev.* C **79** 055207; 2010 *Nucl. Phys.* A **832** 62
[35] Heiselberg H 1994 *Phys. Rev.* D **49** 4739
[36] Boyd G et al 1996 *Nucl. Phys.* B **469** 419
[37] Okamoto M et al 1999 *Phys. Rev.* D **60** 094510
[38] Blaizot J-P, Iancu E and Rebhan E 2001 *Phys. Rev.* D **63** 065003
[39] Gürsoy U et al 2009 Thermal transport and drag force in improved holographic QCD *J. High Energy Phys.* JHEP12(2009)056
[40] Csernai L P, Kapusta J I and McLerran L D 2006 *Phys. Rev. Lett.* **97** 152303

AUTHOR INDEX